X-ray Diffraction of Functional Materials

X-ray Diffraction of Functional Materials

Editors

Thomas Walter Cornelius
Souren Grigorian

MDPI • Basel • Beijing • Wuhan • Barcelona • Belgrade • Manchester • Tokyo • Cluj • Tianjin

Editors
Thomas Walter Cornelius
Université de Toulon
France

Souren Grigorian
Universität Siegen
Siegen, Germany

Editorial Office
MDPI
St. Alban-Anlage 66
4052 Basel, Switzerland

This is a reprint of articles from the Special Issue published online in the open access journal *Materials* (ISSN 1996-1944) (available at: https://www.mdpi.com/journal/materials/special_issues/Ray_Diffr).

For citation purposes, cite each article independently as indicated on the article page online and as indicated below:

LastName, A.A.; LastName, B.B.; LastName, C.C. Article Title. *Journal Name* **Year**, *Volume Number*, Page Range.

ISBN 978-3-0365-3365-0 (Hbk)
ISBN 978-3-0365-3366-7 (PDF)

Contents

About the Editors

Thomas Walter Cornelius, Dr. rer. nat., is CNRS researcher (since 2011) at the Institute for Materials, Microelectronics and Nanoscience of Provence (IM2NP) in Marseille, France. His research focuses on the mechanical properties of nanostructures and the relationship between mechanical strain and physical properties of nanomaterials. For his research, he strongly uses synchrotron X-ray diffraction methods and develops in situ techniques. His previous academic positions were at the European Synchrotron ESRF (Grenoble, 2008-2011) and the Hemlholtz Center for Heavy Ion Research (Darmstadt, 2006-2008). He obtained a PhD in Physics (2006) and diploma degree (2003) from the Ruprecht-Karls University in Heidelberg.

Souren Grigorian completed his PhD in Physics at the Institute of Crystallography, Moscow in 2000. He has since held a variety of research positions at institutions across Europe and became senior scientist at the University of Siegen in 2010. As an internationally recognized researcher, he was offered the role of guest scientist at NIST, Gaithersburg, and invited professor at Aix-Marseille University as well as Sapienza University of Rome. Among his research interests are advanced X-ray techniques for investigating multifunctional soft materials, in situ studies of working organic devices, direct correlation of microstructures and optoelectronic properties, and flexible organic electronics. Outside of his research, Dr Habil Grigorian has actively participated in teaching on an international level, and has coordinated the Volkswagen Foundation's grants for the International Symposium and Young Scientist School over several years.

Preface to "X-ray Diffraction of Functional Materials"

With new functional materials being developed and their properties being strongly related to their microstructure, the characterization of the latter is of paramount interest. Thanks to its non-invasive character, adaptability to various environments, and high sensitivity to crystalline structure and deformation as well as to defects, X-ray diffraction is the ideal tool to investigate the microstructure of these new materials.

The last 15 years have seen the continuous development of X-ray sources, optics, and detectors, making X-ray microscopy of strain and defects a reality with realistic time scales. Thanks to the penetrating power of X-rays, in situ or even operando monitoring of the crystalline structure of materials and devices as a function of mechanical stimuli, temperature, gases, electric fields, etc., is being commonly performed. Such studies are invaluable to investigate the physical mechanisms at work in real or close-to-real conditions. This is the case of elastic and plastic properties, catalytic activity, ferroelectric domain structure, and many others.

Moreover, advances in detectors and computer hardware and software facilitate time-resolved X-ray diffraction studies of the transient behavior of the microstructure, phase transitions, and physical changes caused by external stimuli. The time range covers more than ten orders of magnitude—from sub-picoseconds to kiloseconds. The advent of free electron lasers opens the few femtoseconds range, enabling the study of electronic processes.

This issue is dedicated to the latest advances in X-ray diffraction using both synchrotron radiation as well as laboratory sources for evaluating the microstructure and structure-to-property relation in functional materials (functional oxides, organic and hybrid materials for energy, electronics, etc.). Particular focus will be placed on novel in situ or operando approaches.

Thomas Walter Cornelius, Souren Grigorian

Editors

Article

Structural Characterization and Comparison of Monovalent Cation-Exchanged Zeolite-W

Donghoon Seoung [1], Hyeonsu Kim [1], Pyosang Kim [1], Chihyun Song [2], Suhyeong Lee [2], Sungmin Chae [2], Sihyun Lee [2], Hyunseung Lee [2] and Yongmoon Lee [2,*]

[1] Department of Earth Systems and Environmental Sciences, Chonnam National University, Gwangju 61186, Korea; dseoung@jnu.ac.kr (D.S.); 197942@jnu.ac.kr (H.K.); 197944@jnu.ac.kr (P.K.)
[2] Department of Geological Sciences, Pusan National University, Busan 46241, Korea; autumnleaves@pusan.ac.kr (C.S.); norium@pusan.ac.kr (S.L.); sjk04107@pusan.ac.kr (S.C.); tlgus745@pusan.ac.kr (S.L.); hslee07@pusan.ac.kr (H.L.)
* Correspondence: lym1229@pusan.ac.kr; Tel.: +82-51-510-2254

Received: 24 July 2020; Accepted: 18 August 2020; Published: 20 August 2020

Abstract: We report comparative structural changes of potassium-contained zeolite-W (K-MER, structural analogue of natural zeolite merlinoite) and monovalent extra-framework cation (EFC)-exchanged M-MERs (M = Li^+, Na^+, Ag^+, and Rb^+). High-resolution synchrotron X-ray powder diffraction study precisely determines that crystal symmetry of MERs is tetragonal (*I4/mmm*). Rietveld refinement results reveal that frameworks of all MERs are geometrically composed of disordered Al/Si tetrahedra, bridged by linkage oxygen atoms. We observe a structural relationship between a group of Li-, Na-, and Ag-MER and the group of K- and Rb-MER by EFC radius and position of M(1) site inside double 8-membered ring unit (*d8r*). In the former group, *a*-axes decrease reciprocally, *c*-axes gradually extend by EFC size, and M(1) cations are located at the middle of the *d8r*. In the latter group, *a*- and *c*-axes lengths become longer and shorter, respectively, than axes of the former group, and these axial changes come from middle-to-edge migration of M(1) cations inside the *d8r* channel. Unit cell volumes of the Na-, Ag-, and K-MER are ca. 2005 Å^3, and the volume expansion in the MER series is limited by EFC size, the number of water molecules, and the distribution of extra-framework species inside the MER channel. EFC sites of M(1) and M(2) show disordered and ordered distribution in the former group, and all EFC sites change to disordered distribution after migration of the M(1) site in the latter group. The amount of water molecules and porosities are inversely proportional to EFC size due to the limitation of volume expansion of MERs. The channel opening area of a *pau* composite building unit and the amount of water molecules are universally related as a function of cation size because water molecules are mainly distributed inside a *pau* channel.

Keywords: zeolite-W; cation form; synchrotron X-ray diffraction; Rietveld refinement

1. Introduction

Porous materials have been acknowledged as important specimens due to their pore characteristics, which are dependent on pore morphology, pore size, and pore size distribution [1–3]. Numerous experiments have been conducted to the understand fundamental factors of porous materials, and some have suggested characterization methods, e.g., the work of Liu et al. (2014) [4]. Zeolites are crystalline and are one of the popular functional materials that have been widely studied under various pressure, temperature, and chemical composition conditions for several decades [5–10]. Small-pore zeolites have been utilized in research areas for applications such as water purification, H-storage media, NO_x membrane, and CO_2 removal due to their accessible size, 8-membered ring composition, and selective catalytic reaction [11–14]. Zeolite-W is a synthetic phase whose framework topology is the same as the natural small-pore zeolite, merlinoite (assigned zeolite code: MER) [15,16]. The aspect of crystal structure for zeolite-W is usually determined as a tetragonal (*I4/mmm*) or orthorhombic (*Immm*) space group. Its framework is composed of a secondary building unit (SBU) of Si/Al tetrahedra such as 4-membered rings (4MR), 8-membered rings (8MR), and double 8-membered rings. A channel system usually consists of three-dimensional connections of 8MR pores [16]. Zeolite-W typically has an intermediate Si/Al framework ratio of $2 < Si/Al \leq 5$ and has been synthesized through several different routes, such as in the system of Na_2-K_2O-SiO_2-Al_2O_3, for controlling Si/Al ratio and pore size [17–19]. Many studies have provided information as to synthesis, crystal growth, and physiochemical characterization using methods such as Fourier transform infrared spectroscopy, thermogravimetric analysis, and solid-state NMR; however, there are few reports about the crystal structure of prepared products [19–22]. X-ray diffraction is one of the basic methods used to characterize products in order to understand heterogenetic cation sites, framework topology, and distributions of zeolitic water inside channels.

We have successfully prepared zeolite-W (K-MER) and its fully exchanged monovalent cation forms (Li-, Na-, and Ag-MER) by using conventional hydrothermal synthesis, following the work of Itabashi et al., and the ion-exchange method, respectively [23]. Here, we report comparative and systematic changes in crystal structures and chemical composition of a monovalent M-MER series at ambient conditions using high-resolution synchrotron X-ray powder diffraction, Rietveld structure refinement, energy dispersive X-ray spectroscopy (EDS), and thermogravimetric analysis (TGA).

2. Materials and Methods

2.1. Sample Preparation

The starting material, zeolite-W (K-MER), was synthesized under hydrothermal conditions according to Itabashi et al. [23]. An amount of 50 wt % of aqueous solution of KOH (Sigma-Aldrich) and $Al(OH)_3$ (Sumitomo Chemical Co. Ltd., Tokyo, Japan) was mixed and heated on a hot plate to prepare a potassium aluminate solution. Colloidal silica (Sigma-Aldrich) and a calculated amount of water were then added to the potassium aluminate solution. The batch composition was similar to 3 K_2O:1.5 Al_2O_3:5 SiO_2:100 H_2O. For crystallization, the mixture was put into autoclave and then heated at 150 °C for 48 h. The final product was washed and dried in ambient conditions. Other cation forms were prepared by fully saturated MCl solution (M = Li^+, Na^+, and Ag^+) and K-MER powder in a 100:1 ratio of solution volume to powder weight. The mixture was stirred at 80 °C in a closed system, minimizing the loss of water. After 24 h, the solid was separated from the solution by vacuum filtration. The dried powder was then subjected to two more exchange cycles. The final product was washed with a deionized water and subsequently air-dried at ambient conditions. Stoichiometric analyses were performed on products using EDS, and we confirmed that over 99% of the cation was exchanged. To determine the amount of H_2O molecules, TGA was used with a heating range of 25 to 800 °C and a heating rate of 10 °C/min under a nitrogen atmosphere (Figure S3). Experimental conditions for cation exchange and chemical analysis results are summarized in Tables S1 and S2.

2.2. Synchrotron X-ray Powder Diffraction

High-resolution synchrotron X-ray powder diffraction data were collected for precise determination of crystal symmetry of MERs at the 9B beam line of Pohang Light Source-II (PLS-II) at the Pohang Accelerator Laboratory (PAL). The incident X-rays were vertically collimated by a mirror, and they were monochromatized to a wavelength of 1.4865(1) Å using a double-crystal Si (111) monochromator. The detector arm of the vertical scan diffractometer was composed of six sets of Soller slits, flat Ge (111) crystal analyzers, antiscatter baffles, and scintillation detectors, with each set separated by 21° in 2-theta. Each specimen of approximately 0.2 g powder was prepared using a flat plate side loading method. Step scans were performed at room temperature from 8° in 2-theta with 0.005° increments and 2° overlaps of the detector bank, up to 128.5° in 2-theta. Synchrotron X-ray powder diffraction data were measured in a transmission geometry to avoid preferred orientation at the 3D and 5A beam lines of Pohang Light Source-II (PLS-II) at the Pohang Accelerator Laboratory (PAL) in South Korea. The primary white beam from the bending magnet of 3D or the superconducting insertion device of 5A was directed on a Si (111) crystal, and sets of parallel slits were used to create monochromatic X-rays with wavelengths of 0.6926(1) Å and 0.6199(1) Å for 3D and 5A, respectively. Each sample was loaded into a 0.5 mm diameter SiO_2 glass capillary, and the capillary was subsequently sealed to minimize dehydration by atmosphere during measurement. All samples were measured with sample rotation for 5 min using the MAR345 image plate two-dimensional X-ray detector. The calibrations of the X-ray wavelength and sample-to-detector distance were achieved using a LaB_6 standard (SRM 660c).

2.3. Structural Analysis by Rietveld Refinement

Unit cell length and volume changes were derived from a series of whole-profile fitting procedures using the General Structure Analysis System (GSAS) suite of programs [24]. The background was fixed at selected points, and a pseudo-Voigt profile function proposed by Thompson et al. was used to model the observed Bragg peaks [25]. The structural models of MERs were obtained by Rietveld refinements [24,26,27]. A March–Dollase function was used to account for preferred orientation [28]. In order to reduce the number of parameters, isotropic displacement factors were refined by grouping the framework species and the extra-framework species. Geometrical soft-restraints on the T–O (T = Si, Al) and O–O bond distances of the disordered Si/Al tetrahedra were applied by stoichiometric results of each sample from EDS: the distances of Si–O were restrained to values from 1.642 to 1.648 with esd of ±0.001 Å, and the O–O distances ranged from 2.692 to 2.688 with esd of ±0.005 Å. In the final stages of the refinements, the weights of the soft restraints were gradually reduced. This did not lead to any significant changes in the interatomic distances, and convergence was achieved by simultaneously refining scale factors, lattice constants, 2-theta zero, preferred orientation function, and the atomic positional and thermal displacement parameters. The final refined parameters are summarized in Table S3, and the selected bond distances and angles are listed in Table S4.

3. Results and Discussions

As described in the experimental section, we have successfully prepared K-contained zeolite-W and its series of monovalent cation-exchanged forms by Li^+, Na^+, and Ag^+ (Li-, Na-, and Ag-MER, respectively). High-resolution synchrotron X-ray diffraction patterns of MERs at ambient conditions are shown in Figure 1. Chemical compositions derived from Rietveld refinement are $Li_{6.9}Al_{6.9}Si_{25.0}O_{64.0}·26H_2O$, $Na_{7.5}Al_{7.0}Si_{25.0}O_{64.0}·20.0H_2O$, $Ag_{7.0}Al_{7.0}Si_{25.0}O_{64.0}·22.2H_2O$, and $K_{6.42}Al_{6.5}Si_{25.8}O_{64.0}·15.3H_2O$ for the Li-, Na-, Ag-, and K-MER, respectively (Table S1). Bragg peaks of each phase are indexed to crystal symmetry of a tetragonal system and space group *I4/mmm*. The peak positions such as (101), (200), (220), and (103), related to the *a*- and *c*-axis, in each MER phase seem slightly shifted in 2-theta, while relative peak intensities are obviously changed due to changes in chemical composition. It is therefore shown that distribution and occupancy changes of extra-framework species such as EFCs and water molecules, related to intensity changes, primarily arise from unit cell length changes, related to peak position changes by chemical composition.

Whole-profile fitting analysis reveals that unit cell parameters have two distinguishable relationships in the group containing the Li-, Na-, and Ag-MER and in the group containing the K- and Rb-MER (model from Itabashi et al.) by means of exchanged cation radius and location of M(1) site of each EFC (Figure 2). In the former group (Li-, Na-, and Ag-MER), the *a*-axes linearly decrease from 14.1613(4) to 14.1334(9) Å, and the M(1) (radius < 1.2 Å) site is located at middle of the *d8r* along (010) direction. EFCs of M(1) progressively migrate toward the side wall of the *d8r* as a function of EFC radius (atomic coordinate changes of *x* in Table S3). The bond valence sums (BVS) calculations of MERs show that M(1) cations usually have lower values than M(2) cations, which indicates M(1) cations are under-bonded in comparison with M(2) cations (Table S4) [29]. This helps to understand the reason why M(1) cations migrate easily in the *d8r* channel. The location of the M(1) site is abruptly changed to the side wall of *d8r* when cation radius is larger than the 1.35 Å of K^+, which reflects abrupt increments of *a*-axes in the latter group (K- and Rb-MER). Similarly, the *a*-axes of this group linearly decrease from 14.1927(4) to 14.1798(3) Å by means of EFC size. In the case of *c*-axes lengths, they gradually extend up to 10.040(1) Å of the Ag-MER in the former group; meanwhile, *a*-axes decrease reciprocally by EFC size. The *c*-axis also suddenly collapsed before and after cation radius of 1.35 Å and then linearly decreased identically in the latter group (Figure 2a). Unit cell volume changes due to complex distribution and migration of the EFCs following their radius are shown in Figure 2b. Volumetric changes of M-MERs show gradual expansions and contractions, proportionally for Li^+, Na^+, and Ag^+ and reciprocally for K^+ and Rb^+, as a function of cation radius. Interestingly, volumes of the Na-, Ag-, and K-MER are similar, ca. 2005 $Å^3$, and they constitute ca. 5% volume expansion compared with the volume of Li-MER, ca. 1995.3(1) $Å^3$. Thus, this indicates that the volume expansion in the MER series is limited to ca. 2005 $Å^3$ through alkaline metal cation substitution.

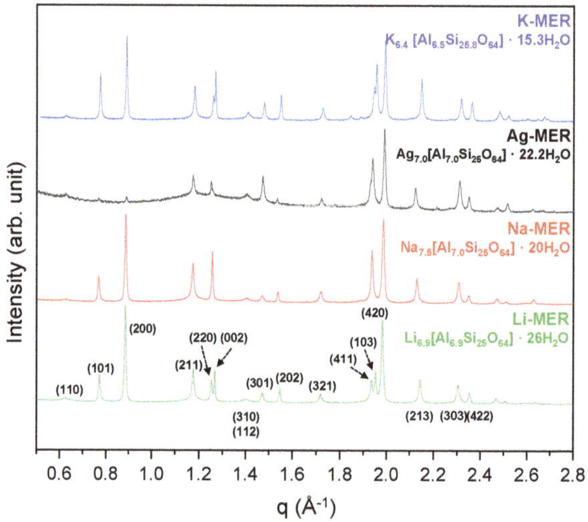

Figure 1. Changes in the synchrotron X-ray powder diffraction patterns of Li-, Na-, Ag-, and K-MER at ambient conditions. Miller indices are shown for the Bragg peaks. Chemical compositions are derived from our refinement results.

Refined crystal structures of MERs are shown in Figure 3 and Figure S1. The EFC site of M(1) (colored beach balls) in the Li-, Na-, and Ag-MER is located at the middle of the *d8r* channel projected to the (010) direction (filled by light-blue color) and has two bonds with one framework oxygen, O(1), and one water molecule, WO(3), whereas the M(1) is near the edge of *d8r* and has three bonds with O(1), O(2), and WO(4) in the K- and Rb-MER. On the other hand, the EFC sites, M(2), are constantly located at the center of the channel along the *c*-axis despite cation changes. Water molecules (red symbols) are designated by five different sites, from WO(1) to WO(5). Sites of WO(1), (2) and (3) (equatorial balls) are located at the middle of the *pau* channel (filled by brown color) and *d8r* along the c-axis, regularly. Both WO(1) and (2) are connected with M(2) cations, and WO(3) is bonded with M(1) cations along the *c*-axis due to a different z-coordinate of WO(3) compared to that of WO(1) and (2). The sites of WO(4) and (5) (hatched and striped balls, respectively) are distributed around the edge of the *pau* channel and have one or two coordinates bonding with M(1) or themselves. Thus, the WO(4) and (5) easily migrate inside the channel, and sites are finally merged to WO(4) during migration of a larger cation, such as K^+ and Rb^+ in the K- and Rb-MER. Changes of occupancies of M(1) and (2) sites are related to cation size (Figure 4). In case of the Li-, Na, and Ag-MER, the M(1) and M(2) sites are partially (less than ca.

0.45) and fully occupied, respectively. As the M(1) cation migrates toward the side of *d8r* due to its size, the occupancy of M(1) gradually increases. The occupancies of M(2) are reduced to less than ca. 0.6 when cation radius is greater than 1.35 Å, in the case of the K- and Rb-MER, and the occupancy of M(1) increases more than that of M(2) in the case of the Rb-MER. These changes of occupancies are related to the number of coordinated bonds with framework oxygens. The M(1) cation in the Li-, Na-, and Ag-MER is connected to one framework oxygen, O(1), with bond distance of ca. 3–3.3 Å. The M(1) site becomes more stable by bonding with two framework oxygens, O(1) and (2), after its migration. From K- to Rb-MER, the M(1) cation is more tightly connected with O(1) than M(2)–O(3) bonding, and therefore occupancy of M(1), in the case of the Rb-MER, can further increase more than the K-MER.

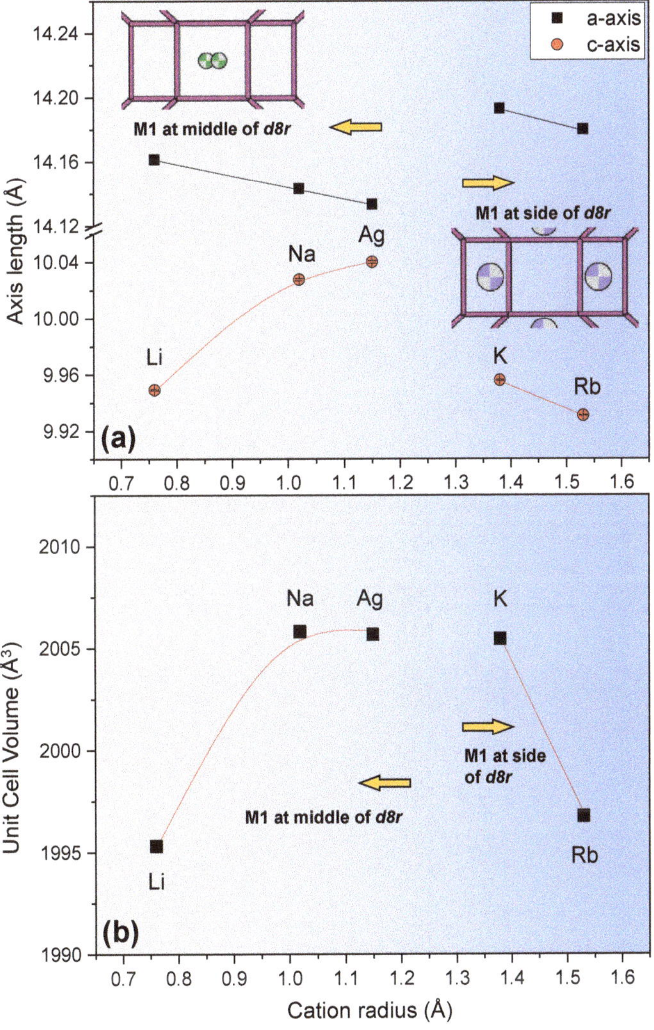

Figure 2. Refined (**a**) *a*- and *c*-axis and (**b**) volume of cation forms of MER as a function of extra-framework cation size.

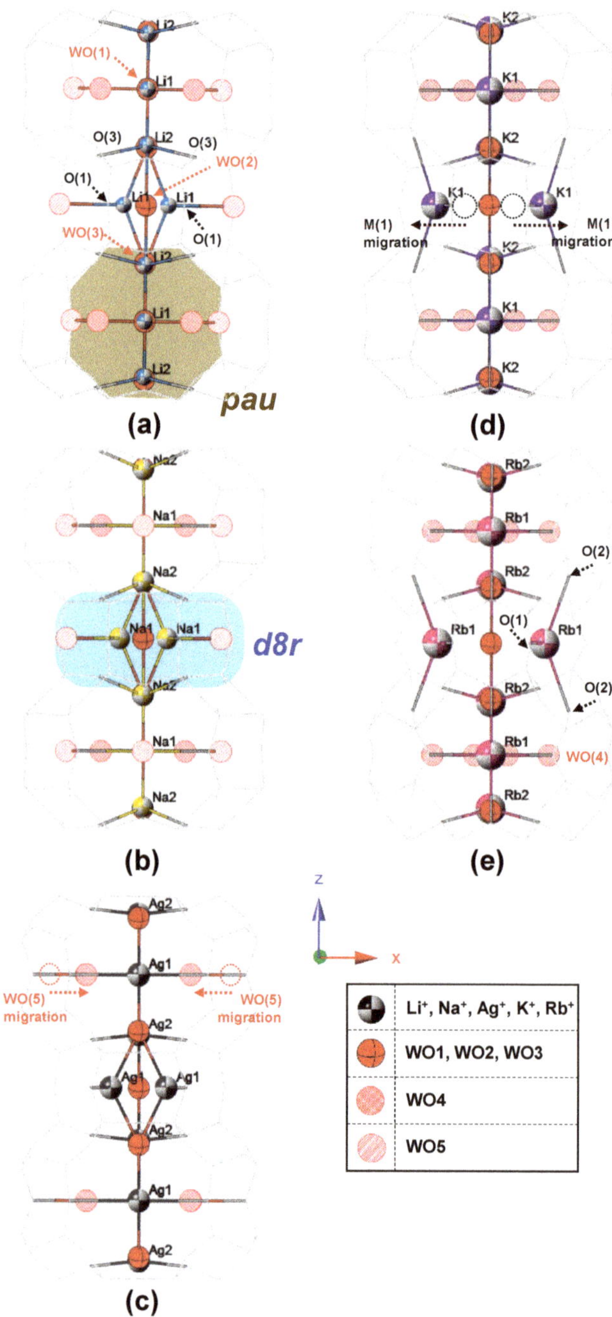

Figure 3. Polyhedral representations of (**a**) Li-MER, (**b**) Na-MER, (**c**) Ag-MER, (**d**) K-MER, and (**e**) Rb-MER along the (010) direction. The grey stick represents disordered Al/Si framework. Each colored beach ball represents an extra-framework cation. Equatorial, hatched, and striped red balls represent oxygens of WO(1)–WO(3), WO(4), and WO(5), respectively.

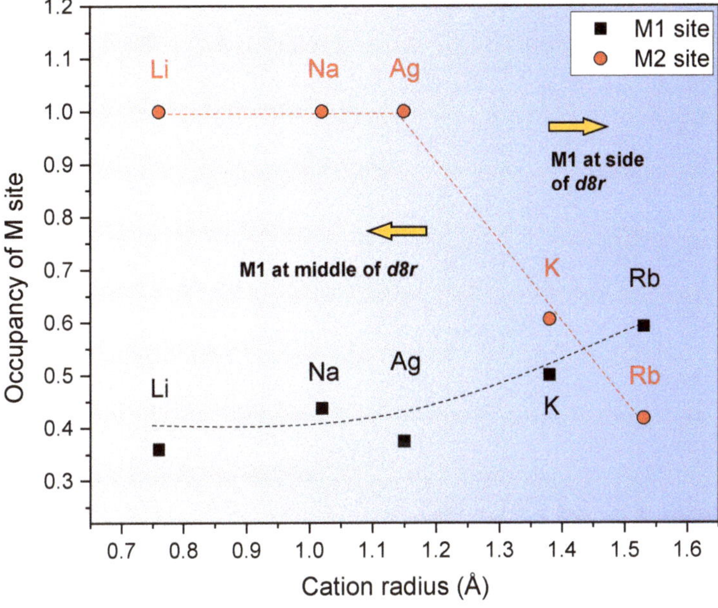

Figure 4. Changes of occupancies of M(1) and M(2) sites (M = Li^+, Na^+, Ag^+, K^+, Rb^+).

We find universal relationships between porosity, channel opening area, and the number of water molecules in the unit cell as a function of cation radius (Figure 5). The porosity of all MERs is calculated from our structure models. The porosity and the number of water molecules are gradually and exponentially decreased by EFC size, respectively. Considering the limitation of the largest volume expansion among MER structures, we expect that the EFC size mainly influences a determination that the porosities and rearrangements of EFCs in the framework are accompanied by cation size in order to absorb water molecules as much as possible. In case of the Ag-MER, the value of water molecules per unit cell is greater than expected value on the trend line (red dotted line in Figure 5a), which might be due to the high electronegativity of the silver cation compared to other cations. The degree of channel opening area, along the (010) direction, of the *pau* composite unit is related to the number of water molecules. In our structure models, all water molecules are distributed inside a *pau* channel along *a*- or *b*-axis; therefore, changes of channel opening area are concomitantly observed in agreement with the number of water molecules along the *a*- or *b*-axis rather than the *c*-axis (Figure 5b and Figure S2).

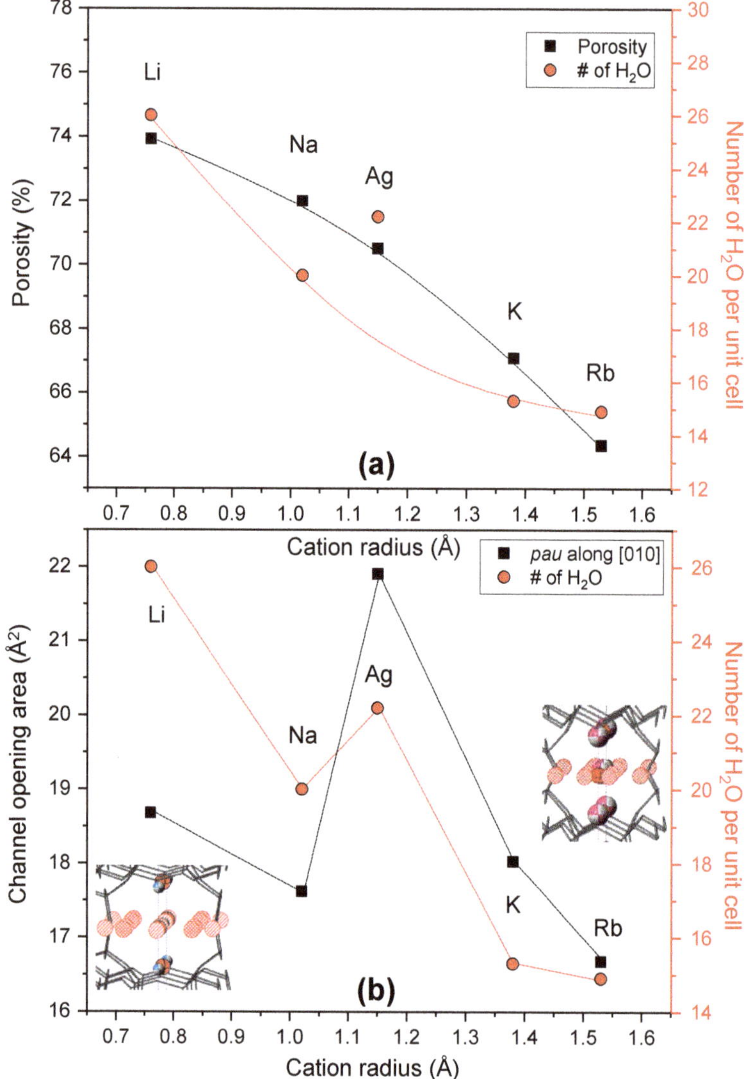

Figure 5. Changes of (**a**) porosity and amount of water molecules and (**b**) channel opening area of the *pau* unit and amount of water molecules as a function of extra-framework cation size.

4. Conclusions

In this work, we have demonstrated comparative crystal chemistry among synthesized K-MER and its monovalent cation forms of Li^+, Na^+, Ag^+, and Rb^+, along with several correlations by means of EFC size. The *a*- and *c*-axis show distinguishable changes between the Li-, Na-, and Ag-MER group and the K- and Rb-MER group depending on whether the cation site of M(1) is located at the middle or the side wall of the *d8r*. The maximal volume expansion is ca. 2005 $Å^3$, and therefore porosity and the amount of water molecules inside the channel decrease with EFC size, accompanying migration of extra-framework species such as EFCs and water molecules. We expect that our results can provide fundamental knowledge to understand the relationship between crystal structure and chemical changes of MER-type zeolites, which can be applied to the enhancement of catalysts and absorbents.

Supplementary Materials: The following are available online at http://www.mdpi.com/1996-1944/13/17/3684/s1, Figure S1: Polyhedral representations of (**a**) Li-MER, (**b**) Na-MER, (**c**) Ag-MER, (**d**) K-MER and (**e**) Rb-MER along (001) direction. Grey stick represents disordered Al/Si framework. Each colored beach ball represents extra-framework cation. Equatorial, hatched, and striped red balls represent oxygens of WO(1)–WO(3), WO(4), and WO(5), respectively, Figure S2: Comparison of channel opening area of *d8r* (black symbol) and *pau* (red symbol) along (001) direction and sum of two areas., Figure S3: Thermogravimetric analysis results of Li-, Na-, Ag-, and K-MER., Table S1: Cation-exchange conditions of MERs and chemical composition from Rietveld refinement and stoichiometric analysis., Table S2. Chemical composition calculated from Energy Dispersive Spectroscopy (EDS) method., Table S3: Refined cell parameters and atomic coordinates of M-MER at ambient condition (M = Li^+, Na^+, Ag^+, K^+ and Rb^+)., Table S4: Selected interatomic distances (Å) and angles (°) for M-MER at ambient condition (M = Li^+, Na^+, Ag^+, K^+, and Rb^+).

Author Contributions: Conceptualization, Y.L.; methodology, H.K., P.K., C.S., S.L. (Suhyeong Lee), S.C., H.L. and S.L. (Sihyun Lee); formal analysis, H.K. and P.K.; data curation, H.K. and P.K.; writing—original draft preparation, Y.L. and D.S.; writing—review and editing, Y.L. and D.S.; visualization, Y.L.; supervision, D.S.; project administration, Y.L.; funding acquisition, Y.L. and D.S. All authors have read and agreed to the published version of the manuscript.

Funding: This research was funded by the National Research Foundation of the Ministry of Science and ICT of Korean Government, grant numbers NRF-2019R1F1A106258, NRF-2017R1D1A1B03035418, NRF-2019K1A3A7A09101574; Pusan National University Research Grant 201902840001; and Chonnam National University Research Grant 2019-0215.

Acknowledgments: Experiments using synchrotron radiation were supported by Pohang Light Source-II (PLS-II) at the Pohang Accelerator Laboratory (PAL). We thank T. Jeon, H.-H. Lee, and D. Ahn for the support at beamlines 3D, 5A, and 9B.

Conflicts of Interest: The authors declare no conflict of interest.

References

1. Liu, P.S.; Chen, G.F. General Introduction to Porous Materials. In *Porous Materials*; Liu, P.S., Chen, G.F., Eds.; Butterworth-Heinemann: Boston, UK, 2014.

2. Tabacchi, G. Supramolecular Organization in Confined Nanospaces. *ChemPhysChem* **2018**, *19*, 1249–1297. [CrossRef] [PubMed]

3. Calzaferri, G. Nanochannels: Hosts for the Supramolecular Organization of Molecules and Complexes. *Langmuir* **2012**, *28*, 6216–6231. [CrossRef] [PubMed]

4. Liu, P.S.; Chen, G.F. Characterization Methods: Basic Factors. In *Porous Materials*; Liu, P.S., Chen, G.F., Eds.; Butterworth-Heinemann: Boston, UK, 2014.

5. Plank, C.J.; Rosinski, E.J.; Hawthorne, W.P. Acidic Crystalline Aluminosilicates. New Superactive, Superselective Cracking Catalysts. *Ind. Eng. Chem. Prod. Res. Dev.* **1964**, *3*, 165–169. [CrossRef]

6. Breck, D.W. *Zeolite Molecular Sieves: Strcuture, Chemistry, and Use*; John Wiley and Sons: New York, NY, USA, 1974.

7. Beyer, H.; Jacobs, P.A.; Uytterhoeven, J.B. Redox behaviour of transition metal ions in zeolites. Part 2—Kinetic study of the reduction and reoxidation of silver-Y zeolites. *J. Chem. Soc. Faraday Trans. 1 Phys. Chem. Condens. Phases* **1976**, *72*, 674–685. [CrossRef]

8. Jacobs, P.A.; Uytterhoeven, J.B.; Beyer, H.K. Redox behaviour of transition metal ions in zeolites. Part 6—Reversibility of the reduction reaction in silver zeolites. *J. Chem. Soc. Faraday Trans. 1 Phys. Chem. Condens. Phases* **1977**, *73*, 1755–1762. [CrossRef]

9. Ackley, M.W. Application of natural zeolites in the purification and separation of gases. *Microporous Mesoporous Mater.* **2003**, *61*, 25–42. [CrossRef]

10. Gatta, G.D.; Lee, Y. Zeolites at high pressure: A review. *Miner. Mag.* **2014**, *78*, 267–291. [CrossRef]

11. Fee, J.P.H.; Murray, J.M.; Luney, S.R. Molecular sieves: An alternative method of carbon dioxide removal which does not generate compound A during simulated low-flow sevoflurane anaesthesia. *Anaesthesia* **1995**, *50*, 841–845. [CrossRef]

12. Segawa, K.; Shimura, T. Effect of dealuminatation of mordenite catalyst for amination reaction of ethanoplamine. In *A New Era of Catecholamines in the Laboratory and Clinic*; Elsevier BV: Amsterdam, The Netherlands, 2000; Volume 130, pp. 2975–2980.

13. Fujimoto, K.; Bischoff, S.; Omata, K.; Yagita, H. Hydrogen effects on nickel-catalyzed vapor-phase methanol carbonylation. *J. Catal.* **1992**, *133*, 370–382. [CrossRef]

14. Altwasser, S.; Welker, C.; Traa, Y.; Weitkamp, J. Catalytic cracking of n-octane on small-pore zeolites. *Microporous Mesoporous Mater.* **2005**, *83*, 345–356. [CrossRef]

15. Baerlocher, C.; McCusker, L.B.; Olson, D.H.; Baerlocher, C.; Burton, A.W.; Baur, W.H.; Broach, R.W.; Dorset, D.L.; Fischer, R.X.; Gies, H.; et al. Preface. In *Atlas of Zeolite Framework Types*; Elsevier Science B.V.: Amsterdam, The Netherlands, 2007.

16. International Zeolite Association. Available online: http://www.iza-structure.org/databases (accessed on 19 August 2020).

17. Barrett, P.A.; Valencia, S.; Camblor, M.A. Synthesis of a merlinoite-type zeolite with an enhanced Si/Al ratioviapore filling with tetraethylammonium cations. *J. Mater. Chem.* **1998**, *8*, 2263–2268. [CrossRef]

18. Quirin, J.C.; Yuen, L.; Zones, S.I. Merlinoite synthesis studies with and without organocations. *J. Mater. Chem.* **1997**, *7*, 2489–2494. [CrossRef]

19. Bieniok, A.; Bornholdt, K.; Brendel, U.; Baur, W.H. Synthesis and crystal structure of zeolite W, resembling the mineral merlinoite. *J. Mater. Chem.* **1996**, *6*, 271. [CrossRef]

20. Belhekar, A.; Chandwadkar, A.; Hegde, S. Physicochemical characterization of a synthetic merlinoite (Linde W-like) zeolite containing Na, K, and Sr cations. *Zeolites* **1995**, *15*, 535–539. [CrossRef]

21. Chen, W.; Guo, Q.; Yang, C.; Hou, J. Preparation of novel functional MER zeolite membrane for potassium continuous extraction from seawater. *J. Porous Mater.* **2017**, *25*, 215–220. [CrossRef]

22. Houlleberghs, M.; Breynaert, E.; Asselman, K.; Vaneeckhaute, E.; Radhakrishnan, S.; Anderson, M.W.; Taulelle, F.; Haouas, M.; Martens, J.A.; Kirschhock, C.E.A. Evolution of the crystal growth mechanism of zeolite W (MER) with temperature. *Microporous Mesoporous Mater.* **2019**, *274*, 379–384. [CrossRef]

23. Itabashi, K.; Ikeda, T.; Matsumoto, A.; Kamioka, K.; Kato, M.; Tsutsumi, K. Syntheses and structural properties of four Rb-aluminosilicate zeolites. *Microporous Mesoporous Mater.* **2008**, *114*, 495–506. [CrossRef]

24. Toby, B.H. EXPGUI, a graphical user interface for GSAS. *J. Appl. Crystallogr.* **2001**, *34*, 210–213. [CrossRef]

25. Thompson, P.; Cox, D.E.; Hastings, J.B. Rietveld refinement of Debye–Scherrer synchrotron X-ray data from Al_2O_3. *J. Appl. Crystallogr.* **1987**, *20*, 79–83. [CrossRef]

26. Larson, A.C.; VonDreele, R.B. *GSAS: General Structure Analysis System Report LAUR*; Los Alamos National Laboratory, LAUR: Los Alamos, NM, USA, 1986; pp. 86–748.

27. Rietveld, H.M. A profile refinement method for nuclear and magnetic structures. *J. Appl. Crystallogr.* **1969**, *2*, 65–71. [CrossRef]

28. Dollase, W.A. Correction of intensities for preferred orientation in powder diffractometry: Application of the March model. *J. Appl. Crystallogr.* **1986**, *19*, 267–272. [CrossRef]

29. Brown, I.D.; Altermatt, D. Bond-valence parameters obtained from a systematic analysis of the Inorganic Crystal Structure Database. *Acta Crystallogr. Sect. B Struct. Sci.* **1985**, *41*, 244–247. [CrossRef]

Article

The Number of Subgrain Boundaries in the Airfoils of Heat-Treated Single-Crystalline Turbine Blades

Jacek Krawczyk [1,*], Włodzimierz Bogdanowicz [1] and Jan Sieniawski [2]

[1] Institute of Materials Engineering, University of Silesia in Katowice, 75 Pułku Piechoty 1a St., 41-500 Chorzów, Poland; wlodzimierz.bogdanowicz@us.edu.pl

[2] Department of Materials Science, Rzeszów University of Technology, Wincentego Pola 2 St., 35-959 Rzeszów, Poland; jansien@prz.edu.pl

* Correspondence: jacek.krawczyk@us.edu.pl; Tel.: +48-32-3497537

Abstract: In the present study, the dendrites deflection mechanism from the mold walls were subjected to verification regarding its heat-treated turbine rotor blades. The number of macroscopic low-angle boundaries created on the cross-section of the blades' airfoil near the tip was experimentally determined and compared to the number of low-angle boundaries calculated from a model based on the dendrites deflection mechanism. Based on the Laue patterns and geometrical parameters of airfoils, the number of low-angle boundaries occurring at the upper part of the blades airfoil after heat treatment was calculated. This number for the analyzed group of blades ranged from 5 to 9.

Keywords: superalloys; low-angle boundaries; X-ray topography; turbine blades; crystal growth

Citation: Krawczyk, J.; Bogdanowicz, W.; Sieniawski, J. The Number of Subgrain Boundaries in the Airfoils of Heat-Treated Single-Crystalline Turbine Blades. *Materials* **2021**, *14*, 8. http://dx.doi.org/10.3390/ma14010008

Received: 6 November 2020
Accepted: 17 December 2020
Published: 22 December 2020

Publisher's Note: MDPI stays neutral with regard to jurisdictional claims in published maps and institutional affiliations.

1. Introduction

The turbine components of aircraft engines are currently most often produced using the CMSX-4 superalloy. The single-crystalline parts made using the CMSX-4 superalloy possess high strength properties, even at high temperatures, which is especially important for turbine rotor blades, as these types of blades are exposed to harsh working conditions [1–5]. The single-crystalline rotor turbine blades, which have a very complex shape, are usually produced by using the Bridgman technique [6–12]. The blades are obtained through directional crystallization using the temperature and withdrawal parameters that allow the formation of an array of dendrites nearly parallel to the blade axis, with the crystal orientation of each dendrite being parallel to the [001] axis. The dendrites and interdendritic areas that are formed during the Bridgman process mainly consist of the Ni-based γ primary solid solution and the Ni_3Al-based γ' secondary solid solution [13–16]. Due to similarity between the structure of both phases and the possibility of obtaining a clear X-ray diffraction pattern, the blades can be recognized as single-crystalline blades [17–19].

The production of the rotor blades includes subjecting them to the heat treatment process to increase chemical and microstructural homogeneity and to obtain a large amount of the γ' phase, as well as to decrease the crystal orientation inhomogeneity through elimination of the low-angle boundaries (LABs) created during crystallization [20–23]. However, it has been stated that not all macroscopic LABs are eliminated by the heat treatment but also that new extra LABs may be created as a result of this treatment. All of the LABs decrease the strength of the rotor blades [24–26].

An airfoil is generally a thin-walled fragment of a blade and has the least durability of all parts of a blade due to its low cross-section and the large complex loads that are acting on it during its operation [27]. Therefore, analysis of the number of defects such as LABs in the treated blades airfoils is extremely important. The number of the LABs with respect to the stretching direction has a strong effect on the creep life of the blades. The LABs also affect the adjacent dislocation density, which is also related to the misorientation angle. It is widely accepted that the boundaries with a misorientation of above 1° have a significant influence on the mechanical properties of the LABs. In a thin-walled airfoil, the interaction of the dendrites with the mold walls may occur more frequently than in the root; therefore, the LABs can be formed more often [28].

During directional crystallization, the dendrites grow directly toward the [001]-type direction, which is only close to the blade axis Z. Therefore, even if the mold walls are parallel to the Z, this means the dendrites can contact them. Additionally, the inclination of the airfoil surfaces in relation to the Z axis of the rotor blade makes the growing dendrites contact the mold walls and interact with them. The character of interactions may depend on the angle between the axis of the dendrite primary arms and the mold surface [28]. For lower angles, the dendrite primary arms may slightly change their growth direction near the surface, thereby creating areas of internal stress, and then stop. The higher angles prevent a further growth of the primary dendrite arms without a change in the growth direction. In both cases, further growth of a dendrite takes place due to the secondary arms that are perpendicular to the primary arms [29]. Both mechanisms can lead to the formation of low-angle boundaries. In the former case, these boundaries may be formed after heat treatment, for example through the process of dislocation polygonization, while in the latter case the boundaries may be formed directly during crystallization. These mechanisms of dendrite interaction with the mold walls were proposed for the first time in Refs. [28,30] and were referred to as the deflection of dendrites on the mold surfaces.

The aim of the research presented here was to check the assumption that in heat-treated single-crystalline blades made of a Ni-based superalloy, each type of dendrite deflection on the mold surface creates LABs. The assessment was performed by using experimental determination of the LABs number in thin-walled blade airfoils where the probability of dendrites interaction with the mold walls is high, and a comparison with the number of LABs could be determined by using a model based on the aforementioned assumption. The kinetics of the dendrites growth may affect the deflection mechanisms; therefore, it was decided to examine blades airfoils that were obtained at different withdrawal rates in the range of 2–5 mm/min, including the most commonly used rate of 3 mm/min.

2. Material and Methods

The Bridgman technique was used for production of two series of single-crystalline turbine rotor blades. The as-cast blades made of CMSX-4 Ni-based superalloy were heat treated and then analyzed. Each series consisted of four blades obtained at different withdrawal rates. The directional crystallization with the use of a spiral selector (S, Figure 1a) and heat treatment process was performed in the Research and Development Laboratory for Aerospace Materials at the Rzeszów University of Technology, Rzeszów, Poland using industrial ALD Vacuum Technologies furnaces. In each series, four crystallization processes were carried out at four different withdrawal rates: 2, 3, 4 and 5 mm/min. The heat treatment, performed by using several steps, consisted of convection heating to 950 °C (in Ar or He), radiation heating to 1350 °C (in vacuum), solution annealing and aging. The temperature-time settings for annealing were: 1277 °C/4 h → 1287 °C/2 h → 1296 °C/3 h → 1304 °C/3 h → 1313 °C/2 h → 1316 °C/5 h → gas furnace quench; for aging, the settings were: 1140 °C/6 h (step 1) and 871 °C/20 h (step 2).

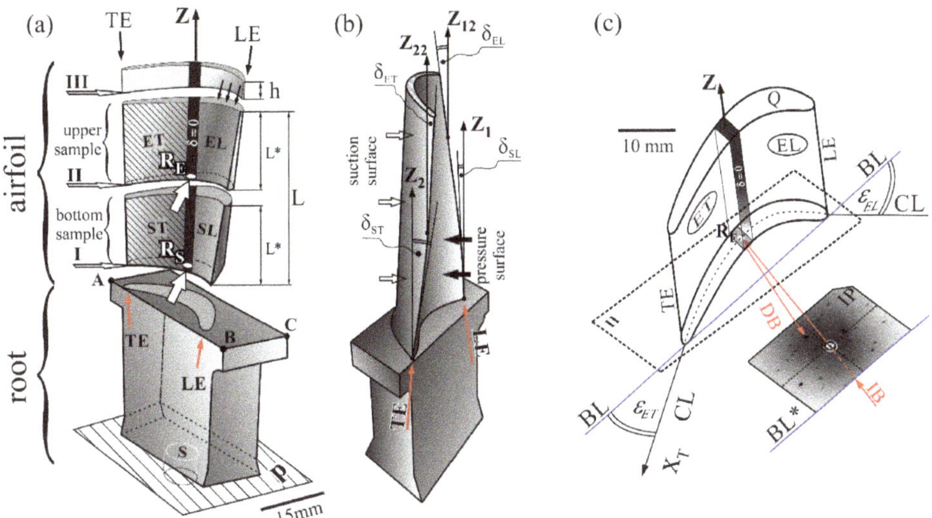

Figure 1. (a) Illustration of a turbine blade with a scheme of cross-sections I, II and III and the location of the ET, ST, EL and SL areas: (b) description of airfoil surface inclinations relative to the axis Z of a blade near the leading edge (LE) and trailing edge (TE); (c) a scheme for obtaining the Laue pattern for the upper sample of an airfoil with an exemplary Lauegram. CL—camber line, BL—base line of cross-section II of an airfoil, δ_{SL}, δ_{EL}, δ_{ST} and δ_{ET}—angles of inclination of LE and TE relative to the axis Z, ε_{EL} and ε_{ET}—angles of airfoil camber, describing the rotation of airfoil surfaces to the BL, IB—incident X-ray beam, DB—diffracted X-ray beam, IP—image plate in which the Laue pattern was obtained. Z is perpendicular to the base plane P; Z_1, Z_{12}, Z_2 and Z_{12} are parallel to the Z.

X-ray diffraction topography and Laue back-reflection diffraction were used to analyze the LABs and define the crystal orientation of airfoil samples. The Laue method is the basic method for defining the crystal orientation which many automatic programs and indexing systems are also based on [31]. The Panalytical X-ray system (Alamelo, The Netherlands) equipped with a microfocus tube (with a quasi-point source of $40 \times 40 \ \mu m^2$) emitting characteristic CuKα divergent beam radiation was used for topography studies. The anode current of 0.3 mA and an anode-cathode voltage of 30 kV were applied. The topograms of the 113 reflection, which is the reflection with the highest intensity for the analyzed sample surface (surfaces of the samples are parallel to the (001) crystal plane), were recorded on the AGFA Structurix D7 X-ray film with a grain size of 7 μm. The oscillations of the coupled sample and film were applied during exposure of the topograms. The source-to-sample distance was 25 mm and the sample-to-film distance was 10 mm. The details of the experiment are presented in Appendix A. The Laue patterns were obtained on the image plates using the X-ray diffractometer of the RIGAKU/EFG XRT-100CCM system provided by EFG Freiberg Instruments (Freiberg, Germany). The accuracy of the angle measurement in the Laue diffraction method was determined by using the spot size and the precision of the sample positioning in the diffractometer holder. The reference plane of the goniometer allowed us to set the sample in the holder with a mean error of about 0.3°. To determine the orientation measurement error related to the size and shape of the Laue spots, a circular envelope of each spot was outlined and the center of the envelope was found. The longest distance from the spot center to the envelope for the Laue pattern was the mean error of 0.5°.

There are several methods that can be used to visualize low-angle boundaries in single crystals. They mainly differ in the type of X-ray source, shape and width of the incident beam, spatial and angular resolution. Table 1 presents some parameters of these X-ray topography methods and their application for different materials. The methods allowing for higher limit resolution with the use of conventional X-ray sources, e.g., the Berg-Barrett or Lang methods, use a highly collimated narrow incident beam. There are two drawbacks to using such methods when examining large engineering elements: the area covered by the X-ray beam is relatively small and may not cover the entire sample; additionally, some areas of the sample with a higher misorientation angle (e.g., several arc degrees) may not be visible on the topogram due to a failure to meet the Bragg condition. This made it impossible to visualize all LABs that may be present in the single-crystalline casts made of the superalloys on the one topogram. Such casts, obtained by using the Bridgman method, consist of a set of almost parallel dendrites and their groups (called subgrains) with a fairly large dispersion of the misorientation angle, including from arc minutes to angular degrees [32]. The X-ray topography with a divergent width beam and oscillation of the coupled sample and film that was applied in this study seemed to meet all of the relevant requirements and was suitable for the visualization of all LABs in the relevant casts.

Table 1. The parameters of different X-ray topography methods and their applications.

The Method	X-ray Source	Incident Beam	Diffraction Geometry	Sample	Limit Resolution	Analyzed Crystals	
						Almost Perfect	Single-Crystalline Superalloys
Berg-Barrett	conventional	narrow parallel	reflection	small	arc seconds [33]	yes	yes
Lang	conventional	narrow parallel	transmission	small, thin	arc seconds [33]	yes	no
Applied in this study	conventional	wide divergent	reflection	large	arc minutes [34]	yes	yes
White beam	synchrotron	parallel	reflection/ transmission	wide range	arc seconds [35]	yes	yes

The rotor blades casts are divided into two main parts—bulk root and fine airfoil (Figure 1a). Three cross-sections I, II and III of the airfoils were made for each analyzed blade. The first section (I) was localized near the platform ABC of the root (Figure 1a). The third cross-section (III) was cut off of the airfoil's tip part with a height h = 3 mm, and the second cross-section (II) divided the remaining fragment of the airfoil with the height L into two parts named the bottom sample and the upper sample. The bottom and the upper samples had the same height $L^* = L/2$. The tip parts of the airfoils were not studied because they contained high internal stresses that made it impossible to create clear diffraction images. The airfoils of rotor blades are bounded by two surfaces, the suction and pressure surface, which are indicated in Figure 1b by white and black arrows, respectively. Both surfaces are twisted around blade' axis Z like a clockwise screw. The twist can be defined by continuous rotation of the chord line called the base line (BL) (Figure 1c) at each transverse section along the axis Z. In addition to this rotation, changes in the cambers of the airfoil may occur, which are related to changes in angles of the airfoil surfaces' rotation to the BL, which can be approximately described by the angles ε, marked for example in Figure 1c as ε_{ET} and ε_{EL} for the upper airfoil samples. The angles ε may be defined as the angles of rotation of camber line fragments located near the leading edge (LE) and the trailing edge (TE).

The values of the angles ε_{EL} and ε_{ET} are different for the cross-sections I and II because the rotation of airfoil surfaces to the BL change continuously along the axis Z (Figure 1a). The suction and pressure surfaces are inclined towards the axis Z. These inclinations lead in turn to inclinations of the LE and TE to the Z axis (Figure 1b). The inclination angles δ of the LE and TE are different to those of the angles ε, and the characteristics of the δ angles' changes along the Z axis are also different. The angle δ for the LE increases along the axis Z ($\delta_{EL} > \delta_{SL}$, Figure 1b) and decreases for the TE ($\delta_{ET} < \delta_{ST}$, Figure 1b). There is a narrow area in the central part of the airfoil, marked in Figure 1a,c in black, for which both the suction and pressure surfaces are parallel to the axis Z. The inclination angle of the area is $\delta = 0$ along the entire height of the airfoil from its beginning near the root to its tip. The geometric dimensions, including the length and thickness of the airfoil and the angles of its edges, were obtained using a 3D scanner. The measurement accuracy was 0.2° for the angles and 0.1 mm for the length dimensions.

In order to experimentally determine the number of LABs, X-ray diffraction topograms were obtained from the upper sample surfaces marked in Figure 1a by thin black downward arrows. To obtain the topograms, the samples were oriented using an additional Laue back-reflection diffraction from the surface Q of section III (Figure 1a,c). The macroscopic LABs are planar defects and their surfaces in airfoils are approximately parallel to the axis Z of the blade [28]. Therefore, it can be assumed that the LABs created during crystallization in the lower part of the airfoil with any distance from the platform ABC would be extended along the axis Z—the axis which indicates the direction of the crystallization process—and would pass through the upper parts of the airfoil. Therefore, in cross-section III near the tip of the airfoil, all of the low-angle boundaries created during crystallization of the airfoil would appear.

3. Results and Discussion

Figure 2 shows X-ray topograms obtained for airfoils of two series of blades obtained at the withdrawal rates of 2 mm/min, 3 mm/min, 4 mm/min and 5 mm/min each. The presented topograms were obtained from the surface Q of the upper sample located near the tip of the airfoil, where the crystallization of the blades was coming to an end. The topograms obtained from the cross-section III contain images of all of the LABs created during the passage of the crystallization front along the axis Z across the entire length of the blade airfoil, as well as images of LABs that were inherited by the airfoil from the root [28]. The topograms consist of the groups of contrast stripes or/and spots marked for example as SG1, SG2, SG3 and SG4 in Figure 2a. The individual contrast spots and stripes visualize the separate dendrites while the groups indicate the individual subgrains (which are SG type subgrains—see Figure 2 for further details). The mutual shift of the images of neighboring dendrites or the images of the neighboring subgrains allows for determination of the angle of their mutual misorientation, which also describes the disorientation angle of the LABs. The mutual misorientation angle calculation method is described in Ref. [34]. The misorientation angle between the single dendrites was found to be up to 0.3°, and it was higher between groups of dendrites. This criterion allowed us to identify the subgrain in nickel-based dendritic superalloys.

There were visible bright bands with a low-contrast (or a lack of contrast) between the images of subgrains. The bands represent LABs between subgrains. The areas of adjacent subgrains in the topograms were spaced and/or shifted relative to each other. Sometimes the low-angle boundaries were visualized in the topograms by using increased contrast, as presented in Figure 2h—LAB no.3. The reason for the above is that the crystal lattices (diffraction planes) of certain neighboring subgrains were inclined toward each other in such a way that their images in the topograms partially overlapped.

The LABs with misorientation angles of above 0.3° are marked by the arrows in Figure 2 and numbered, which allows determination of their number N. The error of the misorientation angle measurement depended on many factors related to the material of the sample and the instrument error. The mean orientation error for the presented results was about 8 arc minutes. The selection of the LAB in the topograms was based on the criteria described in detail later and related to the mechanism of LABs creation.

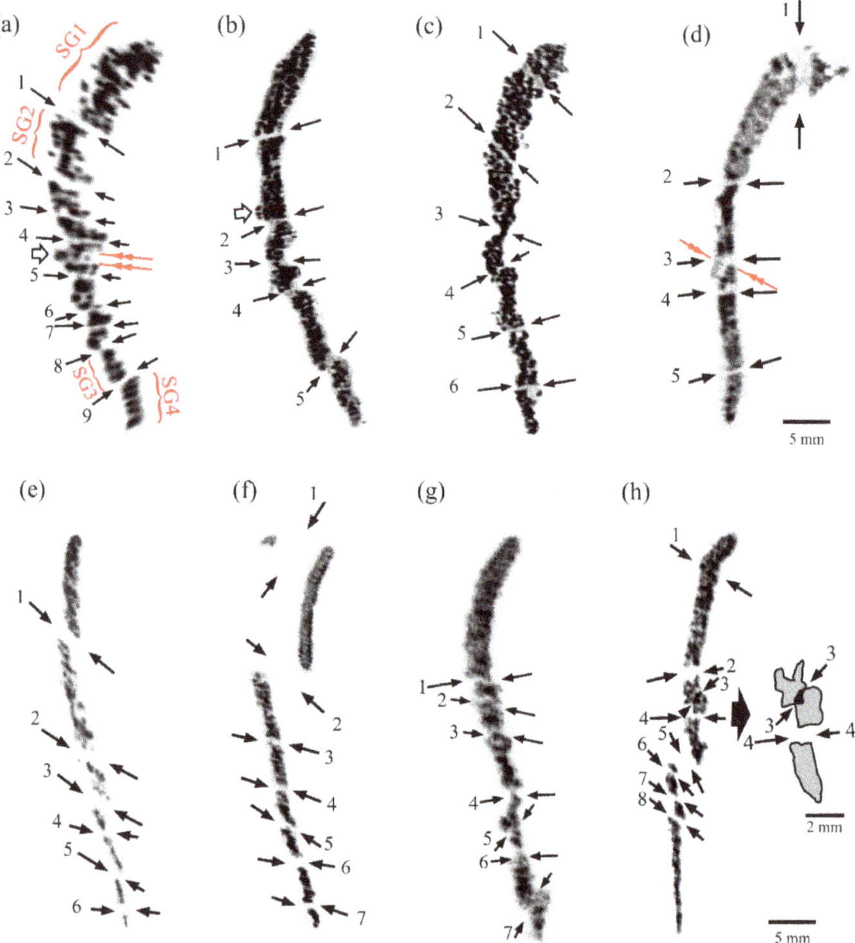

Figure 2. X-ray topograms obtained from micro-section III of an airfoil of I-series (**a–d**) and II-series (**e–h**) obtained at a withdrawal rate of 2 mm/min (**a,e**), 3 mm/min (**b,f**), 4 mm/min (**c,g**) and 5 mm/min (**d,h**). Reflection 113. CuKα radiation.

The boundaries creation may be related to the mechanism of the growing dendrites deflection from the surface of the casting mold, which was first proposed in Ref. [28]. This mechanism was based on stopping the primary arm growth on the mold wall (at the point R, see Figure 3a) and the continuation of crystallization by the secondary arms, and then (at the point T, see Figure 3a) by the tertiary dendrite arms. Because the secondary dendrite arms were arranged in the arrays connecting the sucking and pressure surfaces of the airfoil, their image (and also the image of the LABs) was visible along the entire width of the topogram. Such a growth path occurs when the angle δ between the primary arms and the mold wall is higher than the critical one [30]. At a very low δ angle (Figure 3a) below the arc minute, the dendrites do not deflect on the mold wall—in this case, the dendrites bend and continue to grow parallel to the TE axis, while LABs are not formed either during the dendrite growth from the melt or after the heat treatment. When the angle δ is higher but remains lower than the critical level, then the primary dendrite arm may bend before deflection, which leads to creation of an area of internal stress where extra LABs can also be formed after heat treatment [36]. The surfaces of these LABs are approximately parallel to the blade axis Z [28,30]. Therefore, in the airfoil cross-section III (Figure 1a), all LABs—which formed as a result of deflections that occurred along the entire airfoil height L (Figure 1a)—appeared. It follows that to theoretically calculate the number N of the low-angle boundaries visualized on the cross-section III, it was necessary to calculate the number of acts of dendrites deflection from the surface of the mold walls.

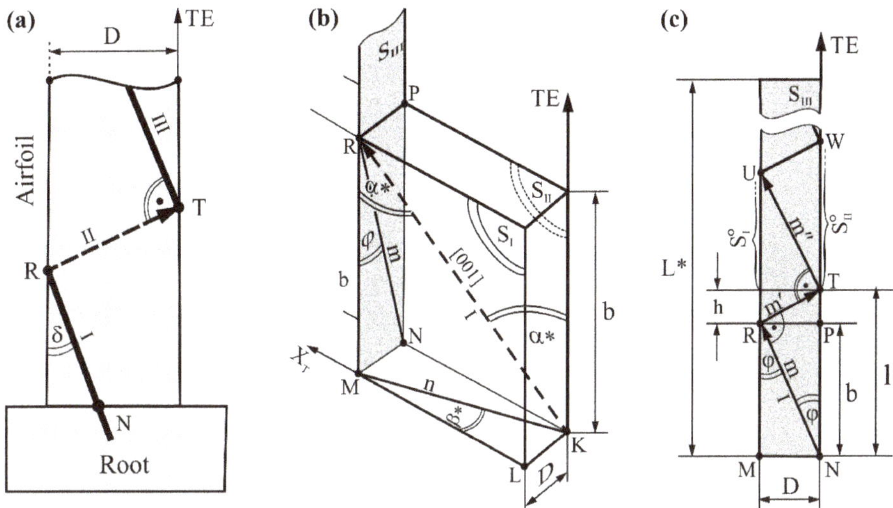

Figure 3. Schemes of dendrite deflection from the mold walls' surfaces parallel to the TE in the ET fragment (**a**) of the blades airfoil. The δ, φ, α^* and β^* angles are enlarged for figure clarity. Section m in (**b,c**) is the projection of the primary arm direction parallel to the [001] on the plane defined by the points M, N, P and R which is perpendicular to the axis X_T of the CL (**c**), and sections m' and m" in Figure 3c are the projections of secondary and tertiary arms deflected from the fragments $S°_I$ and $S°_{II}$ of mold walls planes of the ET area.

The number of deflections is particularly important for fine airfoils. Although the angle δ is usually low (less than 12–15°) because the airfoil dimension D is also low (Figure 3a), the number of deflections is fairly high as a result. The mechanisms for LABs creation in the bulk root connected with the selector provide different results. The main reason for this is a fast, unsteady lateral growth of dendrites near the selector-root connection surface [37]. Usually, two LABs are formed in the root, which are inherited by the airfoil [28]. These macroscopic LABs created in the root are tens of millimeters long. The LABs that are inherited by the airfoil pass through its entire cross-section. Taking the aforementioned mechanisms into account, the criteria used to indicate the LABs image in the topograms can be described as follows: the decreased or increased contrast bands representing the LABs must cross the entire width of the topogram; these contrast bands must be formed between the other contrast bands representing the groups of the dendrites; and the contrast bands representing the dendrite group (subgrains) must be shifted integrally relative to the adjacent bands representing the other group. Typically, the shift of the contrast bands representing subgrains is higher than the shift between the contrast spots and/or the stripes representing the individual dendrite, i.e., a shift greater than 0.3°. The error in determining the misorientation angle is very important when calculating the number of LABs. The LABs may be created as a result of deflection; therefore, the internal stresses may appear, causing blurriness or/and a bend of the contrast line in the topograms (e.g., SG3, Figure 2c). A large misorientation of the single dendrite can also occur, which is visualized in the topogram by a greater shift between the contrast bands visualizing the neighboring dendrites (above LAB no.6 in Figure 2h). As a result of these phenomena, incorrect LAB identification may occur, resulting in inaccurate counting.

The resolution of the method for determining the number of LABs is related to the resolution of the X-ray topography, but it is not crucial in this case. The aim of the experiment was to determine the number of macroscopic LABs formed as a result of the dendrites deflection mechanism using recorded topograms. Theoretically, for almost perfect single-crystals, the resolution of the applied X-ray topography method is several arcmin [34]. However, for the dendritic single-crystalline nickel-based superalloys of the CMSX-4 type, the resolution ranges from a dozen to several dozen arcmin. The outcome depends on the existing internal stresses, the X-ray background level and the arrangement of the diffraction planes in relation to the analyzed surface. For the CMSX-4 containing several alloying elements, it also depends on the heterogeneity of the spatial distribution of these elements. Although the resolution of the applied method is lower than that of the Berg-Barrett or Lang method, it is still useful for visualizing all sample areas with high misorientation angle ranges from a dozen arcmin to several angular degrees on the one topogram [32]. Additionally, the divergence of the X-ray beam allowed us to obtain the topograms from large sample surfaces—even up to approximately 10 cm^2. This is very important for testing engineering products such as turbine blades.

The linear resolution of the applied method was in the order of 100 μm for almost perfect single-crystals. The thickness of the dendrites in the analyzed samples of superalloys obtained by using the Bridgman technique with the withdrawal rate from the high temperature zone of 2–5 mm/min ranged from 300 μm to 100 μm [1], so in this case the images of all dendrites could be seen on the topograms. As the images of the dendrites were visible in the topograms (Figure 2) in the form of stripes or spots, the linear resolution limit ranged from 100–300 μm. In the presented research, it was necessary to visualize all possible existing LABs with a misorientation angle ranging from several arcmins to several degrees of arc. It was not necessary to increase the resolution to arcsec, which could have been achieved by using the Berg-Barrett method. The scheme of the primary dendrite arm arrangement, for example in the fragment ET with the thickness D and height L* (Figure 3b,c), may be used to calculate the number of dendrite deflections. It can be correctly assumed that a dendrite grows directly toward the direction [001], as is commonly believed to occur [38].

S_I and S_{II} in Figure 3b are the side walls of the blade airfoil (and the casting mold) parallel to the TE (Figure 1c). The direction [001] is the direction of the primary dendrite arm that reaches the point R and then deflects. The unit vector KR indicates the direction of the dendrite growth. Then the dendrite growth was continued by the secondary arm up to the point T (Figure 3c). The tertiary arms growth began at the point T and reached the point U where subsequent deflection occurred. In order to simplify considerations, it was assumed that the angle between the subsequent rows of dendrite arms is a right angle (Figure 3c). Since the distance D is small, and the secondary dendrite arm is almost perpendicular to the mold walls, the distance h (Figure 3b) was small in comparison to b, therefore it could be assumed that l ≈ b. Using the above assumptions, the Equation (1) that allowed us to calculate the number of deflections N was obtained (see Appendix B):

$$N = \frac{L^*}{D} \times \tan \alpha^* \times \sin \beta^* \tag{1}$$

where α^* is the angle between the direction [001] and the TE, and β^* is the angle between the projection of the direction [001] on a certain plane (the plane is perpendicular to TE) and the axis X_T (the axis is parallel to the fragment CL of the airfoil area ET) (Figure 1c). The angles could be determined using the Laue patterns of the airfoil cross-sections. The Laue patterns were obtained by arranging the baseline BL* of the image plate parallel to the baseline BL of the airfoil cross-section (Figure 1c). The primary X-ray beam was directed at the airfoil region with δ = 0. The Laue patterns were obtained from the points R_S and R_E (Figure 1a) of the bottom and the upper samples of each blade airfoil, respectively.

An airfoil can be divided into five areas: one area with the angle δ = 0, the two areas SL and EL (Figure 1a) (their inclination to the axis Z can be referred to as the inclination of the LE relative to Z) as well as the two areas ST and ET (Figure 1a) (their inclination can be referred to as the inclination of the TE). The area with $\delta = 0$ is the middle area and was not considered because it is relatively narrow (Figure 1a). The areas SL and ST of the bottom sample, as well as the areas EL and ET of the upper sample, are rotated around the BL of the bottom and the upper samples by the angles ε and are inclined to the blade axis Z by their respective angles δ.

Figure 4 shows a model fragmentation and arrangement of the four airfoil areas with their geometrical parameters. The areas were modeled as plates with parallel surfaces of the average thickness D and height L*. The values of the angles δ and ε, as well as the value of L* = L/2 (Figure 1a) for concerned areas are presented in Table 2.

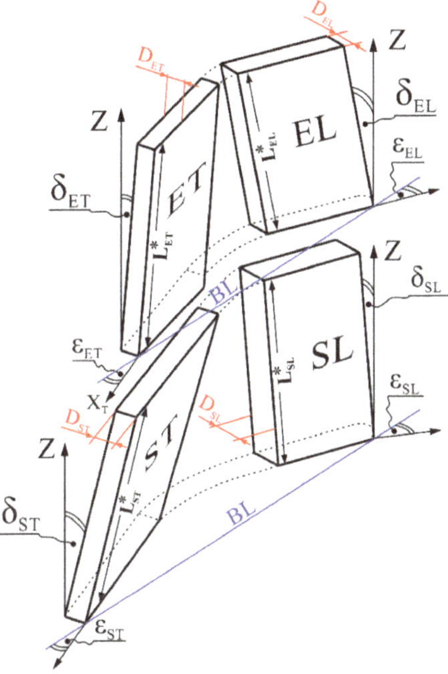

Figure 4. Model scheme of the areas SL and ST of the bottom sample, as well as the areas EL and ET of the upper sample of the blade airfoil with the characteristic parameters of the blades geometry.

Table 2. Characteristic geometric parameters of the SL, ST, EL and ET airfoil areas.

Airfoil Area	L* (mm)	D (mm)	L*/D	δ (°)	ε (°)
SL	15.0	3.7	4.05	7.5	45.0
ST	15.0	1.8	8.33	4.5	48.0
EL	15.0	2.0	7.50	11.0	20.0
ET	15.0	1.2	12.5	1.5	40.0

To determine the number of deflections N of each of the areas SL ST, EL and ET using the Equation (1), it was necessary to experimentally determine the angles α^* and β^* of these areas.

Figure 5 shows a scheme of the areas EL and ET of the upper sample, as well as arrangement of the image plate in which the Laue back-reflection pattern was recorded. During the experiment, both the sample surface and the image plate were positioned vertically, and the X-ray beam was positioned horizontally. The diffracted beam (Figure 5b) was directed in the opposite direction to the dendrites growth direction. The crystallographic orientation—which determines the direction of the growing dendrite—is represented by the spot r_d on the real lauegram. The spot is obtained by using inversion of the 001 reflection (r spot) relative to the center of the lauegram. The position of the Laue spot r is determined by using the vector $\vec{n}^*$, which proceeds in a normal direction towards the diffraction plane (001), and its direction corresponds to the diffracted beam. The direction of the vector $\vec{n}_d$, which also proceeds in a normal direction towards the diffraction plane (001), corresponds to the direction of the growing dendrites. The extension of the diffracted beam up to the intersection with the created virtual lauegram (Figure 5b) determines the location of the spot r_d. To determine the location of the r_d spot, a virtual lauegram was drawn on the other side of the sample where a hypothetical back-reflected Laue pattern could be created. The virtual lauegram was created with the assumption that the incident beam was directed from the top to the point R_E. In this case, the diffracted beam passes through the virtual lauegram in the spot r_d. This spot can be transferred to the real lauegram, which is parallel to the incident beam. As a result of the transfer of the spot r_d from the virtual lauegram to the real lauegram, an extra spot is created. The position of the spot r_d was determined by inversion of the position of the spot r relative to the center of the lauegram.

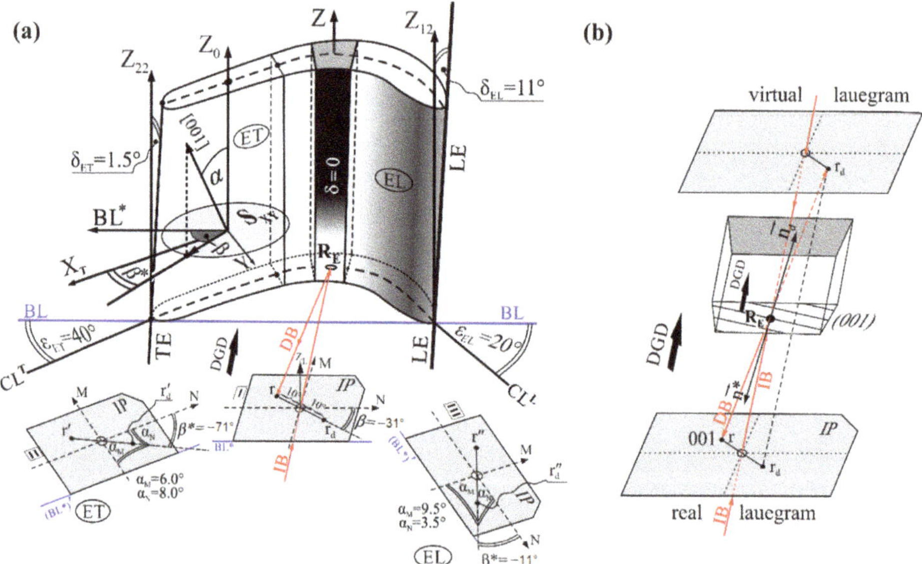

Figure 5. Exemplary illustration of the areas EL and ET of the upper sample of an airfoil obtained at the withdrawal rate of 5 mm/min, with geometric description of angles α, β and β^* (**a**) and a diagram explaining how to determine the position of the spot r_d on the lauegram through inversion of the spot r (**b**). The plane S_{XY} is parallel to the plane p and cross-sections I, II and III in Figure 1a, while the BL* of the lauegram is parallel to the BL of an airfoil cross-section. IP—image plate, IB and DB—incident and diffracted X-ray beam, DGD—dendrite growth direction. $\vec{n}^*$ and $\vec{n}_d$—unit vectors which proceed in a normal direction towards the diffraction plane (001) with the direction corresponding to the diffracted beam and the dendrite growth direction. Z_0, Z_{12} and Z_{22} are parallel to the Z axis of the blade.

The angles ε_{EL} and ε_{ET} between the CL^L and BL, as well as the CL^T and BL, respectively for the areas EL and ET, were determined on the basis of the shape of the micro-section II (Figure 1a). The ε_{EL} was 20° and the ε_{ET} was 40°. Afterwards, using the QLaue software, the Laue patterns recorded from the point R_E were rotated around the axis Z_L by using the angles ε_{ET} and ε_{EL}, until the BL* of the lauegram lined up parallel to the CL^T and CL^L (lauegram II and III, Figure 5a). The reflex r (Laue pattern I), which was obtained from the (001)-type diffraction planes, changed location to r' and r'' on the rotated Laue patterns II and III, respectively. However, the transformed r_d spot was taken into account in further analysis because it is related to the direction of growing dendrites. As an example, the area ET of the upper sample was considered. In the cubic system, the directions of [001]-type are perpendicular to the crystal planes of (001)-type, which allowed us to define the inclination of the crystallographic direction [001] and the dendrites growth direction to the pressure and suction surfaces, meaning in relation to the walls of the mold. The primary arms growth direction may be defined by the position of the spot r_d' on the rotated lauegram II. Based on the rotated Laue pattern II and using the QLaue software, the rotation angles components α_M and α_N of the primary arms growth direction relative to the axes M and N, which are respectively perpendicular and parallel to the CL^T (X^T), may be determined (Figure 5). When considering the blades obtained at a withdrawal rate of 5 mm/min, the angles were found to be $\alpha_M = 6.0°$ and $\alpha_N = 8.0°$ for the exemplary area ET (Figure 5a). It is important to note that for the angle δ_{ET}, the component δ_{ET} of the rotation around the axis N of the primary arms growth may be calculated by using the equation $\gamma_{ET} = \alpha_N + \delta_{ET}$ and in a more general case by the equation $\gamma_{ET} = \alpha_N \pm \delta_{ET}$. Additionally, the Laue pattern II allowed us to determine the angle β^* using the equation $\beta^* = \beta - \varepsilon_{ET}$ ($\beta^* = -31 - 40 = -71°$). The angle γ_{EL} may be determined similarly, using the rotated Laue pattern III.

As an example, for the upper sample, the angles of the surfaces inclination relative to the axis Z and rotation around the axis Z relative to the BL are as follows: for the area ET — $\delta_{ET} = 1.5°$, $\varepsilon_{ET} = 40.0°$, for the area EL — $\delta_{EL} = 11.0°$, $\varepsilon_{EL} = 20.0°$ (see Table 2).

Because the angle β^* can take negative values (Laue pattern II, Figure 5), the absolute value of $\sin \beta^*$ must be used in Equation (1). Therefore, the following Equation (2) should be used for the calculation of N:

$$N = \frac{L^*}{D} \tan(\alpha^*) \times |\sin(\beta^*)| \tag{2}$$

Taking into consideration the components α_M and α_N of the area ET and the fact that $(\alpha^*) = \sqrt{\alpha_M^2 + \gamma_{ET}^2} = \sqrt{\alpha_M^2 + (\alpha_N \pm \delta_{ET})^2}$, Equation (2) can be denoted into the following Equation (3):

$$N = \frac{L^*}{D} \times \tan\left[\sqrt{\alpha_M^2 + (\alpha_N \pm \delta_{ET})^2}\right] \times |\sin(\beta \pm \varepsilon_{ET})| \tag{3}$$

The choice of the $+/-$ sign was made on the basis of the localization of the image plate quadrant in which the spot r_d' is present.

As an example, calculations of the number of deflections N_{ET} for the area ET are presented below. In this case, there is the sign "$-$" in front of the δ_{ET} because the inclination $\delta_{ET} = 1.5°$ of the ET was consistent with the inclination of the dendrite relative to the axis N. The sign "$-$" also appears in front of the ε_{ET} because the rotation of the image plate was counterclockwise. Given the above, the Equation (3) takes the Equation (4).

$$N_{ET} = \frac{L^*_{ET}}{D_{ET}} \times \tan\left[\sqrt{\alpha_M^2 + (\alpha_N - \delta_{ET})^2}\right] \times |\sin(\beta - \varepsilon_{ET})| \tag{4}$$

Putting these values into Equation (4), we obtained:

$$N_{ET} = 12.5 \times \tan\left[\sqrt{6^2 + (8 - 1.5)^2}\right] \times \left|\sin\left(\overbrace{-31° - 40°}^{-71°}\right)\right| = 1.92 \approx 2 \tag{5}$$

Similar calculations can be made for the area EL of the upper sample. In this case, there is the sign "$+$" in front of the δ_{EL} because the inclination $\delta_{EL} = 11°$ (Figure 5a) of the leading edge and the surfaces' EL was opposite to the inclination of the component α_N of the dendrite. The sign "+" also appears in front of the ε_{EL} because the rotation of the image plate was clockwise. Given the above, the equation for the N_{EL} takes the Equation (5).

$$N_{EL} = \frac{L^{*}_{EL}}{D_{EL}} \times \tan\left[\sqrt{\alpha_M^2 + (\alpha_N + \delta_{EL})^2}\right] \times |\sin(\beta + \varepsilon_{EL})| \tag{6}$$

Putting these values into Equation (5), the N_{EL} was obtained as follows:

$$N_{EL} = 7.5 \times \tan\left[\sqrt{9.5^2 + (3.5 + 11)^2}\right] \times \left|\sin\left(\overbrace{-31° + 20°}^{-11}\right)\right| = 0.41 \approx 0 \tag{7}$$

The sinus and the tangent functions can take real numbers, therefore the calculated values were rounded to integers. The number N of the low-angle boundaries, which can be determined experimentally from the topograms obtained from the airfoil cross-section III (Figure 1a), is equal to the sum of the numbers of the low-angle boundaries created by deflection in the whole airfoil:

$$N = N_{ST} + N_{ET} + N_{SL} + N_{EL} \tag{8}$$

In addition, two LABs from the root are usually inherited to the airfoil [28], which results in the equation:

$$N = N_{ST} + N_{ET} + N_{SL} + N_{EL} + 2 \tag{9}$$

To verify the aforementioned mechanism of deflection, the data obtained experimentally in topograms were compared to the data calculated using Equation (7). Table 3 shows the experimental results and the calculations results of the number of LABs in the cross-section III for two series of blade casts obtained by using the Bridgman technique at the withdrawal rate of 2, 3, 4 and 5 mm/min and then heat treated.

Table 3. The number of LABs defined by calculations and from X-ray topograms for airfoils of blades obtained in two series at the withdrawal rates of 2, 3, 4 and 5 mm/min.

	W (mm/min)	Number of Subgrains								Sum	Total (+2 from the Root)	Identified in Topograms
		ET		ST		EL		SL				
Series I	2	2.34	2	1.23	1	2.90	3	0.81	1	7	9	9
	3	1.20	1	0.16	0	2.21	2	1.12	1	4	6	5
	4	0.37	0	0.63	1	1.12	1	0.71	1	3	5	6
	5	1.92	2	1.22	1	0.41	0	0.03	0	3	5	5
Series II	2	1.47	2	1.7	2	0.18	0	0.20	0	3	6	6
	3	1.7	2	1.1	1	1.04	1	0.38	0	4	6	7
	4	2.06	2	1.36	1	1.97	2	0.38	0	5	7	7
	5	2.3	2	0.71	1	2.40	2	0.75	1	6	8	8

The analysis of the data presented in Table 3 shows that there is good compatibility of the model and the experimental data. The differences between them are of the order of one LAB. However, for the airfoil of the series I of the blades cast at the withdrawal rate of 2 mm/min, the obtained topograms did not allow us to accurately determine the number of LABs. In the case of the subgrains marked as SG1, SG2, SG3 and SG4, one can state that the diffraction images of the dendrites (in the form of strips and spots) clearly formed groups with the same common crystallographic orientation. The other areas of the topograms were strongly and irregularly fragmented, therefore localization of subgrains and determination of the number of LABs was difficult. As a result, we could identify two more LABs (Table 3) in those areas, which are marked by two double red arrows in Figure 2a. The additional subgrain marked by the arrow with the envelope could be visualized in the topogram by using three contrast spots. In Figure 2b, all 5 marked LABs are clearly visible due to the lack of contrast bands that cross the entire width of the topogram. In Figure 2c, only the contrast band visualizing LAB no. 3 may be related to the macro-stress rather than to the misoriented subgrains; therefore, the number of boundaries counted may be one less than it otherwise would be. Analysis of the topogram presented in Figure 2d allowed us to indicate one additional LAB (marked by the red arrow). In the case of the topograms shown in Figure 2e–g, the determination of the LAB number was quite unambiguous. In the Figure 2h only the contrast band visualizing LAB no.6 could be questionable, therefore the counted number of boundaries may be one less than it otherwise would be because the contrast spot above LAB no.6 may belong to the group of dendrites located below LAB no.6. This is likely because the misorientation angle of LAB no.6 was close to $0.3°$.

In the present study, the mechanism of dendrites deflection from the mold walls was verified regarding the heat-treated turbine rotor blades. The number of macroscopic LABs created on the cross-section of the blades airfoil near the tip was experimentally determined and compared to the LABs number calculated from the model based on the dendrites deflection mechanism.

Based on the Laue patterns and geometrical parameters of the airfoils, the number of the low-angle boundaries occurring at the upper part of the blades airfoil after heat treatment could be calculated. The number for the analyzed group of blades ranged from 5 to 9 (Table 3).

Due to the complex shape of the analyzed blades airfoil, which is similar in shape to the applied rotor blades, the aforementioned geometric assumptions based on the division into four flat areas were significantly simplified, but it is possible to verify this model. All the low-angle boundaries, which were formed during crystallization and inherited by the airfoil, appear near the blade tip. In the proposed model, to calculate the number of LABs in blades airfoils, it is necessary to experimentally determine the α^* and β^* angles used in Equation (1) and the values of the angles δ and ε describing the airfoil geometry. The values α^* and β^* can be determined in a non-destructive way by obtaining two Laue patterns directly from the pressure or suction surface, using, for example, the diffractometric method called Ω-scan and described in Ref. [39]. Determination of the number of LABs for blades airfoil allows for a non-destructive quality control process during blades production.

4. Conclusions

The dendrites deflection mechanism effectively describes the process of the low-angle boundaries creation in the airfoil of heat-treated rotor blades and allowed us to calculate the LABs number. It also confirms the proposed "deflection" mechanisms of the LABs creation in other thin-walled parts of single-crystalline casts. The dendrites deflection mechanism was verified for blades airfoils obtained by using directional crystallization in the direction [001] of the CMSX-4 superalloy at the withdrawal rates of 2 to 5 mm/min. The simplification of the geometric assumptions in the model could limit its use. However, the model can be the basis for more precise calculations. To increase the accuracy of the model, the complex shape of the blades should be taken into account, for example by using the finite element method. The proposed model is the basis for a non-destructive technique for determining the number of LABs in the blades airfoils, based only on two Laue patterns. Application of the method, without the need for cutting the blades and preparing metallographic sections, allows it to be used for quality control on the production lines.

Author Contributions: Conceptualization: W.B. and J.K.; methodology: J.K.; software: J.K.; investigation: J.K.; data curation: W.B. and J.K.; writing—original draft preparation: W.B. and J.K.; writing—review and editing: J.K.; visualization: J.K.; supervision: J.S.; project administration: J.S. All authors have read and agreed to the published version of the manuscript.

Funding: This research received no external funding.

Data Availability Statement: Data sharing is not applicable to this article.

Conflicts of Interest: The authors declare no conflict of interest.

Appendix A

The X-ray diffraction topography is based on the following principles. First, the upper sample is oriented using the Laue diffraction pattern obtained from its surface Q (marked by thin black downward arrows in Figure 1a). The sample is mounted in the holder so that the zone line defined by the reflections from planes (001) and (113) is arranged horizontally with the simultaneous vertical arrangement of the oscillation axis T of the sample and X-ray film (Figure A1). The sample is mounted in the center of the horizontal divergent incident beam (S1) and inclined at the Bragg angle $\theta^{113}_{CuK\alpha} = 31°$ to the plane (113). The α, which is the angle between the sample surface and the plane (113), is determined from the Laue pattern. During exposure, the sample is oscillated about the T axis around the $\theta^{113}_{CuK\alpha}$ angle within $\pm 4°$. The X-ray film is arranged parallel to the Q sample plane and to the T axis.

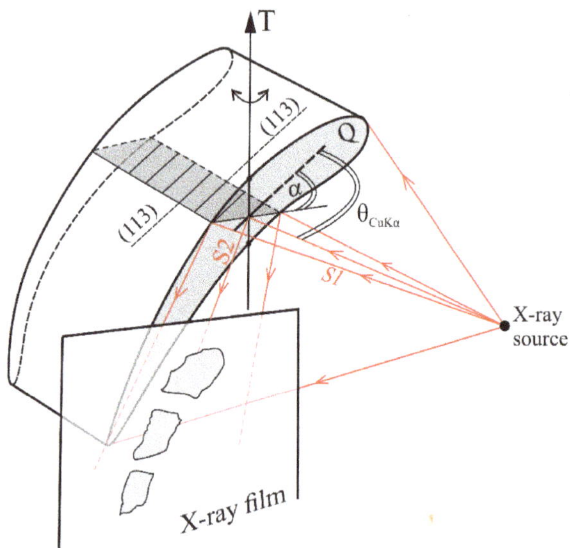

Figure A1. Diagram presenting the geometric arrangement of the X-ray source, sample and X-ray film for the X-ray topography method with the oscillation. S1—Incident divergent beam, S2—diffracted beam.

Appendix B

The number N of dendrite deflections can be calculated for the area ET of the upper sample in the way presented below. Let's calculate the number of dendrite arms deflections from the mold surface, based on the schemes from Figure 3c.

From Figure 3c and the assumption that h = 0 and b $\approx$ l, it follows that

$$N = \frac{L^*}{l} \tag{A1}$$

where L* is the height of the area ET.

From the triangle ΔNRT (Figure 3c) it follows that $l = \frac{m}{\cos \varphi}$, and from the triangle ΔNRM it follows that $m = \frac{D}{\sin \varphi}$.

The comparison of the above equations gives

$$l = \frac{D}{\sin \varphi \times \cos \varphi} \tag{A2}$$

By putting Equation (A2) into Equation (A1), $N = \frac{L^*}{D} \sin \varphi \times \cos \varphi$ is obtained, and the next the Equation (A3) is created

$$N = \frac{L^*}{D} \sin \varphi \sqrt{1 - \sin^2 \varphi} \tag{A3}$$

Considering the triangle ΔKRM in Figure 3b it can be determined that $\tan(\alpha^*) = \frac{n}{b}$ and considering the triangle ΔKLM it can be determined that $\sin(\beta^*) = \frac{D}{n}$.

The comparison of the above equations gives that the $\tan(\alpha^*) \times \sin(\beta^*) = \frac{D}{b}$. From Figure 3b,c it follows that $\frac{D}{b} = \tan \varphi$, so:

$$\tan(\alpha^*) \times \sin(\beta^*) = \tan \varphi \tag{A4}$$

Taking into account that $\tan \varphi = \frac{\sin \varphi}{\cos \varphi}$ and putting it into the Equation (A4) it follows that

$$\tan \varphi = \frac{\sin \varphi}{\sqrt{1 - \sin^2 \varphi}} = \tan \alpha^* \times \sin \beta^* \tag{A5}$$

Treating the right side of the Equation (A5) as comparable to $\sin^2 \varphi$, the equations (A6) are obtained as follow:

$$\sin^2 \varphi = \frac{\tan^2(\alpha^*) \times \sin^2(\beta^*)}{1 + \tan^2(\alpha^*) \times \sin^2(\beta^*)}, \ \sin \varphi = \frac{\tan(\alpha^*) \times \sin(\beta^*)}{\sqrt{1 - \tan^2(\alpha^*) \times \sin^2(\beta^*)}} \tag{A6}$$

Putting the $\sin \varphi$ and $\sin^2 \varphi$ from Equation (A6) into Equation (A3), after simple transformation, Equation (A7) is obtained as follow:

$$N = \frac{L^*}{D} \times \frac{\tan \alpha^* \times \sin \beta^*}{\left(1 + \tan^2 \alpha^* \sin^2 \beta^*\right)} \tag{A7}$$

Due to the low value of $\alpha^* - \tan^2(\alpha^*) \times \sin^2(\beta^*) \ll 1$.

Considering the above, the simplified equation, from Equation (A7), is given:

$$N = \frac{L^*}{D} \times \tan(\alpha^*) \times \sin(\beta^*) \tag{A8}$$

References

1. Reed, R.C. *The Superalloys Fundamentals and Applications*; Cambridge University Press: Cambridge, UK, 2006.
2. Donachie, M.J.; Donachie, S.J. *Superalloys—A Technical Guide*, 2nd ed.; ASM International: Geauga, OH, USA, 2002.
3. Long, H.; Mao, S.; Liu, Y.; Zhang, Z.; Han, X. Microstructural and compositional design of Ni-based single crystalline superalloys—A review. *J. Alloys Compd.* **2018**, *743*, 203–220. [CrossRef]
4. Williams, J.C.; Starke, J.E.A. Progress in structural materials for aerospace systems. *Acta Mater.* **2003**, *51*, 5775–5799. [CrossRef]
5. Muktinutalapati, N.R. Materials for gas turbines—An overview. In *Advances in Gas Turbine Technology*; Benini, E., Ed.; InTech Open: Rijeka, Croatia, 2011; pp. 293–314.
6. Hong, J.; Ma, D.; Wang, J.; Wang, F.; Dong, A.; Sun, B.; Bührig-Polaczek, A. Geometrical effect of freckle formation on directionally solidified superalloy CM247 LC components. *J. Alloys Compd.* **2015**, *648*, 1076–1082. [CrossRef]
7. Zou, Z.; Wang, S.; Liu, H.; Zhang, W. *Axial Turbine Aerodynamics for Aero-Engines*; Springer: Singapore, 2018.
8. Szybicki, D.; Burghardt, A.; Kurc, K.; Pietruś, P. Calibration and verification of an original module measuring turbojet engine blades geometric parameters. *Arch. Mech. Eng.* **2019**, *66*, 97–109. [CrossRef]
9. Wang, X.; Zou, Z. Uncertainty analysis of impact of geometric variations on turbine blade performance. *Energy* **2019**, *176*, 67–80. [CrossRef]
10. Wang, F.; Ma, D.; Zhang, J.; Bogner, S.; Bührig-Polaczek, A. Solidification behavior of a Ni-based single crystal CMSX-4 superalloy solidified by downward directional solidification process. *Mater. Charact.* **2015**, *101*, 20–25. [CrossRef]
11. Ma, D.; Wang, F.; Wu, Q.; Bogner, S.; Bührig-Polaczek, A. Innovations in casting techniques for single crystal turbine blades of superalloys. In Proceedings of the 13th International Symposium on Superalloys, Pittsburgh, PA, USA, 11–15 September 2016; pp. 237–246.
12. Ma, D. Novel casting processes for single-crystal turbine blades of superalloys. *Front. Mech. Eng.* **2018**, *13*, 3–16. [CrossRef]

13. Strickland, J.; Nenchev, B.; Dong, H. On directional dendritic growth and primary spacing—A review. *Crystals* **2020**, *10*, 627. [CrossRef]

14. Wang, W.; Lee, P.D.; McLean, M. A model of solidification microstructures in nickel-based superalloys: Predicting primary dendrite spacing selection. *Acta Mater.* **2003**, *51*, 2971–2987. [CrossRef]

15. Van Sluytman, J.S.; Pollock, T.M. Optimal precipitate shapes in nickel-base γ/γ' alloys. *Acta Mater.* **2012**, *60*, 1771–1783. [CrossRef]

16. Singh, A.R.P.; Nag, S.; Chattopadhyay, S.; Ren, Y.; Tiley, J.; Viswanathan, G.B.; Fraser, H.L.; Banerjee, R. Mechanisms related to different generations of γ' precipitation during continuous cooling of a nickel base superalloy. *Acta Mater.* **2013**, *61*, 280–293. [CrossRef]

17. Grosdidier, T.; Hazotte, A.; Simon, A. Precipitation and dissolution processes in γ/γ' single crystal nickel-based superalloys. *Mater. Sci. Eng. A* **1998**, *256*, 183–196. [CrossRef]

18. Long, H.; Wei, H.; Liu, Y.; Mao, S.; Zhang, J.; Xiang, S.; Chen, Y.; Gui, W.; Li, Q.; Zhang, Z.; et al. Effect of lattice misfit on the evolution of the dislocation structure in Ni-based single crystal superalloys during thermal exposure. *Acta Mater.* **2016**, *120*, 95–107. [CrossRef]

19. Protasova, N.A.; Svetlov, I.L.; Bronfin, M.B.; Petrushin, N.V. Lattice-parameter misfits between the γ and γ' phases in single crystals of nickel superalloys. *Phys. Met. Metall.* **2008**, *106*, 495–502. [CrossRef]

20. Wang, A.; Lv, J.; Chen, C.; Xu, W.; Zhang, L.; Mao, Y.; Zhao, Y. Effects of heat treatment on microstructure and high-temperature tensile properties of nickel-based single-crystal superalloys. *Mater. Res. Express* **2019**, *6*, 126527. [CrossRef]

21. Su, X.; Xu, Q.; Wang, R.; Xu, Z.; Liu, S.; Liu, B. Microstructural evolution and compositional homogenization of a low Re-bearing Ni-based single crystal superalloy during through progression of heat treatment. *Mater. Des.* **2018**, *141*, 296–322. [CrossRef]

22. D'Souza, N.; Newell, M.; Devendra, K.; Jennings, P.A.; Ardakani, M.G.; Shollock, B.A. Formation of low angle boundaries in Ni-based superalloys. *Mater. Sci. Eng. A* **2005**, *413*, 567–570. [CrossRef]

23. Hallensleben, P.; Scholz, F.; Thome, P.; Schaar, H.; Steinbach, I.; Eggeler, G.; Frenzel, J. On crystal mosaicity in single crystal Ni-based superalloys. *Crystals* **2019**, *9*, 149. [CrossRef]

24. Condruz, M.R.; Matache, G.; Paraschiv, A.; Pușcașu, C. Homogenization heat treatment and segregation analysis of equiaxed CMSX-4 superalloy for gas turbine components. *J. Anal. Calorim.* **2018**, *134*, 443–453. [CrossRef]

25. Huo, M.; Liu, L.; Yang, W.; Li, Y.; Hu, S.; Su, H.; Zhang, J.; Fu, H. Formation of low-angle grain boundaries under different solidification conditions in the rejoined platforms of Ni-based single crystal superalloys. *J. Mater. Res.* **2019**, *34*, 251–260. [CrossRef]

26. Jiang, S.; Sun, D.; Zhang, Y.; Yan, B. Influence of heat treatment on microstructures and mechanical properties of NiCuCrMoTiAlNb nickel-based alloy. *Metals* **2018**, *8*, 217. [CrossRef]

27. Gross, J.; Buhl, P.; Weber, U.; Schuler, X.; Krack, M. Effect of creep on the nonlinear vibration characteristics of blades with interlocked shrouds. *Int. J. Non-Linear Mech.* **2018**, *99*, 240–246. [CrossRef]

28. Bogdanowicz, W.; Krawczyk, J.; Tondos, A.; Sieniawski, J. Subgrain boundaries in single crystal blade airfoil of aircraft engine. *Cryst. Res. Technol.* **2017**, *52*, 1600372. [CrossRef]

29. Bogdanowicz, W.; Krawczyk, J.; Paszkowski, R.; Sieniawski, J. Primary crystal orientation of the thin-walled area of single-crystalline turbine blade airfoils. *Materials* **2019**, *12*, 2699. [CrossRef]

30. Bogdanowicz, W.; Tondos, A.; Krawczyk, J.; Albrecht, R.; Sieniawski, J. Dendrite growth in selector-root area of single crystal CMSX-4 turbine blades. *Acta Phys. Pol. A* **2016**, *130*, 1107. [CrossRef]

31. Chung, J.-S.; Ice, G.E. Automated indexing for texture and strain measurement with broad-bandpass X-ray microbeams. *J. Appl. Phys.* **1999**, *86*, 5249–5255. [CrossRef]

32. Husseini, N.S.; Kumah, D.P.; Yi, J.Z.; Torbet, C.J.; Arms, D.A.; Dufresne, E.M.; Pollock, T.M.; Jones, J.W.; Clarke, R. Mapping single-crystal dendritic microstructure and defects in nickel-base superalloys with synchrotron radiation. *Acta Mater.* **2008**, *56*, 4715–4723. [CrossRef]

33. Bowen, D.K.; Tanner, B.K. *High Resolution X-ray Diffractometry and Topography*; Taylor & Francis Ltd.: London, UK, 2005.

34. Bogdanowicz, W. Martensitic transformation in β1-CuZnAl single crystals studied by X-ray topography method. *Scr. Mater.* **1997**, *37*, 829–835. [CrossRef]

35. Colli, A.; Attenkofer, K.; Raghothamachar, B.; Dudley, M. Synchrotron X-ray Topography for encapsulation stress/strain and crack detection in crystalline silicon modules. *IEEE J. Photovolt.* **2016**, *6*, 1387–1389. [CrossRef]

36. Krawczyk, J.; Bogdanowicz, W.; Hanc-Kuczkowska, A.; Tondos, A.; Sieniawski, J. Influence of heat treatment on defect structures in single-crystalline blade roots studied by X-ray topography and positron annihilation lifetime spectroscopy. *Met. Mater. Trans. A* **2018**, *49*, 4353–4361. [CrossRef]

37. Krawczyk, J.; Paszkowski, R.; Bogdanowicz, W.; Hanc-Kuczkowska, A.; Sieniawski, J.; Terlecki, B. Defect creation in the root of single-crystalline turbine blades made of Ni-based superalloy. *Materials* **2019**, *12*, 870. [CrossRef] [PubMed]

38. Bürger, D.; Parsa, A.B.; Ramsperger, M.; Körner, C.; Eggeler, G. Creep properties of single crystal Ni-base superalloys (SX): A comparison between conventionally cast and additive manufactured CMSX-4 materials. *Mater. Sci. Eng. A* **2019**, *762*, 138098. [CrossRef]

39. Berger, H.; Bradaczek, H.A.; Bradaczek, H. Omega-Scan: An X-ray tool for the characterization of crystal properties. *J. Mater. Sci. Mater. Electron.* **2008**, *19*, 351–355. [CrossRef]

Article

Optimization of Residual Stress Measurement Conditions for a 2D Method Using X-ray Diffraction and Its Application for Stainless Steel Treated by Laser Cavitation Peening

Hitoshi Soyama [1,*], Chieko Kuji [2], Tsunemoto Kuriyagawa [2], Christopher R. Chighizola [3] and Michael R. Hill [3]

1 Department of Finemechanics, Tohoku University, Sendai 980-8579, Japan
2 Department of Mechanical Systems Engineering, Tohoku University, Sendai 980-8579, Japan; kuji.shinkou@gmail.com (C.K.); tkuri@tohoku.ac.jp (T.K.)
3 Department of Mechanical and Aerospace Engineering, University of California Davis, Davis, CA 95616, USA; crchighizola@ucdavis.edu (C.R.C.); mrhill@ucdavis.edu (M.R.H.)
* Correspondence: soyama@mm.mech.tohoku.ac.jp; Tel.: +81-22-795-6891

Abstract: As the fatigue strength of metallic components may be affected by residual stress variation at small length scales, an evaluation method for studying residual stress at sub-mm scale is needed. The $\sin^2\psi$ method using X-ray diffraction (XRD) is a common method to measure residual stress. However, this method has a lower limit on length scale. In the present study, a method using at a 2D XRD detector with ω-oscillation is proposed, and the measured residual stress obtained by the 2D method is compared to results obtained from the $\sin^2\psi$ method and the slitting method. The results show that the 2D method can evaluate residual stress in areas with a diameter of 0.2 mm or less in a stainless steel with average grain size of 7 μm. The 2D method was further applied to assess residual stress in the stainless steel after treatment by laser cavitation peening (LCP). The diameter of the laser spot used for LCP was about 0.5 mm, and the stainless steel was treated with evenly spaced laser spots at 4 pulses/mm^2. The 2D method revealed fluctuations of LCP-induced residual stress at sub-mm scale that are consistent with fluctuations in the height of the peened surface.

Keywords: residual stress; X-ray diffraction; laser cavitation peening; pulse laser

Citation: Soyama, H.; Kuji, C.; Kuriyagawa, T.; Chighizola, C.R.; Hill, M.R. Optimization of Residual Stress Measurement Conditions for a 2D Method Using X-ray Diffraction and Its Application for Stainless Steel Treated by Laser Cavitation Peening. *Materials* **2021**, *14*, 2772. https://doi.org/10.3390/ma14112772

Academic Editors: Thomas Walter Cornelius and Souren Grigorian

Received: 23 April 2021
Accepted: 20 May 2021
Published: 24 May 2021

Publisher's Note: MDPI stays neutral with regard to jurisdictional claims in published maps and institutional affiliations.

1. Introduction

As residual stress is one of the most important factors related to the fatigue strength of metallic materials [1–8], it is worth measuring the residual stress in local areas subject to fatigue crack nucleation. It is well known that conventional welding causes tensile residual stress near the welded line due to the heat-affected zone (HAZ) [9–12]. Friction stir welding (FSW) also generates tensile residual stress near the FSW region [13–18], as FSW produces stirring and a thermo-mechanically affected zone. Residual stress is one of the key factors for mechanical surface treatments such as shot peening (SP) [19]. Laser peening can also improve fatigue properties by introducing compressive residual stress [20–23]. As the distribution of the residual stress of conventional welding and the FSWed part drastically changes with distance from the welding line, the residual stress of the laser-peened surface is also distributed with a laser spot size of the mm-order. One of the conventional methods used to evaluate the residual stress is X-ray diffraction. As the size of the measured area using X-ray diffraction is similar to that of the distribution of the residual stress of the welding part and/or the laser-peened surface in sub-millimeter order, it is necessary to improve the accuracy of residual stress measurements by using X-ray diffraction. Note that the most important factor of the stresses measurement accuracy in local area using X-ray diffraction is the number of the grains.

The $\sin^2\psi$ method is the most popular method for evaluating the residual stress of polycrystal metals using X-ray diffraction [24], and a 2D method using a two-dimensional detector has been developed [25]. Regarding JSMS standard, 3×10^5 to 6×10^5 grains is required for the $\sin^2\psi$ method. Each method is based on its own theory, and each has advantages and disadvantages. For example, in the case of the $\sin^2\psi$ method, a simple goniometer is sufficient to evaluate the residual stress. On the other hand, the 2D method can evaluate the 3D stress state, but a highly accurate multi-axis goniometer is needed. The great advantage of using a 2D detector is that the Debye ring can be evaluated by interpolation and extrapolation. Namely, the 2D method using a 2D detector could be used to evaluate residual stress of very small area and/or coarse grain. A 2D method with specimen oscillation by moving the detector in the direction orthogonal to θ-direction was proposed to obtain better Debye ring in the reference [26], but the obtained result was not compared with the result obtained using the other method. One way to evaluate residual stress measurements using X-ray diffraction is to compare them with mechanical relaxation method such as a slitting method [27] and/or a hole drilling method [28]. The slitting method is relatively easy to perform, can be performed quickly, and provides excellent repeatability, which makes it very useful for actual laboratory residual stress measurements [27]. As the experimental deviation of the slitting method was smaller than that of the hole drilling method [29], the slitting method was chosen in the present experiment.

As mentioned above, laser peening introduces compressive residual stress and enhances the fatigue properties [20–23]. Y. Sano et al. measured the residual stress distribution with depth for stainless steel SUS304 and demonstrated an improvement in fatigue strength by laser peening without protective coating [21]. In the case of laser peening with coating [20,22], coating or tapes such as black polymer tape or metal foil is pasted on material to control laser energy absorption and prevent the surface from melting. The distributions of residual stress were precisely measured, but fluctuations due to the laser spot were not observed [30,31]. The distribution of changes in residual stress with depth was precisely measured, but there was no information provided for the distribution on the surface [32,33]. On the other hand, in the case of the numerical simulation, residual stress distributions due to laser spots were clearly observed [6,34–36]. Recently, the patterns of residual stress on the surface due to laser spots were also observed [37,38]. G. Xu et al. measured the residual stress of SUS316L by the $\sin^2\psi$ method, in which the diameter of the measured area was 2 mm with a 0.5 mm step; the laser spot was 3×3 mm^2; the overlapping rates were 30%, 50% and 70%; and the cyclic pattern of the residual stress was obtained [37]. X. Pan et al. measured the residual stress of Ti6Al4V by the $\sin^2\psi$ method, in which the diameter of the measuring area was 2 mm with a 1 mm step, the laser spot was 2.4 mm, the overlapping rate was 40% and a cyclic pattern due to laser spots was not observed in the distribution of residual stress [39]. Using a synchrotron, Y. Sano et al. measured the distribution of the residual stress crossing over a single laser spot with 1D line irradiation by measuring an area of 0.2 mm in diameter; the laser spot was about 1 mm in diameter and the authors reported the tensile residual stress due to the laser spot [38]. It was determined by numerical simulation that the crack propagation was affected by the residual stress distribution due to laser peening [40,41]; therefore, the precise distribution of the residual stress had to be determined. Figure 1 shows the typical pattern of a fractured fatigue specimen of a stainless steel (a) non-peened specimen and (b) laser-peened specimen [23]. As shown in Figure 1b, the fatigue crack of the laser-peened specimen propagated nearly straight compared with that of the non-peened one, in which several cracks were propagated in parallel due to the increase of the hardness near the crack tips by plastic deformation. The distribution of residual stress of laser peened specimen. The distribution of residual stress of the laser peened specimen could be one of the reasons for crack propagation. Considering a previous report [38], the measurement of the residual stress in a submillimeter-order area is required. Thus, a method that can measure residual stress at a submillimeter level using a conventional X-ray diffraction apparatus is needed.

This paper consists of two parts. The first half reveals the optimization of measuring the condition of the 2D method using a two-dimensional detector for residual stress measurements to evaluate the surface modification layer compared with the mechanical method, i.e., the slitting method [27], and the $\sin^2\psi$ method [24]. The second half demonstrates the residual stress measurement of the peened specimen by laser cavitation peening using the proposed 2D method.

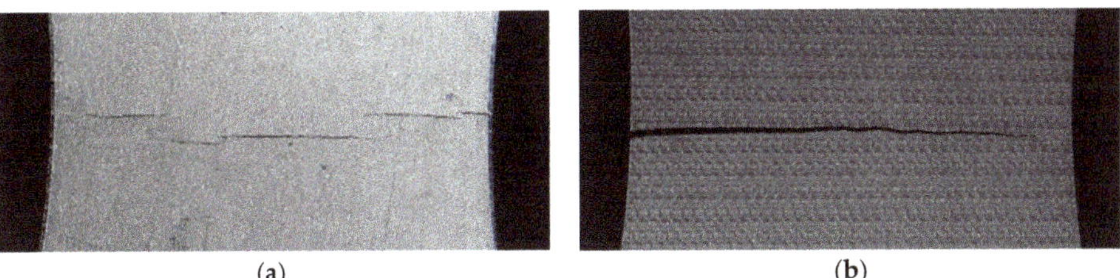

(a)   (b)

Figure 1. Aspect of a fractured fatigue specimen of stainless steel. (**a**) Non-peened specimen (σ_a = 301 MPa, N_f = 7.8 × 10^5); (**b**) specimen treated by laser cavitation peening (σ_a = 308 MPa, N_f = 4.8 × 10^6).

2. Experimental Apparatus and Procedures

2.1. Peening Systems

To prepare specimens with compressive residual stress, cavitation peening (CP) using a cavitating jet (see Figure 2) and laser cavitation peening (LCP) using a pulse laser (see Figure 3) were applied. In the case of CP, a high-speed water jet was injected into a water-filled chamber. The cavitating conditions were the same as those in a previous paper [8]; the injection pressure of the jet was 30 MPa, the diameter of the nozzle d was 2 mm and the standoff distance was 222 mm. To enhance the peening intensity, the nozzle had a cavitator with a diameter d_c of 3 mm [42] and an outlet bore with a diameter D and length L of 16 and 16 mm, respectively [43]. The specimen was installed in the recess. In the case of LCP, a Nd:YAG (Nd:$Y_3Al_5O_{12}$) laser with Q-switch was used to generate laser cavitation [12]. The repetition frequency of the pulse laser was 10 Hz. The used wavelength was 1064 nm. The pulse laser was focused onto the specimen, which was placed in a water-filled glass chamber. The standoff distance in the air s_a and in water s_w was optimized by measuring the peening intensity [12]. The specimen based the stage was moved perpendicularly to the direction of the laser axis by the stepping motors.

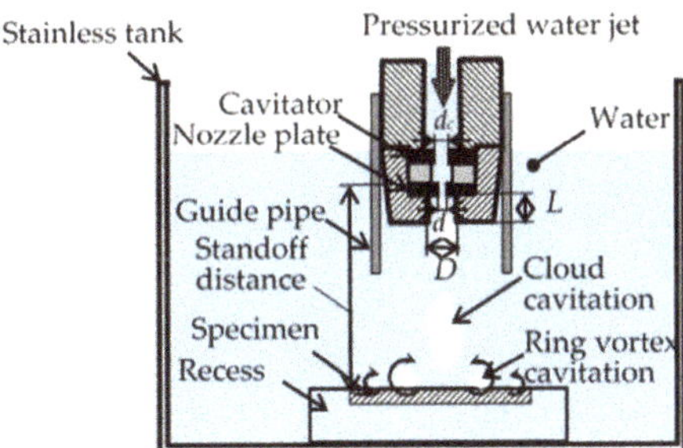

Figure 2. Schematic diagram of cavitation peening (CP) using a cavitating jet.

As the backside surface of the peened plate had compressive residual stress [44], a recirculating shot peening (SP) system accelerated by a water jet [45] was used for the peening. Note that compressive residual stress was introduced onto the backside surface, but the grain size on the back side was not affected, as the backside surface was not peened. At SP, the shots were installed a chamber, whose diameter was 54 mm, and accelerated by the water jet through three holes with a diameter of 0.8 mm. The diameter and the number of the shots were 3.2 mm and 500, respectively. The injection pressure of the water jet was

12 MPa. The distance from the nozzle to the specimen surface was 50 mm. To avoid a loss of shots, the specimen was set in the recess.

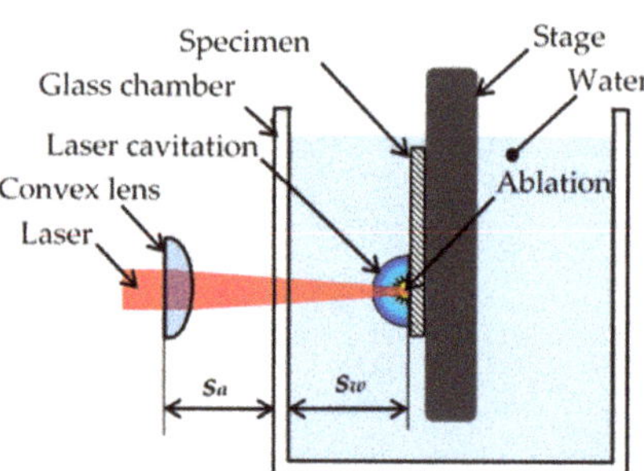

Figure 3. Schematic diagram of cavitation peening using a pulse laser, i.e., laser cavitation peening (LCP).

2.2. Material

The tested material was austenitic stainless steel, Japanese Industrial Standards JIS SUS316L. Four different specimens were used, as shown in Table 1. All specimens were made from plates 2, 3 and 6 mm in thickness, respectively, and all the plates featured a No. 2B surface finish accomplished by temper rolling. Specimen A was used to measure the residual stress of the peened side by the 2D method compared with the slitting method and $\sin^2\psi$ method. Specimens A, B and C were used to optimize the measuring conditions of the 2D method. Specimen D was used to demonstrate the effect of the laser spot on the residual stress distribution. During LCP, the surface was laser and the outermost surface showed tensile residual stress; the surface was then removed by electrochemical polishing. The peening intensity of CP, SP and LCP was controlled based on processing time per unit length, processing time and pulse density, respectively. The processing time per unit length of specimen A was chosen based on the values in the reference [8]. To introduce large compressive residual stress into the specimen, 100 pulses/mm^2 pulse density and 6 mm thickness were chosen for specimen B. Considering the results of the preliminary experiment by measuring the residual stress, 30 s and 3 mm thickness were chosen for specimen C. For specimen D, the pulse density of 4 pulses/mm^2 was optimized by measuring the fatigue life changes based on pulse density.

Table 1. Specimens for residual stress measurements.

Symbol	Peening Method	Peening Intensity	Thickness	Measured Side	Electrochemical Polishing
A	Cavitation peening CP	8 s/mm	2 mm	Peened side	None
B	Laser cavitation peening LCP	100 pulses/mm^2	6 mm	Peened side	39 μm
C	Shot peening SP	30 s	3 mm	Back side	None
D	Laser cavitation peening LCP	4 pulses/mm^2	2 mm	Peened side	33 μm

As mentioned above, the number of grains in the measurement area is important factor for the accuracy of the stresses using X-ray diffraction method. The average grain size, i.e., spatial diameter [46] and the grain size that occupied 50% of the area d_N were measured. The d_N was obtained by the following procedure. The area A_i of each grain was measured and they were sorted from small value to large value, then the cumulated area A_C was calculated as follows.

$$A_C = \sum_{i=1}^{N} A_i \tag{1}$$

when the number N at $A_C/A = 50\%$ was obtained, $d_N = 2\sqrt{A_N/\pi}$ was defined as the grain size that occupied 50%. Here, A was test area. 300 grains were measured in the present experiment.

Figure 4 illustrates a schematic diagram of specimen D and the coordinates of the residual stress with the scanning direction of the laser. The specimen was moved at 5 mm/s, and the reputation frequency of the laser was 10 Hz. Then, the specimen was stepped at 0.5 mm, as shown in Figure 4. The positional relationship of the laser spots in each row was different for each row as shown in Figure 4, as the stepping motors and pulse laser were not synchronized. Note that the rolling direction was y-direction in Figure 4.

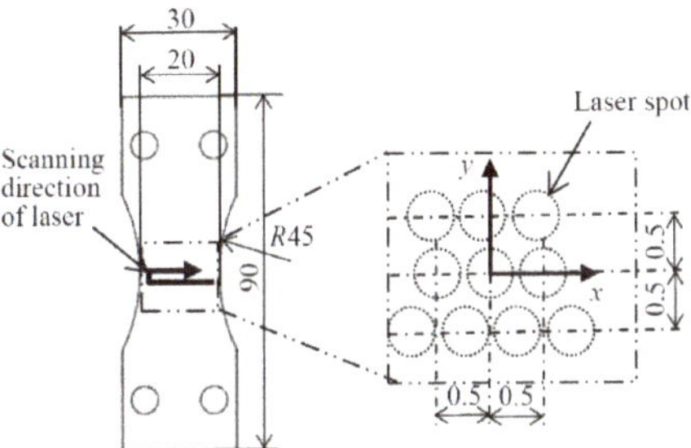

Figure 4. Schematic diagram of specimen D treated by laser cavitation peening and the coordinates for residual stress measurement.

2.3. Residual Stress Measurement

To confirm the compressive residual stress of specimen A introduced by cavitation peening, the residual stress was measured using the slitting method [27]. The slitting was done using a wire electric discharge machine (EDM, Sodick, Chicago, IL, USA), and the residual stress was evaluated from the strain obtained by the strain gage. The diameter of the wire was 0.254 mm and the gage length of the strain gage was 0.81 mm. The distribution of the residual stress with depth under the surface was obtained by using the recorded strain and solving an inverse problem following the procedure in the reference [15]. In the present paper, slitting of 0.0254 mm in depth was used as the reference value. The distribution of the residual stress with depth and more details on the slitting method can be found in the reference [8].

To measure the residual stress by X-ray diffraction, an XRD system (Bruker Japan K. K., Tokyo, Japan) with a two-dimensional position sensitive proportional counter (2D-PSPC) was used. The same system was used for the $\sin^2\psi$ method [24] and 2D method [47]. The schematic diagram and coordinates θ, ψ, ω, χ, φ of the XRD system with 2D-PSPC are shown in Figure 5. The Kα X-rays from a Cr-tube operating at 35 kV and 40 mA were used. The used diffracted plane was γ-Fe (2 2 0), and the diffraction angle without strain was $128°$. In the residual stress analysis for both the $\sin^2\psi$ method and the 2D method, Relevant software (Leptos ver 7.9, Bruker Japan K. K., Tokyo, Japan) was used. The used Young's modulus and Poisson ratio were 191.975 GPa and 0.3, respectively. To investigate the effects of the measuring area, five different collimators with diameters of 0.1460, 0.3, 0.5, 0.5724 and 0.8 mm were used. The 0.1460 and 0.5724 mm collimators were of the total reflection type, and the other collimators were of the double-holed type. Tables 2 and 3 show the measuring conditions and analyzed areas of X-ray diffraction obtained using each method based on the standard method [24] and the previous report [26]. Under both the $\sin^2\psi$ method and the 2D method, 24 frames were measured. In the case of the $\sin^2\psi$ method, the X-ray diffraction profile obtained an accumulating X-ray diffraction of $\chi = 85$–$95°$, and diffracted peaks 2θ were obtained at each ψ. Then, the residual stress was calculated from the $\sin^2\psi$–2θ diagram. To eliminate the ψ split, $+\psi$ and $-\psi$ were measured for each x and y direction. At the $\sin^2\psi$ method, σ_{Rx} was obained by $\varphi = 90°$ and $270°$, σ_{Ry} was obtaied by $\varphi = 0$ and $180°$, respectively. Namely, 12 frames in Table 2 were used to obtain σ_{Rx} and σ_{Ry}, respectively.

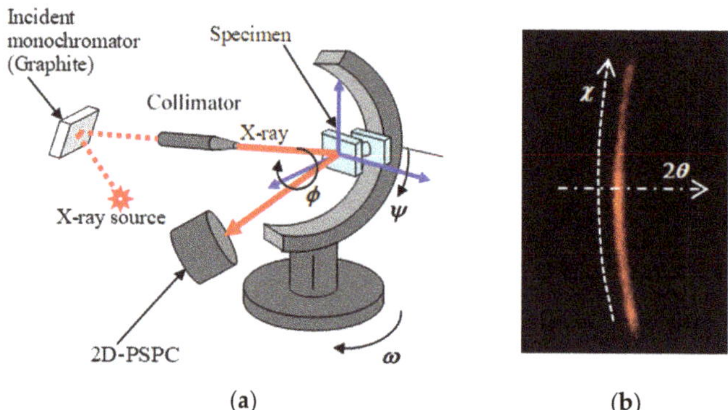

Figure 5. Schematic diagram and coordinates of XRD system with 2D-PSPC. (**a**) Coordinates of the system; (**b**) coordinates on the 2D-PSPC.

Table 2. Measuring conditions for each method.

Method	$\psi°$	$\varphi°$
	0	0, 90, 180, 270
	20.268	0, 90, 180, 270
$\sin^2\psi$ method	29.334	0, 90, 180, 270
	36.870	0, 90, 180, 270
	43.854	0, 90, 180, 270
	50.768	0, 90, 180, 270
	0	0, 45, 90, 135, 180, 225, 270, 315
2D method	30	0, 45, 90, 135, 180, 225, 270, 315
	60	0, 45, 90, 135, 180, 225, 270, 315

Table 3. Analyzed area of obtained X-ray diffraction.

Method	$2\theta°$	$\chi°$
$\sin^2\psi$ method	125–132	85–95
2D method	125–132	70–110

2.4. Observation of Specimen Surface

To investigate the grain size of the tested material, the surface was observed using a scanning electron microscope (SEM; JCM-7000, JEOL Ltd., Tokyo, Japan). The aspect of the laser-cavitation-peened specimen was also observed using a laser confocal microscope (VK-100, Keyence Corporation, Osaka, Japan) to obtain the surface profile.

3. Results

3.1. Comparison of Measured Residual Stress between the Slitting Method, $\sin^2\psi$ Method and 2D Method

In order to compare the residual stress measured by the slitting method, the $\sin^2\psi$ method and the 2D method, Figure 6 illustrates the residual stress σ_{Ry} of specimen A. For the $\sin^2\psi$ method and the 2D method, the effect of the measuring area was investigated by changing diameter of the collimator d_{col}. As shown in Figure 6, the exposure time at each frame t_{exp} was also changed based on the area of the collimator. In the case of $d_{col} = 0.8$ mm and $t_{exp} = 20$ s, the specimen was moved in both directions, $x = \pm 2$ mm and $y = \pm 2$ mm, to minimize the exposure time. As shown in Figure 6, in the case of d_{col} = 0.8 mm and t_{exp} = 40 s at $x = \pm0$ mm and $y = \pm0$ mm, the residual stress σ_R measured using the $\sin^2\psi$ method and the 2D method was -220 ± 74 and -220 ± 14 MPa; these results are similar to -251 ± 16 MPa, which was measured by the slitting method. For the $\sin^2\psi$ method, the residual stress measured using $d_{col} \geq 0.5724$ mm was similar to that of the slitting method. However, at $d_{col} \leq 0.5$ mm, the residual stress was too small and the standard deviation of the residual stress was too large. For the reference, Appendix A reveals the diagram of $\sin^2\psi - 2\theta$ for d_{col} = 0.146 mm, t_{exp} = 20 min and d_{col} = 0.8 mm, t_{exp} = 40 s. On the other hand, in the case of the 2D method, the residual stress measured using d_{col} = 0.146 mm was -187 ± 29 MPa. Thus, it can be concluded that the 2D method can evaluate the residual stress by using d_{col} = 0.146 mm. Specifically, the 2D method can measure the residual stress in the 1/15 region of the $\sin^2\psi$ method under the presented conditions.

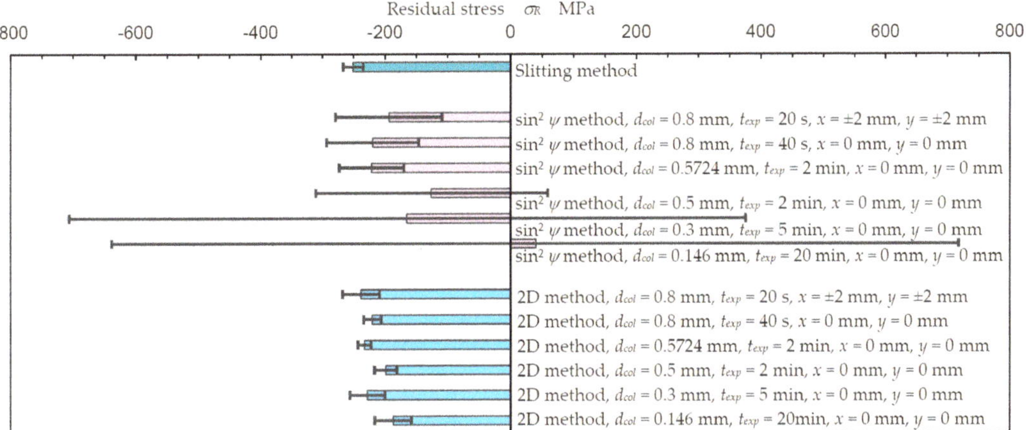

Figure 6. Comparison of the slitting method, $\sin^2\psi$ method, and the 2D method for the residual stress of stainless steel treated by cavitation peening (specimen A).

In order to investigate the difference in the measurement of residual stress between the $\sin^2\psi$ method and the 2D method for the austenitic stainless steel tested using a collimator of d_{col} = 0.146 mm, Figure 7 shows the aspects of the surface of the measured sample observed using a scanning electron microscope (SEM). The average grain size, i.e., spatial diameter [46], was 6.6 ± 4.0 μm in diameter, and the grain size that occupied 50% of the area was about 11 μm. As shown in Figure 7, specific anisotropy is not observed. Thus, at the present condition, the rolling direction did not affect the results of residual stress measurement.

Figure 8 illustrates a typical X-ray diffraction pattern that was a part of the Debye ring, as detected by 2D-PSPC from the stainless steel treated by cavitation peening—i.e., specimen A treated using d_{col} = 0.146 mm—and the analyzed area for (a) the $\sin^2\psi$ method and (b) the 2D method. As illustrated in Figure 7, the grain size was about 1/10 of the diameter of the collimator, and the X-ray diffraction pattern was a mottled pattern, as shown in Figure 8. In the case of the $\sin^2\psi$ method, the diffraction pattern located near $\chi \approx 90°$ should be used due to the theory of the $\sin^2\psi$ method; the standard deviations of the $\sin^2\psi$ method at $d_{col} \leq 0.5$ mm were remarkably large, as the diffraction pattern at $\chi \approx 90°$ was weak or not obtained. For the present residual stress analysis, $\chi = 85–95°$ was used for the $\sin^2\psi$ method. The residual stress obtained by the $\sin^2\psi$ method for d_{col} = 0.8 mm, t_{exp} = 40 s, $x = 0$ mm and $y = 0$ mm was -300 ± 46 MPa based on analysis using $2\theta = 125–132°$ and $\chi = 70–110°$. These values were too large compared to the value of -251 ± 16 MPa measured by the slitting method. Namely, when large area, i.e., $\chi = 70–110°$. was used, the number of counts of the X-ray diffraction was increased. However, $\chi = 70–110°$ was too large for the $\sin^2\psi$ method.

On the other hand, $\chi = 70\text{–}110°$ was used for the 2D method, as the 2D method obtained the residual stress from the distortion of the Debye ring. For the 2D method, the Debye ring was obtained by interpolation and extrapolation of the patchy patterns of the diffraction spots. It was concluded that the 2D method could evaluate the residual stress in a smaller area compared to the $\sin^2\psi$ method, as the Debye ring was used in the relatively large area of χ. Thus, it can be said that the key point for evaluating the residual stress in a small area is to obtain a more uniform Debye ring.

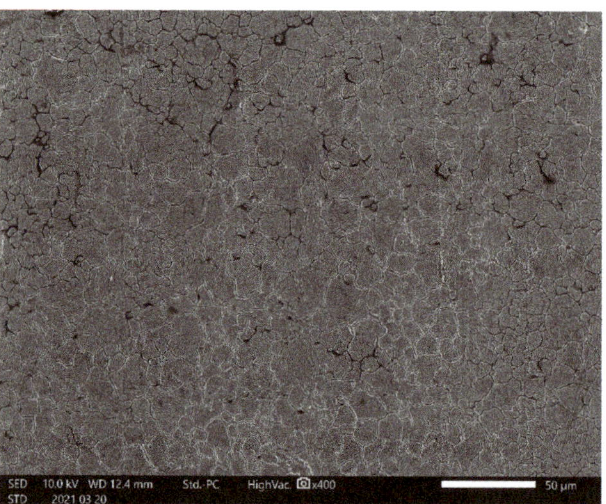

Figure 7. Aspect of the surface of specimen A observed by scanning electron microscope (SEM).

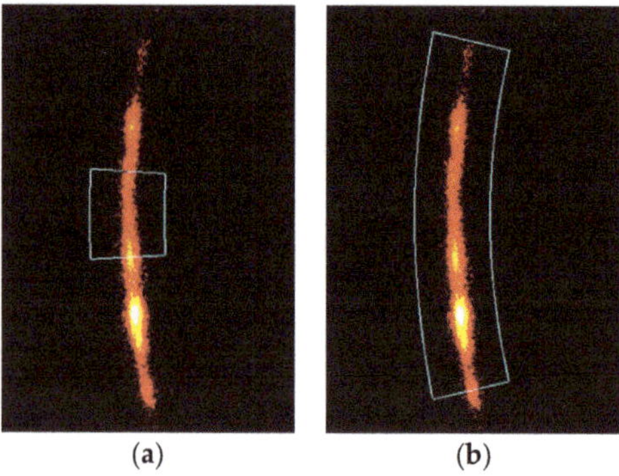

(a)   (b)

Figure 8. Typical X-ray diffraction pattern detected by 2D-PSPC from stainless steel treated using cavitation peening (Specimen A) and the analyzed area ($d_{col} = 0.146$ mm, $\psi = 0°$, $\varphi = 0°$, $\Delta\omega = 0°$, $t_{exp} = 20$ min, $x = 0$ mm, $y = 0$ mm); (a) analyzed area for the $\sin^2\psi$ method ($2\theta = 125\text{–}132°$, $\chi = 85\text{–}95°$); (b) analyzed area for the 2D method ($2\theta = 125\text{–}132°$, $\chi = 70\text{–}110°$).

3.2. Optimum Condition for the 2D Method to Evaluate Residual Stress

In the present paper, to obtain a better Debye ring, specimen oscillation in the ω direction—i.e., ω-oscillation—was proposed. Figure 9 shows the X-ray diffraction pattern (a) without ω-oscillation (i.e., $\Delta\omega = 0°$) and (b) with ω-oscillation at $\Delta\omega = 10°$. As shown in Figure 9a ($\psi = 0$, $\varphi = 0$) and (b) ($\psi = 0$, $\varphi = 0$), the diffraction spot at $\chi \approx 102°$ became a streak by ω-oscillation. Precisely, the diffraction spot became a streak in the χ direction by ω-oscillation of the specimen. The ω-oscillation helped to achieve a better

Debye ring. Note that, in the case of 2D method, there should be an optimum value of $\Delta\omega$, as the residual stress was obtained by the distortion of Debye ring. The optimum value of $\Delta\omega$ is discussed in the following.

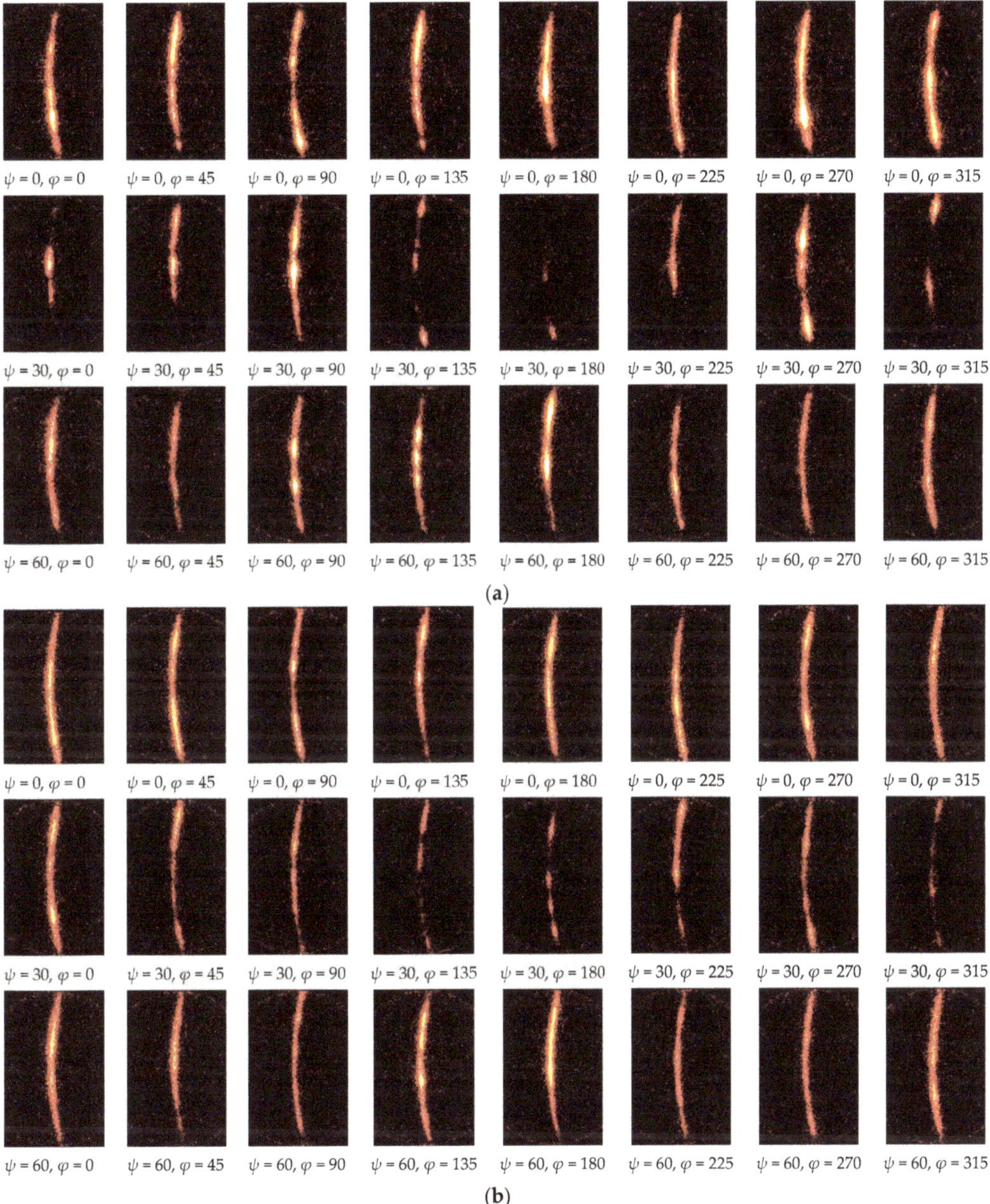

Figure 9. X-ray diffraction patterns detected by 2D-PSPC from the stainless steel treated by cavitation peening (Specimen A). (a) $\Delta\omega = 0°$; (b) $\Delta\omega = 10°$.

In order to investigate the ω-oscillation of the specimen both qualitatively and quantitatively, Figure 10 shows the typical X-ray diffraction pattern of specimen A at $\psi = 0°$

and $\varphi = 0$ changing with (a) the collimator diameter d_{col}, (b) the exposure time t_{exp} and (c) the ω-oscillation angle $\Delta\omega$. Figure 11 illustrates the relationship between the total count of X-ray diffraction and the standard deviation of the residual stress $\Delta\sigma_R$. As shown in Figure 10a, the X-ray diffraction pattern became a mottled pattern with a decrease in the diameter of the collimator d_{col}. Then, the $\Delta\sigma_R$ increased with a decrease of d_{col}, as shown in Figure 11. When the exposure time t_{exp} was increased, the Debye ring became clear, as shown in Figure 10b, and then $\Delta\sigma_R$ decreased with an increase of t_{exp}. As shown in Figure 11, $\Delta\sigma_R$ decreased with an increase in the total count of X-ray diffraction for both d_{col} and t_{exp}. As shown in Figure 10c, the diffraction pattern changed from a mottled pattern to a streak-like pattern with an increase of $\Delta\omega$. The $\Delta\sigma_R$ was 64 MPa for $\Delta\omega = 0°$, 49 MPa for $\Delta\omega = 2°$, 49 MPa for $\Delta\omega = 4°$, 37 MPa for $\Delta\omega = 6°$, 29 MPa for $\Delta\omega = 8°$ and 32 MPa for $\Delta\omega = 10°$. Specifically, $\Delta\sigma_R$ decreased with an increase of $\Delta\omega$ at $\Delta\omega = 0$–$8°$ and presented its minimum at $\Delta\omega = 8°$; then, $\Delta\sigma_R$ slightly increased at $\Delta\omega = 10°$. As the 2D method evaluates the residual stress caused by distortion of the Debye ring, a $\Delta\omega$ that is too large dims the distortion by averaging too large an area in the χ direction. Thus, it can be concluded that the ω-oscillation of the specimen was effective, and the optimum value of $\Delta\omega$ was $8°$. In Figure 11, $\Delta\sigma_R = 64$ MPa. In Figure 11, $\Delta\sigma_R = 64$ MPa for $\Delta\omega = 0°$ corresponds to $t_{exp} = 4$ or 5 min and $\Delta\sigma_R = 29$ MPa for $\Delta\omega = 8°$ corresponds to $t_{exp} = 20$ min. Thus, ω-oscillation of the specimen has the effect of shortening the measurement time to $1/4$–$1/5$.

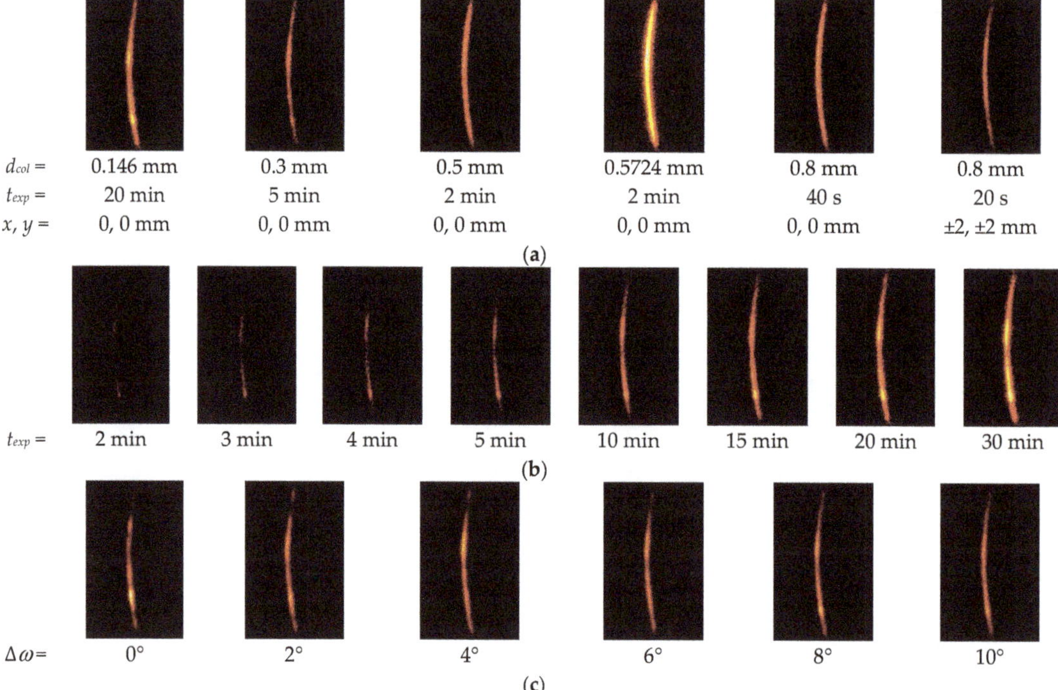

Figure 10. Typical X-ray diffraction patterns detected by 2D-PSPC at $\psi = 0°$ and $\varphi = 0°$ from stainless steel treated by cavitation peening (Specimen A). (**a**) effect of the collimator diameter d_{col} at $\Delta\omega = 8°$; (**b**) effect of exposure time t_{exp} at $d_{col} = 0.146$ mm and $\Delta\omega = 8°$; (**c**) effect of the ω-oscillation angle $\Delta\omega$ at $d_{col} = 0.146$ mm and $t_{exp} = 20$ min.

In order to confirm that the 2D method can evaluate the residual stress of the austenitic stainless steel by using the 0.146 mm-diameter collimator, Figure 12a reveals the residual stress (σ_{Rx}, σ_{Ry}) of specimen A, B and C as a function of the diameter of the collimator d_{col}, and Figure 12b shows the standard deviation of the residual stress. The data for specimen A in Figure 6 were used as the σ_{Ry} in Figure 12. For all three specimens, A, B and C, as both σ_{Rx} and σ_{Ry} at $d_{col} = 0.146$ mm were nearly equal to the values at $d_{col} = 0.8$ mm, the residual stresses of $d_{col} = 0.146$ mm for specimens A, B and C were able to be evaluated.

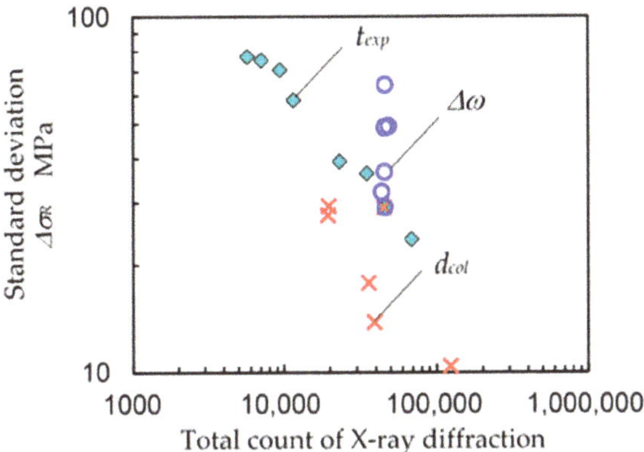

Figure 11. Relationship between the counts at $\psi = 0°$ and $\varphi = 0°$ detected by 2D-PSPC and the standard deviation of the residual stress of stainless steel treated by cavitation peening (specimen A) changing with exposure time t_{exp}, collimator diameter d_{col} and ω-oscillation angle $\Delta\omega$.

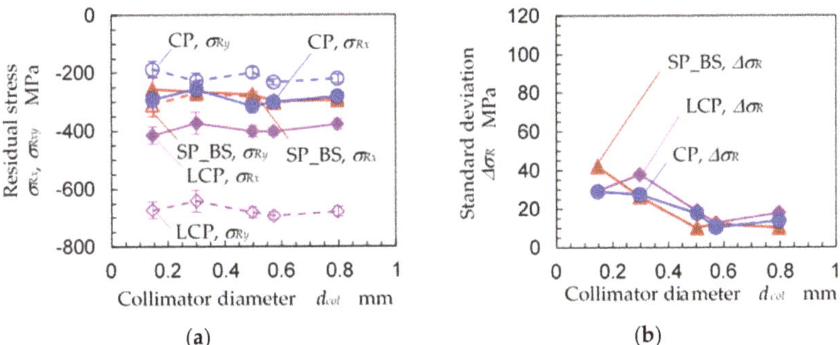

Figure 12. Effect of the collimator dimeter on the residual stress measurement of the stainless steel by the 2D method ($\Delta\omega = 8°$); CP: specimen A, LCP: specimen B, SP_BS: specimen C. (**a**) residual stress; (**b**) standard deviation of residual stress.

In the case of specimen B, i.e., LCP, the specimen was treated with 100 pulses/mm^2, as mentioned in Table 1. The specimen was moved in the x-direction at 1 mm/s. As the repetition frequency of the pulse laser was 10 Hz, the distance of each laser spot was 0.1 mm. After each specimen was treated in the x-direction, it was moved stepwise at 0.1 mm in the y-direction. As shown in Figure 12a, σ_{Rx} and σ_{Ry} were about -400 MPa and -670 MPa, respectively. Specifically, the compressive residual stress introduced in y-direction, i.e., the stepwise direction, under laser cavitation peening was 270 MPa larger than that in the x-direction. Even though the distances between the laser spots in both x- and y-direction were the same, the compressive residual stress introduced in the y-direction was larger than that in the x-direction. This tendency was similar to the results in the reference [38].

In the present experiment, SP was applied, then the residual stress on the back side of shot peened specimen was measured to avoid the effects of the change of the grain size etc.

If the treated surfaces by SP, CP and LCP were evaluated, the characteristics of the peened surfaces have different features. It was reported that the dislocation density of CP and LCP of austenitic stainless steel SUS316L was lower than that of SP at the equivalent peening condition, i.e., the equivalent arc height condition [48].

In order to investigate the effects of ω-oscillation of the specimen, Figure 13 reveals the residual stress (σ_{Rx}, σ_{Ry}) and the standard deviation of the residual stress $\Delta\sigma_R$ as a function of the ω-oscillation angle $\Delta\omega$ for specimens A, B and C. As shown in Figure 13, the residual stress of specimen A, B and C was roughly constant for all $\Delta\omega$ values, and $\Delta\sigma_R$ roughly decreased with an increase of $\Delta\omega$. In the case of LCP, i.e., specimen C, the compressive residual stress increased with an increase of $\Delta\omega$ at $\Delta\omega \leq 8°$; then, the compressive residual stress slightly decreased at $\Delta\omega = 10°$. Further, the $\Delta\sigma_R$ of specimen C had its minimum at $\Delta\omega = 8°$, and the $\Delta\sigma_R$ at $\Delta\omega = 10°$ was larger than the $\Delta\sigma_R$ of $\Delta\omega = 8°$. As the 2D method obtained the residual stress from the distortion of the Debye ring, a $\Delta\omega$ value that was too large caused a decrease of the residual stress and an increase of $\Delta\sigma_R$, just as with specimen A. It can be concluded that the ω-oscillation of the specimen is effective for evaluating the residual stress and that the optimum value of $\Delta\omega$ is $8°$.

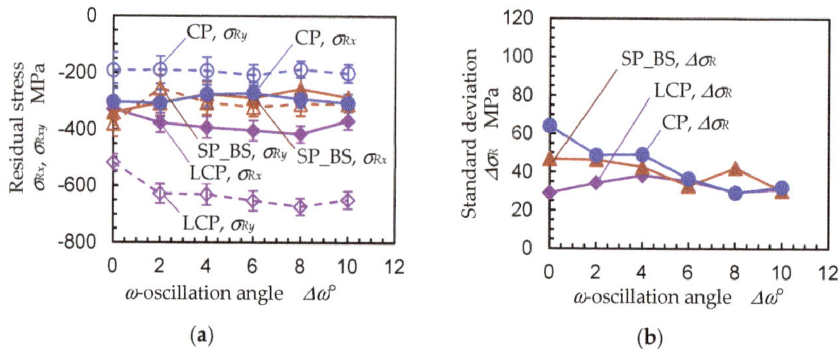

Figure 13. Effect of the ω-oscillation angle $\Delta\omega$ on the residual stress measurement of stainless steel using the 2D method (d_{col} = 0.146 mm, t_{exp} = 20 min); CP: specimen A, LCP: specimen B, SP_BS: specimen C; (**a**) residual stress; (**b**) standard deviation of residual stress.

In order to determine the optimum exposure time needed to obtain the X-ray diffraction pattern, Figure 14 shows (a) the residual stress σ_{Rx}, σ_{Ry} and (b) standard deviation of the residual stress $\Delta\sigma_R$ as a function of exposure time t_{exp} for specimens A, B and C. Under all measurement conditions in Figure 14, the specimens were oscillated at $\Delta\omega = 8°$, and the diameter of the collimator was 0.146 mm. For specimens A and B, the σ_{Rx} and σ_{Ry} were nearly constant at all t_{exp} values. For specimen C, σ_{Rx} and σ_{Ry} decreased and became saturated at t_{exp} = 15 and 20 min. The $\Delta\sigma_R$ of specimens A, B and C decreased with an increase of t_{exp} and became saturated at t_{exp} = 15 or 20 min. Under the studied conditions, t_{exp} = 20 min is the optimum exposure time to obtain residual stress. The X-ray diffraction totaled about 4.6×10^4 counts.

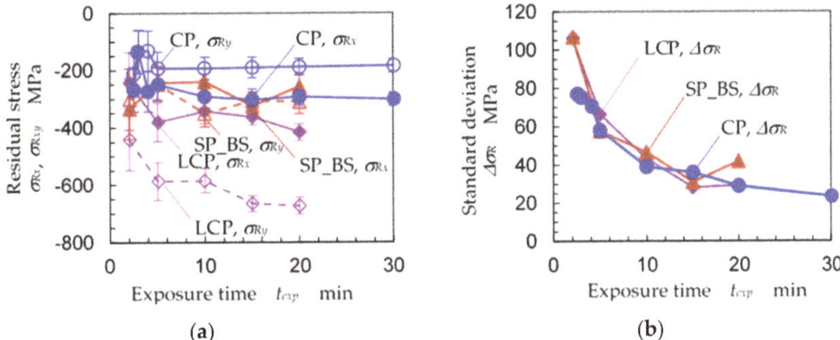

Figure 14. Effect of exposure time t_{exp} on the residual stress measurements of stainless steel under the 2D method (d_{col} = 0.146 mm, $\Delta\omega$ = 8°); CP: specimen A, LCP: specimen B, SP_BS: specimen C. (**a**) Residual stress; (**b**) standard deviation of residual stress.

3.3. Residual Stress Distribution of Specimen Treated by Laser Cavitation Peening

To determine the typical results for the residual stress of austenitic stainless steel JIS SUS316L in the local area measured by the 2D method, the residual stress of the specimen treated by laser cavitation peening, i.e., that of specimen D, was evaluated using the 2D method. Considering the results in Sections 3.1 and 3.2, the measuring conditions were as follows: The diameter of the collimator d_{col} was 0.146 mm, the ω-oscillation angle $\Delta\omega$ was 8° and the exposure time of each frame t_{exp} was 20 min. The pulse density of the laser cavitation peening d_L was 4 pulses/mm^2, as the fatigue life was greatest at d_L = 4 pulses/mm^2 and changed with the pulse density [23]. Under these conditions, the laser spot distances in the x- and y-directions were 0.5 and 0.5 mm, respectively. As mentioned above, the top surface revealed tensile residual stress, and then the surface of 33 μm was removed by electrochemical polishing. Note that the residual stress at 30 μm accurately corresponded to the fatigue life, as the fatigue life estimated by the residual stress at 30 μm, the surface hardness and the surface roughness was proportional to the experimental value [23].

Figure 15 shows (a) the aspects of the laser-cavitation-peened specimen observed using a CCD camera on the XRD system and (b) the aspects of the same specimen observed using a laser confocal microscope. As the specimen was treated with d_L = 4 pulses/mm^2, the distances of the x- and y-directions between the laser spots were 0.5 and 0.5 mm. Under the presented conditions, the depth of the laser spot was about 10 μm. As shown in Figure 15, the laser spot diameter was about 0.5 mm after the 33 μm electrochemical polishing. The vertical positional relationship of the laser spots was not aligned, as the stepping motor and laser system were not synchronized.

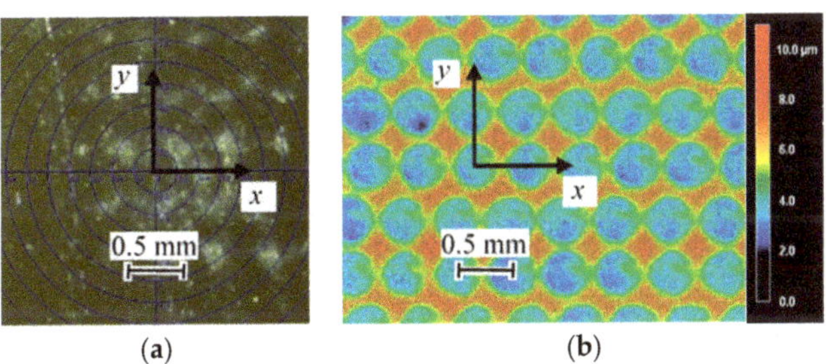

(**a**) (**b**)

Figure 15. Aspects of the surface of the stainless-steel specimen treated by laser cavitation peening (specimen D); (**a**) observation using the CCD camera on the XRD system; (**b**) observations from the laser confocal microscope.

Figure 16 shows the residual stress distribution as a function of y at $x = 0$, 0.125, 0.25, 0.375 and 0.5 mm for (a) σ_{Rx} and (b) σ_{Ry}. The standard deviation at the residual stress was about 70 MPa. It was difficult to recognize the 0.5 mm interval period at each position of x, as shown in Figure 16a,b, because the positional relationships of the laser spots at $y = 0$, 0.5, 1, 1.5 and 2 mm were different, as shown in Figure 15. As σ_{Rx} and σ_{Ry} varied from 0 to -150 MPa, there was a difference of about 150 MPa depending on the location for both σ_{Rx} and σ_{Ry}.

Figure 17 reveals the residual stress σ_{Rx} and σ_{Ry} changing with distance x at $y = 0$ with the laser spot aligned in the x-direction. The σ_{Rx} was about -100 MPa at $x = 0$ mm; it had a peak of 0 MPa at $x = 0.125$ mm and then decreased with an increase in x. Then, σ_{Rx} had a minimum of 150 MPa at $x = 0.375$ mm and increased to 0 MPa at x = 0.5 mm. On the other hand, σ_{Ry} had a minimum at x = 0.125 mm and a maximum at x = 0.375 mm. Even though the standard deviations were somewhat large, a 0.5 mm cycle was observed for both σ_{Rx} and σ_{Ry}. It can be concluded from Figures 16 and 17 that the residual stress may differ by about 150 MPa depending on the location when austenitic stainless steel JIS SUS316L is treated using laser cavitation peening at 4 pulses/mm². As shown in Figures 12 and 13, when the residual stress was relatively uniform, the experimental deviation at the present condition using the 0.146 mm collimator was about ±40 MPa. At the measurement of LCP specimen treated at 4 pulse/mm², the residual stress was changed from 0 MPa to -150 MPa within 0.25 mm in length, thus the experimental deviation using the 0.146 mm collimator was about ±70 MPa due to the spatial distribution.

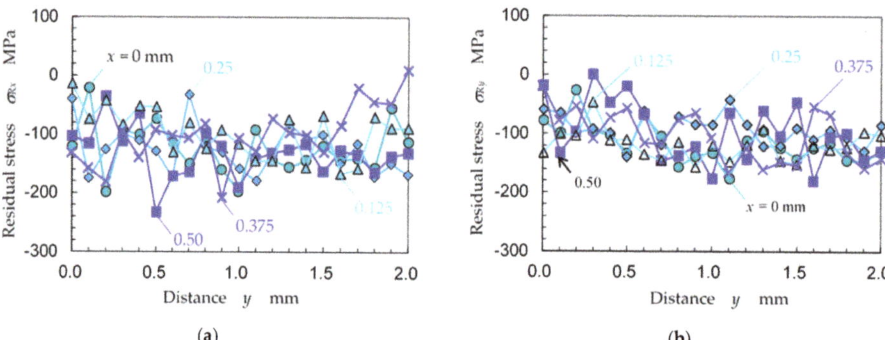

Figure 16. Distribution of the residual stress changing with distance y at various positions of x (specimen D); (a) residual stress in the x-direction σ_{Rx}; (b) residual stress in the y-direction σ_{Ry}.

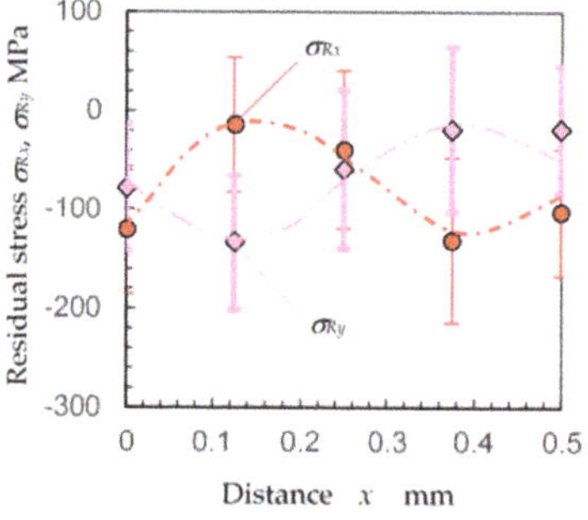

Figure 17. Distribution of the residual stress σ_{Rx} and σ_{Ry} changing with distance x (specimen D).

4. Conclusions

To clarify the possibilities and measure the conditions of residual stress in a mechanical-surface-modified layer with a small area by the 2D method using X-ray diffraction, the residual stress of the austenitic stainless steel JIS SUS316L treated by cavitation peening was measured by the 2D method changing with the diameter of the collimator comparing with the $\sin^2\psi$ method and the slitting method. The measured sample was austenitic stainless steel with temper rolling. The average diameter and 50% area of the grain size of the tested specimen were 6.6 ± 4.0 μm and about 11 μm, respectively. The specimens were treated by cavitation peening using a cavitating jet and a pulse laser, i.e., laser cavitation peening. The results obtained can be summarized as follows:

(1) Compared to the $\sin^2\psi$ method, the 2D method can evaluate the residual stress in a small area, which is 1/15 of the area ratio of the $\sin^2\psi$ method. In the present experiment, the measurable areas of the $\sin^2\psi$ method and 2D method were 0.5724 mm in diameter and 0.146 mm in diameter, respectively.

(2) The ω-oscillation of the specimen using the 2D method had the effect of reducing the measurement error to 1/2. This result is equivalent to the effect of reducing the measurement time to 1/5–1/4. The optimum ω-oscillation angle $\Delta\omega$ was $8°$.

(3) The 2D method using optimized conditions can evaluate the residual stress distribution for a laser spot with a diameter of 0.5 mm.

(4) The compressive residual stress under laser cavitation peening at 100 pulses/mm^2 was larger in the stepwise direction than in the orthogonal direction.

Author Contributions: Conceptualization, H.S., C.K. and T.K.; methodology, H.S., C.K. and M.R.H.; validation, H.S. and C.R.C.; formal analysis, H.S. and C.R.C.; investigation, H.S. and C.R.C.; resources, H.S.; data curation, H.S. and C.R.C.; writing—original draft preparation, H.S.; writing—review and editing, H.S. and C.K.; funding acquisition, H.S.; supervision, T.K. and M.R.H. All authors have read and agreed to the published version of the manuscript.

Funding: This research was partly supported by JSPS KAKENHI Grant Number 18KK0103 and 20H02021.

Institutional Review Board Statement: Not applicable.

Informed Consent Statement: Not applicable.

Data Availability Statement: The data presented in this study are available on request from the author.

Conflicts of Interest: The authors declare no conflict of interest.

Appendix A

For the reference, Figure A1 shows the diagram of $\sin^2\psi - 2\theta$ for d_{col} = 0.146 mm, t_{exp} = 20 min and d_{col} = 0.8 mm, t_{exp} = 40 s. As shown in Figure A1, the peak of the profile of the X-ray diffraction, i.e., the diffraction angle of d_{col} = 0.146 mm was scattered, therefore the experimental deviation in Figure 6 of d_{col} = 0.146 mm was large.

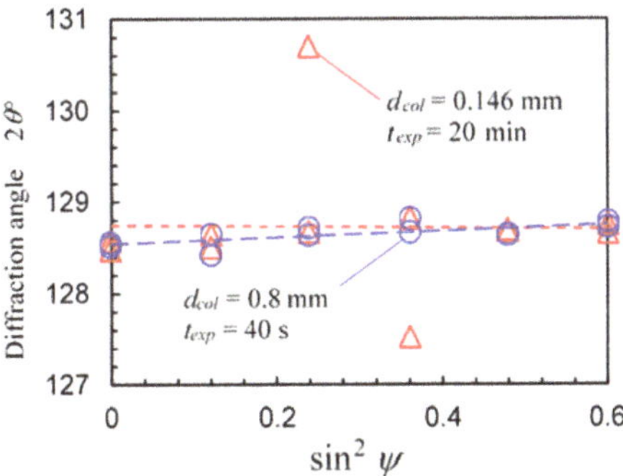

Figure A1. Relationship between $\sin^2\psi - 2\theta$ (d_{col} = 0.146 mm, t_{exp} = 20 min and d_{col} = 0.8 mm, t_{exp} = 40 s).

References

1. Nikitin, I.; Scholtes, B.; Maier, H.J.; Altenberger, I. High temperature fatigue behavior and residual stress stability of laser-shock peened and deep rolled austenitic steel aisi 304. *Scr. Mater.* **2004**, *50*, 1345–1350. [CrossRef]
2. Withers, P.J. Residual stress and its role in failure. *Rep. Prog. Phys.* **2007**, *70*, 2211–2264. [CrossRef]
3. Gujba, A.K.; Medraj, M. Laser peening process and its impact on materials properties in comparison with shot peening and ultrasonic impact peening. *Materials* **2014**, *7*, 7925–7974. [CrossRef] [PubMed]
4. Soyama, H. Comparison between shot peening, cavitation peening and laser peening by observation of crack initiation and crack growth in stainless steel. *Metals* **2020**, *10*, 63. [CrossRef]
5. Arakawa, J.; Hanaki, T.; Hayashi, Y.; Akebono, H.; Sugeta, A. Evaluating the fatigue limit of metals having surface compressive residual stress and exhibiting shakedown. *Fatigue Fract. Eng. Mater. Struct.* **2020**, *43*, 211–220. [CrossRef]
6. Bikdeloo, R.; Farrahi, G.H.; Mehmanparast, A.; Mahdavi, S.M. Multiple laser shock peening effects on residual stress distribution and fatigue crack growth behaviour of 316L stainless steel. *Theor. Appl. Fract. Mech.* **2020**, *105*, 11. [CrossRef]
7. Tang, L.Q.; Ince, A.; Zheng, J. Numerical modeling of residual stresses and fatigue damage assessment of ultrasonic impact treated 304L stainless steel welded joints. *Eng. Fail. Anal.* **2020**, *108*, 23. [CrossRef]
8. Soyama, H.; Chighizola, C.R.; Hill, M.R. Effect of compressive residual stress introduced by cavitation peening and shot peening on the improvement of fatigue strength of stainless steel. *J. Mater. Process. Technol.* **2021**, *288*, 116877. [CrossRef]
9. Edwards, L.; Bouchard, P.J.; Dutta, M.; Wang, D.Q.; Santisteban, J.R.; Hiller, S.; Fitzpatrick, M.E. Direct measurement of the residual stresses near a 'boat-shaped' repair in a 20 mm thick stainless steel tube butt weld. *Int. J. Press. Vessels Pip.* **2005**, *82*, 288–298. [CrossRef]
10. Zhang, W.Y.; Jiang, W.C.; Zhao, X.; Tu, S.T. Fatigue life of a dissimilar welded joint considering the weld residual stress: Experimental and finite element simulation. *Int. J. Fatigue* **2018**, *109*, 182–190. [CrossRef]
11. Luo, Y.; Gu, W.B.; Peng, W.; Jin, Q.; Qin, Q.L.; Yi, C.M. A study on microstructure, residual stresses and stress corrosion cracking of repair welding on 304 stainless steel: Part i-effects of heat input. *Materials* **2020**, *13*, 2416. [CrossRef]
12. Soyama, H. Laser cavitation peening and its application for improving the fatigue strength of welded parts. *Metals* **2021**, *11*, 531. [CrossRef]
13. Webster, P.J.; Oosterkamp, L.D.; Browne, P.A.; Hughes, D.J.; Kang, W.P.; Withers, P.J.; Vaughan, G.B.M. Synchrotron X-ray residual strain scanning of a friction stir weld. *J. Strain Anal. Eng. Des.* **2001**, *36*, 61–70. [CrossRef]
14. Peel, M.; Steuwer, A.; Preuss, M.; Withers, P.J. Microstructure, mechanical properties and residual stresses as a function of welding speed in aluminium AA5083 friction stir welds. *Acta Mater.* **2003**, *51*, 4791–4801. [CrossRef]
15. Prime, M.B.; Gnaupel-Herold, T.; Baumann, J.A.; Lederich, R.J.; Bowden, D.M.; Sebring, R.J. Residual stress measurements in a thick, dissimilar aluminum alloy friction stir weld. *Acta Mater.* **2006**, *54*, 4013–4021. [CrossRef]
16. Haghshenas, M.; Gharghouri, M.A.; Bhakhri, V.; Klassen, R.J.; Gerlich, A.P. Assessing residual stresses in friction stir welding: Neutron diffraction and nanoindentation methods. *Int. J. Adv. Manuf. Technol.* **2017**, *93*, 3733–3747. [CrossRef]
17. Zhang, L.; Zhong, H.L.; Li, S.C.; Zhao, H.J.; Chen, J.Q.; Qi, L. Microstructure, mechanical properties and fatigue crack growth behavior of friction stir welded joint of 6061-T6 aluminum alloy. *Int. J. Fatigue* **2020**, *135*, 11. [CrossRef]
18. Soyama, H.; Simoncini, M.; Cabibbo, M. Effect of cavitation peening on fatigue properties in friction stir welded aluminum alloy AA5754. *Metals* **2021**, *11*, 59. [CrossRef]
19. Delosrios, E.R.; Walley, A.; Milan, M.T.; Hammersley, G. Fatigue-crack initiation and propagation on shot-peened surfaces in A316 stainless-steel. *Int. J. Fatigue* **1995**, *17*, 493–499. [CrossRef]
20. Peyre, P.; Fabbro, R.; Merrien, P.; Lieurade, H.P. Laser shock processing of aluminium alloys. Application to high cycle fatigue behaviour. *Mater. Sci. Eng. A* **1996**, *210*, 102–113. [CrossRef]
21. Sano, Y.; Obata, M.; Kubo, T.; Mukai, N.; Yoda, M.; Masaki, K.; Ochi, Y. Retardation of crack initiation and growth in austenitic stainless steels by laser peening without protective coating. *Mater. Sci. Eng. A* **2006**, *417*, 334–340. [CrossRef]
22. Gill, A.; Telang, A.; Mannava, S.R.; Qian, D.; Pyoun, Y.-S.; Soyama, H.; Vasudevan, V.K. Comparison of mechanisms of advanced mechanical surface treatments in nickel-based superalloy. *Mater. Sci. Eng. A* **2013**, *576*, 346–355. [CrossRef]
23. Soyama, H. Comparison between the improvements made to the fatigue strength of stainless steel by cavitation peening, water jet peening, shot peening and laser peening. *J. Mater. Process. Technol.* **2019**, *269*, 65–78. [CrossRef]
24. SMS Committee on X-ray Study of Mechanical Behavior of Materials. Standard method for X-ray stress measurement. *JSMS Stand.* **2005**, *JSMS-SD-10-05*, 1–21.
25. He, B.B.; Preckwinkel, U.; Smith, K. Advantage of using 2D detectors for residual stress measurements. *Adv. X-ray Anal.* **2000**, *42*, 429–438.
26. Takakuwa, O.; Soyama, H. Optimizing the conditions for residual stress measurement using a two-dimensional XRD method with specimen oscillation. *Adv. Mater. Phys. Chem.* **2013**, *03*, 8–18. [CrossRef]
27. Hill, M. The slitting method. In *Practical Residual Stress Measurement Methods*; Schajer, G.S., Ed.; John Wiley & Sons, Inc.: Hoboken, NJ, USA, 2013; pp. 89–108.
28. ASTM E837-13a. *Standard Test Method for Determining Residual Stresses by the Hole-Drilling Strain-Gage Method*; ASTM: West Conshohocken, PA, USA, 2020.
29. Chighizola, C.R.; D'Elia, C.R.; Weber, D.; Kirsch, B.; Aurich, J.C.; Linke, B.S.; Hill, M.R. Intermethod comparison and evaluation of measured near surface residual stress in milled aluminum. *Exp. Mech.* **2021**, in press.

30. Hatamleh, O.; Rivero, I.V.; Swain, S.E. An investigation of the residual stress characterization and relaxation in peened friction stir welded aluminum-lithium alloy joints. *Mater. Des.* **2009**, *30*, 3367–3373. [CrossRef]

31. Correa, C.; de Lara, L.R.; Diaz, M.; Gil-Santos, A.; Porro, J.A.; Ocana, J.L. Effect of advancing direction on fatigue life of 316L stainless steel specimens treated by double-sided laser shock peening. *Int. J. Fatigue* **2015**, *79*, 1–9. [CrossRef]

32. Kallien, Z.; Keller, S.; Ventzke, V.; Kashaev, N.; Klusemann, B. Effect of laser peening process parameters and sequences on residual stress profiles. *Metals* **2019**, *9*, 655. [CrossRef]

33. Wang, Z.D.; Sun, G.F.; Lu, Y.; Chen, M.Z.; Bi, K.D.; Ni, Z.H. Microstructural characterization and mechanical behavior of ultrasonic impact peened and laser shock peened AISI 316L stainless steel. *Surf. Coat. Technol.* **2020**, *385*, 19. [CrossRef]

34. Bhamare, S.; Ramakrishnan, G.; Mannava, S.R.; Langer, K.; Vasudevan, V.K.; Qian, D. Simulation-based optimization of laser shock peening process for improved bending fatigue life of Ti-6Al-2Sn-4Zr-2Mo alloy. *Surf. Coat. Technol.* **2013**, *232*, 464–474. [CrossRef]

35. Correa, C.; de Lara, L.R.; Diaz, M.; Porro, J.A.; Garcia-Beltran, A.; Ocana, J.L. Influence of pulse sequence and edge material effect on fatigue life of Al2024-T351 specimens treated by laser shock processing. *Int. J. Fatigue* **2015**, *70*, 196–204. [CrossRef]

36. Keller, S.; Chupakhin, S.; Staron, P.; Maawad, E.; Kashaev, N.; Klusemann, B. Experimental and numerical investigation of residual stresses in laser shock peened AA2198. *J. Mater. Process. Technol.* **2018**, *255*, 294–307. [CrossRef]

37. Xu, G.; Luo, K.Y.; Dai, F.Z.; Lu, J.Z. Effects of scanning path and overlapping rate on residual stress of 316L stainless steel blade subjected to massive laser shock peening treatment with square spots. *Appl. Surf. Sci.* **2019**, *481*, 1053–1063. [CrossRef]

38. Sano, Y.; Akita, K.; Sano, T. A mechanism for inducing compressive residual stresses on a surface by laser peening without coating. *Metals* **2020**, *10*, 816. [CrossRef]

39. Pan, X.L.; Li, X.; Zhou, L.C.; Feng, X.T.; Luo, S.H.; He, W.F. Effect of residual stress on S-N curves and fracture morphology of Ti6Al4V titanium alloy after laser shock peening without protective coating. *Materials* **2019**, *12*, 3799. [CrossRef]

40. Busse, D.; Ganguly, S.; Furfari, D.; Irving, P.E. Optimised laser peening strategies for damage tolerant aircraft structures. *Int. J. Fatigue* **2020**, *141*, 12. [CrossRef]

41. Sun, R.J.; Keller, S.; Zhu, Y.; Guo, W.; Kashaev, N.; Klusemann, B. Experimental-numerical study of laser-shock-peening-induced retardation of fatigue crack propagation in Ti-17 titanium alloy. *Int. J. Fatigue* **2021**, *145*, 13. [CrossRef]

42. Soyama, H. Enhancing the aggressive intensity of a cavitating jet by introducing a cavitator and a guide pipe. *J. Fluid Sci. Technol.* **2014**, *9*, 1–12. [CrossRef]

43. Soyama, H. Enhancing the aggressive intensity of a cavitating jet by means of the nozzle outlet geometry. *J. Fluids Eng.* **2011**, *133*, 1–11. [CrossRef]

44. Nishikawa, M.; Soyama, H. Two-step method to evaluate equibiaxial residual stress of metal surface based on micro-indentation tests. *Mater. Des.* **2011**, *32*, 3240–3247. [CrossRef]

45. Naito, A.; Takakuwa, O.; Soyama, H. Development of peening technique using recirculating shot accelerated by water jet. *Mater. Sci. Technol.* **2012**, *28*, 234–239. [CrossRef]

46. ASTM E112-13. *Standard Test Methods for Determining Average Grain Size*; ASTM: West Conshohocken, PA, USA, 2020.

47. He, B.B. *Two-Dimensional X-Ray Diffraction*; John Wiley & Sons, Inc.: Hoboken, NJ, USA, 2009; pp. 249–328.

48. Kumagai, M.; Curd, M.E.; Soyama, H.; Ungár, T.; Ribárik, G.; Withers, P.J. Depth-profiling of residual stress and microstructure for austenitic stainless steel surface treated by cavitation, shot and laser peening. *Mater. Sci. Eng. A* **2021**, *813*, 141037. [CrossRef]

 *materials*

Article

Relationship between Young's Modulus and Planar Density of Unit Cell, Super Cells (2 × 2 × 2), Symmetry Cells of Perovskite (CaTiO$_3$) Lattice

Marzieh Rabiei [1], Arvydas Palevicius [1,*], Sohrab Nasiri [1,*], Amir Dashti [2], Andrius Vilkauskas [1] and Giedrius Janusas [1,*]

[1] Faculty of Mechanical Engineering and Design, Kaunas University of Technology, LT-51424 Kaunas, Lithuania; marzieh.rabiei@ktu.edu (M.R.); andrius.vilkauskas@ktu.lt (A.V.)

[2] Department of Materials Science and Engineering, Sharif University of Technology, Tehran 11365-9466, Iran; a.dashty@merc.ac.ir

* Correspondence: arvydas.palevicius@ktu.lt (A.P.); sohrab.nasiri@ktu.edu (S.N.); giedrius.janusas@ktu.lt (G.J.); Tel.: +370-618-422-04 (A.P.); +370-655-863-29 (S.N.); +370-670-473-37 (G.J.)

Abstract: Calcium titanate-CaTiO$_3$ (perovskite) has been used in various industrial applications due to its dopant/doping mechanisms. Manipulation of defective grain boundaries in the structure of perovskite is essential to maximize mechanical properties and stability; therefore, the structure of perovskite has attracted attention, because without fully understanding the perovskite structure and diffracted planes, dopant/doping mechanisms cannot be understood. In this study, the areas and locations of atoms and diffracted planes were designed and investigated. In this research, the relationship between Young's modulus and planar density of unit cell, super cells (2 × 2 × 2) and symmetry cells of nano CaTiO$_3$ is investigated. Elastic constant, elastic compliance and Young's modulus value were recorded with the ultrasonic pulse-echo technique. The results were C$_{11}$ = 330.89 GPa, C$_{12}$ = 93.03 GPa, C$_{44}$ = 94.91 GPa and E = 153.87 GPa respectively. Young's modulus values of CaTiO$_3$ extracted by planar density were calculated 162.62 GPa, 151.71 GPa and 152.21 GPa for unit cell, super cells (2 × 2 × 2) and symmetry cells, respectively. Young's modulus value extracted by planar density of symmetry cells was in good agreement with Young's modulus value measured via ultrasonic pulse-echo.

Keywords: nano-perovskite (CaTiO$_3$); X-ray diffraction; Young's modulus; ultrasonic-pulse echo; planar density

Citation: Rabiei, M.; Palevicius, A.; Nasiri, S.; Dashti, A.; Vilkauskas, A.; Janusas, G. Relationship between Young's Modulus and Planar Density of Unit Cell, Super Cells (2 × 2 × 2), Symmetry Cells of Perovskite (CaTiO$_3$) Lattice. *Materials* **2021**, *14*, 1258. http://doi.org/10.3390/ma14051258

Academic Editor: Thomas Walter Cornelius and Souren Grigorian

Received: 5 January 2021
Accepted: 2 March 2021
Published: 6 March 2021

Publisher's Note: MDPI stays neutral with regard to jurisdictional claims in published maps and institutional affiliations.

1. Introduction

Perovskites have a general formula of ABO$_3$. In these structures, the A site cation is a typical lanthanide, alkaline or alkaline-earth metal with 12-fold oxygen coordination, and the B-site is any one of a variety of transition metal cations [1]. Calcium titanate (CaTiO$_3$) was established in 1839 by a Russian mineralogist Perovski, and materials with the same type of CaTiO$_3$ were introduced as the perovskite structure. CaTiO$_3$ has ionic bonds, as well as the ionic radii of Ca^{2+}, O^{2-} and Ti^{4+} are 1 Å, 1.40 Å and 0.6 Å, respectively [2]. In recent years, researchers have focused on developing perovskites and their mechanical properties in order to obtain a high yield. Furthermore, CaTiO$_3$ is a well-known component in ferroelectric perovskite category, which has been considerably utilized as a dopant/doping in electronic materials due to its dielectric manner and flexibility in structural transformations [3,4]. The modulus of elasticity (E) or Young's modulus is defined as the proportion of the stress to the strain, created by the stress on the body when the body is in the elastic region [5]. The elastic constants are specified from the lattice crystal deformation against force. These elastic moduli are: Young's modulus, shear modulus and volumetric modulus. These modules are registered via inherent elastic properties of materials and their resistance to deformation due to loading. Elastic behavior of materials is described by models such

45

as Cauchy elastic, hypo-elastic and hyper-elastic. A hyper-elastic is a constitutive model for ideally elastic material that responds against stress gain from a strain energy density function, while for hypo-elastic material, their governing equation is independent of finite strain quantity except in the linearized state [6]. The elastic properties are intimately connected to the crystal structure, the intrinsic character of bonding between the atoms and the anisotropic nature of materials [7,8]; therefore, elastic constants can be derived from crystal lattice calculations [9]. There are several studies on the relationship between elastic constants and planes/directions in a lattice structure, for example, in [10–12]. One of the most accurate methods to measure the elastic stiffness constants and Young's modulus is to determine the velocity of long-wavelength acoustic waves through the ultrasonic pulse-echo technique [13]. In a crystal structure, points, directions and planes are described with an indexing scheme, and planar density is obtained as the number of atoms per unit area, which are centered on a specific crystallographic plane with a defined index [14]. Since the discovery of X-rays at the end of the 19th century, this method has been often used for material characterization [15]. It is used to identify the atomic-scale structure of different materials in a variety of states [16]. X-ray diffraction is the only method that provides the specification of both the mechanical and microstructural character of each diffracted plane. These planes are used as a strain to quantify Young's modulus in one or more planes/directions of the diffraction vector [17]. In forming, designing and manufacturing equipment industries, the use of non-destructive, accurate and convenient methods to determine the mechanical properties of materials is particularly important. Mechanical tests, such as tensile, strike and collision tests, are destructive. Ultrasonic methods are very time-consuming and require operator expertise in this area, and theoretical methods require time-consuming density functional theory (DFT) calculation and may need verification with experimental tests. Our proposed method only needs the XRD analysis, which is a routine test and calculation of planar density; therefore, it can be very significant in terms of industrial application. In this study, the effects of cell size on the accuracy of Young's modulus calculation were considered. Locations of atoms and diffracted planes of unit cell, super cells ($2 \times 2 \times 2$) and symmetry cells of $CaTiO_3$ are designed and investigated. The super cell is a cell that describes the same crystal but has a larger volume than a unit cell. By extension of a unit cell proportional to the lattice vectors, the super cells are generated. In super cells ($2 \times 2 \times 2$), the extension is twice of unit cell length in each direction; likewise, for super cells ($8 \times 8 \times 8$), the extension is 8 times. The result extracted by symmetry cells was in good agreement with results recorded via ultrasonic technique. Therefore, this new approach of exploration of reliable Young's modulus quantity based on XRD is proposed for either single crystal or polycrystalline of $CaTiO_3$.

2. Experimental

2.1. Materials

In this study, for synthesis $CaTiO_3$, titanium (IV) butoxide, calcium chloride dehydrate, sodium hydroxide and ethanol reagents were purchased from Sigma Aldrich (Taufkirchen, Germany) and deionized water as the solvent for dispersions was prepared.

2.2. Instrumentation

In this research, a Bruker D8 Advance X-ray diffractometer (Kaunas, Lithuania) with $Cu_{K\alpha}$ radiation was used. The powder X-ray diffraction was taken at 40 kV and 40 mA, and it was registered at a scanning rate of 2.5 degrees/minute and a step size of 0.02 degrees. The XRD peaks were interpreted by High Score X'Pert software (4.9.0) analysis to get the output ASC type files. The pulse-echo technique was applied for the determination of sound velocity for both transverse and longitudinal ultrasonic signals. For ultrasonic measurement, the model of pulser receiver and oscilloscope were Panametrics Co. (waltham, MA, USA) and Iwatsu (Tokyo, Japan) (100 MHz), respectively. For powder pressing, the model of mechanical machines was CD04-Z and CIP (CP 360). Additionally, the specific surface area of the sample was investigated by desorption isotherms of nitrogen (N_2) gas via using a Brunauer-Emmett-Teller (BET) apparatus Gemini V analyzer, micrometrics GmbH (Tehran, Iran). Moreover, transmission electron microscopy (TEM) CM 10-Philips (Tehran, Iran) with acceleration voltage from 50 to 80 KV was utilized.

2.3. Methods

2.3.1. Synthesis of Nano-Powder $CaTiO_3$

Calcium titanate ($CaTiO_3$) was synthesized by solvothermal method. A simple procedure, namely the solvothermal method, was performed for the synthesis of $CaTiO_3$ (Figure S1). In the first step, (1) calcium chloride dehydrate was stirred with ethanol and deionized water. (2) Titanium (IV) butoxide and ethanol were added to the system drop by drop, under stirring for around 10 min (750 rpm). The molar ratio of ingredients was achieved to calcium chloride dehydrate = 1, ethanol = 5, Titanium (IV) butoxide = 1 and deionized water = 100 respectively. (3) To create pH = 14, sodium hydroxide solution was utilized. (4) The produced solution was placed into the autoclave and the temperature was ~250 °C for 5 h. (5) Afterward, the product was under the drying conditions involved at 110 °C and 0.76 bar, respectively. (6) After a day, the mixture was washed, (7) filtered and dried (110 °C for 4 h), respectively. This method was used in previous studies [18,19].

2.3.2. X-ray Diffraction of $CaTiO_3$ and Planar Density Calculations

Combining X-ray diffraction of crystalline $CaTiO_3$ and calculation of planar density values of each diffracted plane was performed. In our study, the atomic density of each plane was considered as the planar density, which was determined as the area of atoms with the center positioned at the plane divided by the total area of the plane, and it is a determinant factor for mechanical properties of each plane. Planar density is a unitless parameter, and its value is less than 1 in each cell. Furthermore, the values of planar density are related to the positions and situations of atoms in the planes. For determination of atomic area, the Crystal Maker, Version 10.2.2 software was performed. First of all, the three-dimensional (3D) geometry of crystal structures was designed, and then, from the intersection area of each diffracted plane with atoms located at the plane, the atomic area was calculated. When an atom with diameter D was involved completely, the atomic area will be $A = \pi\left(\frac{D}{2}\right)^2$; otherwise, it will be a percentage of this amount.

2.3.3. Ultrasonic Pulse-Echo Technique of $CaTiO_3$

An ultrasonic wave is a type of elastic wave spread in the medium with high frequency to obtain the Young's modulus value of samples. Mastering the ultrasonic parameters can be used to acquire more accurate values of mechanical properties [20]. Recently, different studies on mechanical properties have been done by ultrasonic techniques. Basically, the crossing of longitudinal and transverse waves in nano- or microstructures is performed at different velocities. Each returned velocity is considered as the represented properties.

For ultrasonic measurements based on the Christoffel procedure, the first cubic specimen of $CaTiO_3$ was prepared by cold isostatic press. The schematics of ultrasonic measurement are depicted in Figure 1a. The main part of the ultrasonic system is the pulser-receiver, which creates an electric pulse and stimulates the probe. Furthermore, the produced pulses enter the specimen, and after a sweep, they can be received via a probe. In this measurement, some drops of water were utilized to prevent the depreciation of waves in the air, and the effect of hand pressure on the probe was decreased [21].

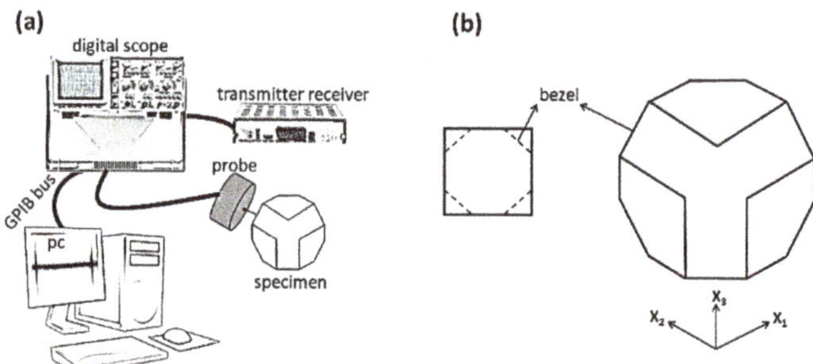

Figure 1. Schematic of (**a**) ultrasonic pulse instrument and (**b**) a sketch of prepared CaTiO$_3$ sample.

At any position in the sample, a local coordinate is adjusted, such as X$_1$, the radial coordinate; X$_2$, the circumferential coordinate; and X$_3$, the axial coordinate. V$_{i/j}$ denotes the velocity of an ultrasound wave propagating in the X$_i$ direction with particle displacements in the X$_j$ direction. V$_{i/j}$ with the same i and j is longitudinal, and with i $\neq$ j is related to the transverse waves. For the measurement of quasi-longitudinal or quasi-transverse velocity (V$_{ij/ij}$), the specimen should be cut (bezel) on the edges of the surfaces perpendicular to the X directions. A sketch of the sample is represented in Figure 1b.

3. Results

3.1. X-ray Diffraction of CaTiO$_3$ and Planar Density Calculations

The XRD pattern of CaTiO$_3$ is presented in Figure 2. The characteristic peaks of CaTiO$_3$ correspond to the report in Ref [22]. The crystal structure of CaTiO$_3$ is cubic, the atomic positions of Ti are at (000), Ca at $(\frac{1}{2},\frac{1}{2},\frac{1}{2})$ and O at $(\frac{1}{2},0,0)$, $(0,\frac{1}{2},0)$, $(0,0,\frac{1}{2})$. According to X-ray powder diffraction results, the lattice parameter is 3.79 $\pm$ 0.02 Å, which is in good corresponds with the amount recorded in the Ref [23]. In addition, crystallographic parameters (Table S1) of CaTiO$_3$ and analyzed data by X′Pert [24] nasiri are recorded as the cell volume = 54.44 $\pm$ 0.01 Å^3 and crystal density = 4.14 $\pm$ 0.01 g/cm^3, and the space group is Pm-3m. In addition, the crystal size of CaTiO$_3$ was calculated by the Monshi–Scherrer equation (Figure S2) [25] and BET analysis. The crystal size values were registered at ~59.10 and 63.02 nm, respectively. The Monshi–Scherrer method is described in Section 2 of the supporting information. Furthermore, a TEM image of CaTiO$_3$ is shown in Figure S3. According to the images shown in Figure S3, the size of CaTiO$_3$ particles basically corresponds to the crystallite size, and it is clear that particles of powder have nanoscale and size can be reported almost ±50 nm.

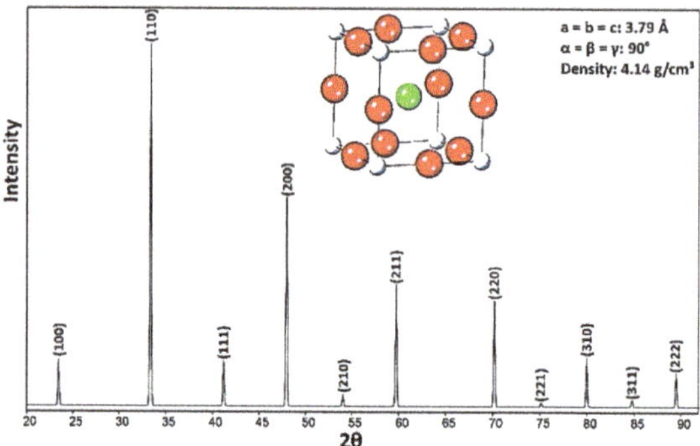

Figure 2. X-ray diffraction of CaTiO$_3$ (powder sample).

For the evaluation of cells as the results, the comprehensive calculations of the planar density of diffracted planes in the unit cell, super cells ($2 \times 2 \times 2$) and super cells ($8 \times 8 \times 8$) of $CaTiO_3$ lattice are presented in Figures S4–S6 respectively. In addition, the locations of atoms, geometry of planes and calculations of planar density of (211) super cell ($4 \times 4 \times 4$), (211) super cell ($8 \times 8 \times 8$), (221) super cell ($4 \times 4 \times 4$), (221) super cell ($8 \times 8 \times 8$), (311) super cell ($3 \times 3 \times 3$), (311) super cell ($4 \times 4 \times 4$), (311) super cell ($8 \times 8 \times 8$), (222) super cell ($3 \times 3 \times 3$) and (222) super cell ($8 \times 8 \times 8$) are depicted briefly in Figures 3–6 respectively. Furthermore, the completed calculations with their figures are shown in Figures S7–S10.

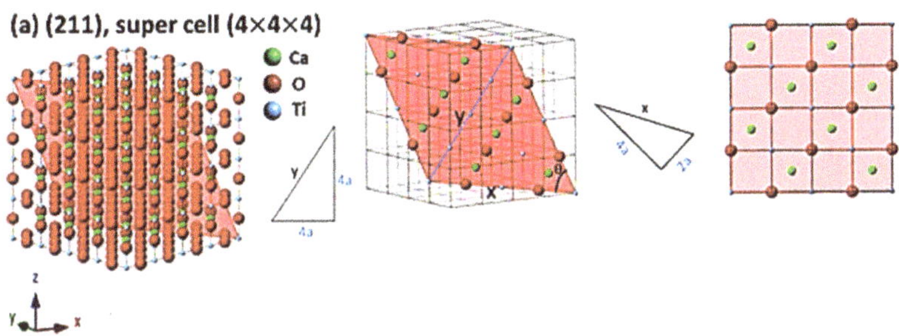

number of atoms in the plane (211) × area of each atom in the plane (211) =

$$\left[\left(\left(2 \times \frac{78.52}{360} + 2 \times \frac{101.48}{360} + 4 \times \frac{1}{2} + 5\right) \times \pi \left(r_{Ti^{4+}}\right)^2\right) + \left(\left(8 \times \frac{1}{2} + 4\right) \times \pi \left(r_{O^{2-}}\right)^2\right)\right] + \left[\left(8 \times \pi \left(r_{Ca+}\right)^2\right)\right] =$$

$$\left[\left(8 \times \pi (0.60)^2\right) + \left(8 \times \pi (1.40)^2\right) + \left(8 \times \pi (1)^2\right)\right] = 83.44$$

$$\text{Planar density} = \frac{\text{number of atoms in the plane (211)} \times \text{area of each atom in the plane (211)}}{\text{area of the plane (211)}} = \frac{83.44}{334.1} = 0.25$$

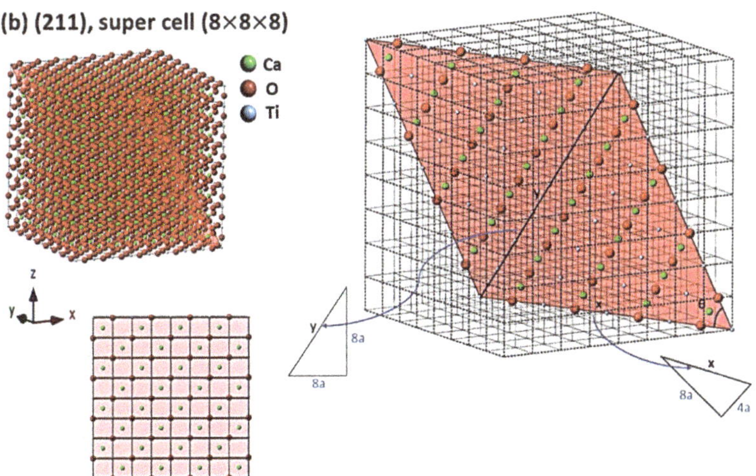

Figure 3. *Cont.*

number of atoms in the plane (211) × area of each atom in the plane (211) =

$$\left[\left(\left(2 \times \frac{78.52}{360} + 2 \times \frac{101.48}{360} + 12 \times \frac{1}{2} + 25\right) \times \pi \left(r_{Ti^{4+}}\right)^2\right) + \left(\left(16 \times \frac{1}{2} + 24\right) \times \pi \left(r_{O^{2-}}\right)^2\right)\right] + \left[\left(32 \times \pi \left(r_{Ca+}\right)^2\right)\right] =$$

$$[(32 \times \pi (0.60)^2) + (32 \times \pi (1.40)^2) + (32 \times \pi (1)^2)] = 333.76$$

$$\text{Planar density} = \frac{\text{number of atoms in the plane (211)} \times \text{area of each atom in the plane (211)}}{\text{area of the plane (211)}} = \frac{333.76}{1335.24} = 0.25$$

Figure 3. Geometry of planes and calculations of planar density of (**a**) (211) super cell (4 × 4 × 4) and (**b**) (211) super cell (8 × 8 × 8) (which shows and emphasizes the symmetry of (8 × 8 × 8) super cells).

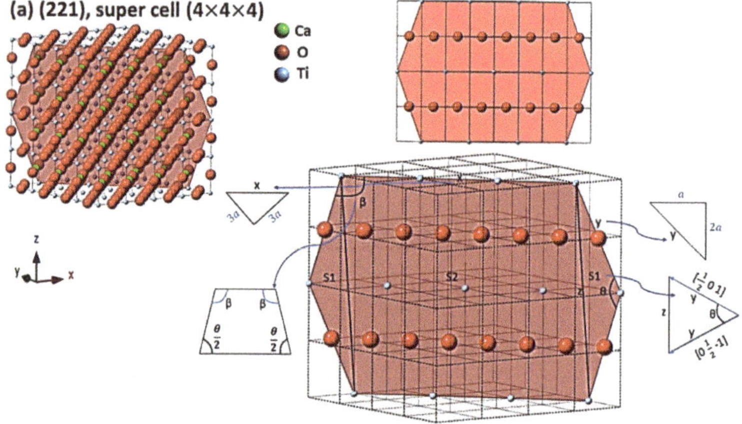

number of atoms in the plane (221) × area of each atom in the plane (221) =

$$\left[\left(\left(2 \times \frac{143.13}{360} + 4 \times \frac{108.44}{360} + 4 \times \frac{1}{2} + 3\right) \times \pi \left(r_{Ti^{4+}}\right)^2\right) + \left(\left(12 + 4 \times \frac{1}{2}\right) \times \pi \left(r_{O^{2-}}\right)^2\right)\right] = [(7 \times \pi (0.60)^2) + (14 \times \pi (1.40)^2)] = 94.12$$

$$\text{Planar density} = \frac{\text{number of atoms in the plane (221)} \times \text{area of each atom in the plane (221)}}{\text{area of the plane (221)}} = \frac{94.12}{301.71} = 0.31$$

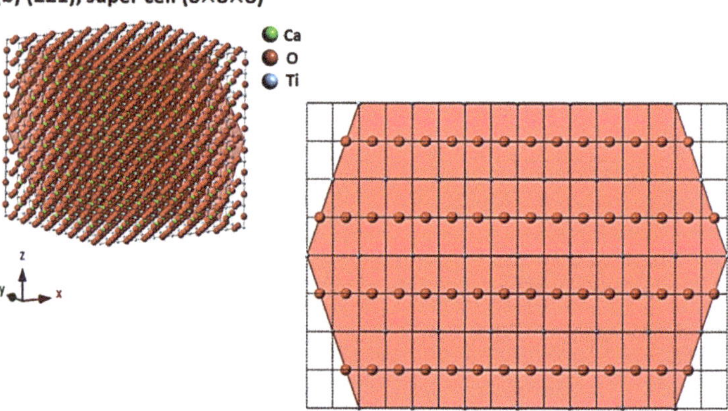

Figure 4. *Cont.*

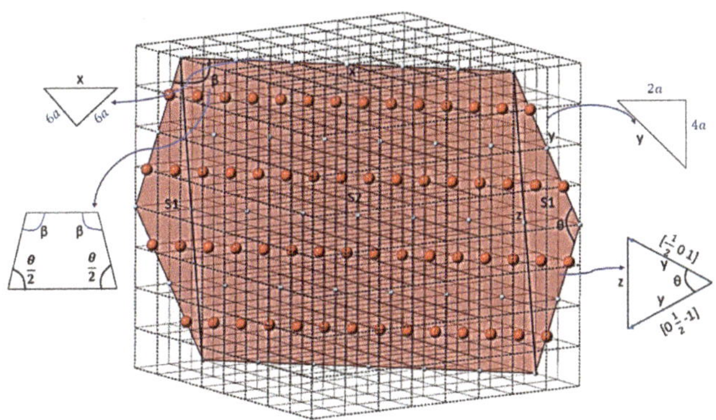

number of atoms in the plane (221) × area of each atom in the plane (221) =

$$\left[\left(\left(2\times\tfrac{143.13}{360}+4\times\tfrac{109.44}{360}+14\times\tfrac{1}{2}+19\right)\times\pi\left(r_{Ti^{4+}}\right)^2\right)+\left(\left(52+8\times\tfrac{1}{2}\right)\times\pi\left(r_{O^{2-}}\right)^2\right)\right]=\left[\left(28\times\pi\,(0.60)^2\right)+\left(56\times\pi\,(1.40)^2\right)\right]=376.49$$

$$\text{Planar density}\ =\frac{\text{number of atoms in the plane (221)}\times\text{area of each atom in the plane (221)}}{\text{area of the plane (221)}}=\frac{376.49}{1206.65}=0.31$$

Figure 4. Geometry of planes and calculations of planar density of (**a**) (221) super cell (4 × 4 × 4) and (**b**) (221) super cell (8 × 8 × 8) (which shows and emphasizes the symmetry of (8 × 8 × 8) super cells).

(a) (311), super cell (3×3×3)

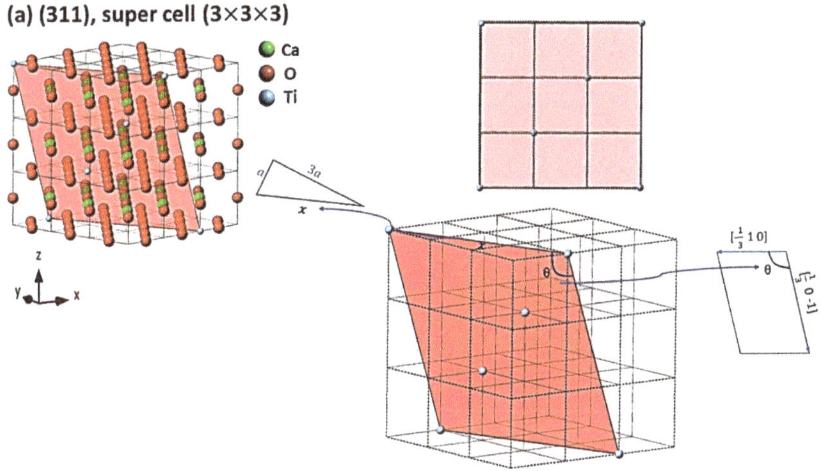

number of atoms in the plane (311) × area of each atom in the plane (311) =

$$\left[\left(\left(2\times\frac{95.74}{360}+2\times\frac{84.26}{360}+2\right)\times\pi\left(r_{Ti^{4+}}\right)^2\right)\right]=\left(3\times\pi\times(0.6)^2\right)=3.39$$

$$\text{Planar density}\ =\frac{\text{number of atoms in the plane (311)}\times\text{area of each atom in the plane (311)}}{\text{area of the plane (311)}}=\frac{3.39}{143.04}=0.02$$

Figure 5. *Cont.*

(b) (311), super cell (4×4×4)

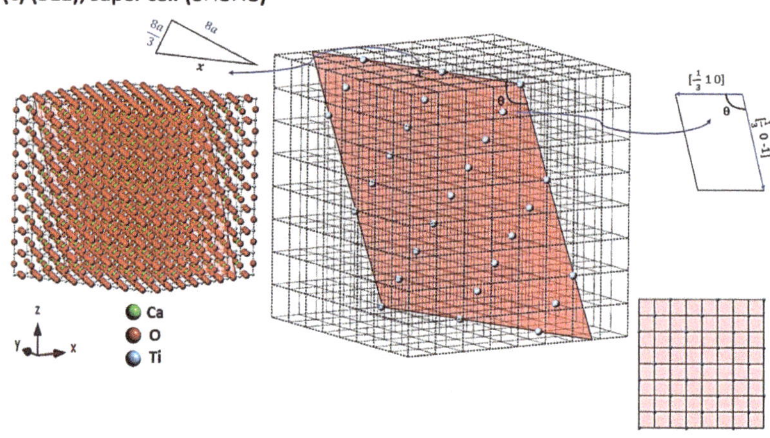

number of atoms in the plane (311) × area of each atom in the plane (311) =

$$\left[\left(\left(2\times\frac{95.74}{360}+4\times\frac{1}{2}+3\right)\times\pi\left(r_{Ti^{4+}}\right)^2\right)\right]=(5.53\times\pi\times(0.6)^2)=6.25$$

$$\text{Planar density}\;=\frac{\text{number of atoms in the plane (311)}\times\text{ area of each atom in the plane (311)}}{\text{area of the plane (311)}}=\frac{6.25}{254.40}=0.02$$

(c) (311), super cell (8×8×8)

number of atoms in the plane (311) × area of each atom in the plane (311) =

$$\left[\left(\left(2\times\frac{95.74}{360}+8\times\frac{1}{2}+17\right)\times\pi\left(r_{Ti^{4+}}\right)^2\right)\right]=(21.53\times\pi\times(0.6)^2)=24.35$$

$$\text{Planar density}\;=\frac{\text{number of atoms in the plane (311)}\times\text{ area of each atom in the plane (311)}}{\text{area of the plane (311)}}=\frac{24.35}{1016.32}=0.02$$

Figure 5. The concept of a symmetry cell; geometry of planes and calculations of planar density of (**a**) (311) super cell (3 × 3 × 3), (**b**) (311) super cell (4 × 4 × 4) and (**c**) (311) super cell (8 × 8 × 8).

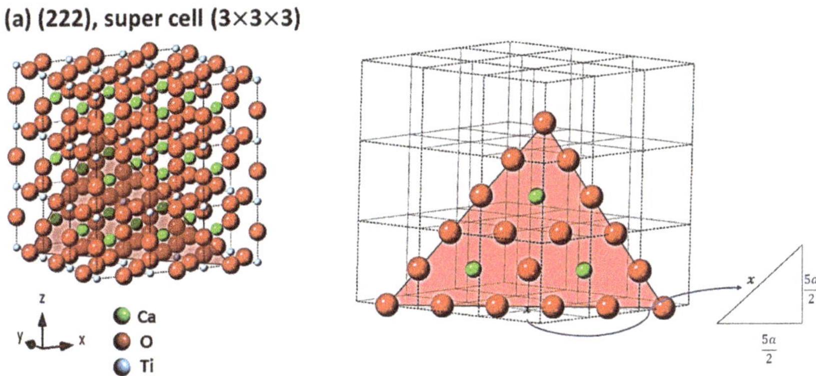

number of atoms in the plane (222) × area of each atom in the plane (222) $= \left[\left(\left(3 \times \frac{1}{6} + 12 \times \frac{1}{2} + 3\right) \times \pi \left(r_{O^{2-}}\right)^2\right)\right] + \left[\left(3 \times \pi \times \left(r_{Ca^+}\right)^2\right)\right] =$

$[(9.5 \times \pi (1.40)^2)] + [(3 \times \pi (1)^2)] = 67.92$

Planar density $= \dfrac{\text{number of atoms in the plane (222)} \times \text{area of each atom in the plane (222)}}{\text{area of the plane (222)}} = \dfrac{67.92}{77.28} = 0.88$

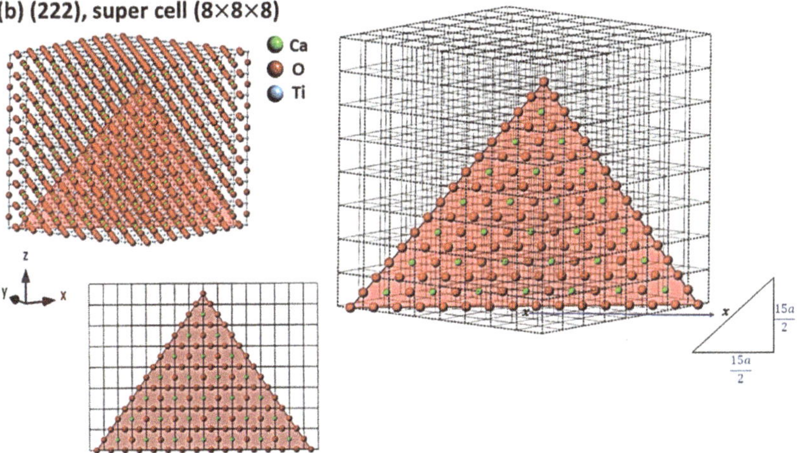

number of atoms in the plane (222) × area of each atom in the plane (222) $= \left[\left(\left(3 \times \frac{1}{6} + 42 \times \frac{1}{2} + 63\right) \times \pi \left(r_{O^{2-}}\right)^2\right)\right] + \left[\left(28 \times \pi \times \left(r_{Ca^+}\right)^2\right)\right] =$

$[(84.5 \times \pi (1.40)^2)] + [(28 \times \pi (1)^2)] = 608.28$

Planar density $= \dfrac{\text{number of atoms in the plane (222)} \times \text{area of each atom in the plane (222)}}{\text{area of the plane (222)}} = \dfrac{608.28}{695.92} = 0.88$

Figure 6. Geometry of planes and calculations of planar density of (**a**) (222) super cell ($3 \times 3 \times 3$) and (**b**) (222) super cell ($8 \times 8 \times 8$).

3.2. Investigation of Results Obtained from Ultrasonic Pulse-Echo Technique of CaTiO₃

Taking into account the Christoffel equation, the connection between ultrasonic phase velocity and the stiffness matrix is given as follows:

$$\left(C_{ijkl}l_jl_l - \rho V^2 \delta_{ik}\right)\alpha_k = 0$$

where V is the ultrasonic phase velocity, C_{ijkl} is the general stiffness matrix, ρ is the material density, l is the orientation of propagation, α_k is the polarization direction and δ_{ik} is the Kronecker delta (note that i, j, k, I = 1 to 3). For the extraction and calculation of elastic constants from ultrasonic measurements based on the Christoffel equation, with the propagation in X_1, X_2 and X_3 directions, all of the diagonal elements of the stiffness matrix are obtained. For the determination of whole constants, we cut the specimen on the edges of the surfaces perpendicular to principal directions (bezel) and the velocity was measured from the propagation of ultrasound wave normal to these planes.

Based on Equations (1)–(5) [26,27] and the measured velocity according to the Table 1, stiffness constants values were obtained. C_{11} is in the agreement with longitudinal distortion and longitudinal compression/tension, so C_{11} can be described as the hardness. Moreover, the transverse distortion is connected to the C_{12}, and C_{12} is obtained from the transverse expansion correlated to the Poisson's ratio. C_{44} is based on the shear modulus, as well as C_{44} is in the settlement with C_{11} and C_{12} [26].

$$C_{11} = \rho V_{\bar{1}}^2 \tag{1}$$

$$C_{22} = \rho V_{\bar{2}}^2 \tag{2}$$

$$C_{66} = \rho V_{\bar{1}}^2 = \rho V_{\bar{2}}^2 \tag{3}$$

$$C_{12} = \sqrt{(C_{11} + C_{66} - 2\rho V_{\overline{12}}^2)(C_{22} + C_{66} - 2\rho V_{\overline{12}}^2)} - C_{66} \tag{4}$$

$$C_{44} = \rho V_{\bar{3}}^2 = \rho V_{\bar{2}}^2 \tag{5}$$

Table 1. The values of longitudinal and transverse velocity of the sample.

Longitudinal Velocity (m/s)	Transverse Velocity (m/s)	Quasi Longitudinal or Quasi Transverse (m/s)
$V_{1/1}$ = 9261.85 $V_{2/2}$ = 8013.51	$V_{2/3}$ = 4960.5 $V_{1/2}$ = 4283.65	$V_{12/12}$ = 4976.63

After substitution and calculation, C_{11}, C_{12} and C_{44} were registered at 330.89, 93.03 and 94.91 GPa respectively. These values of CaTiO₃ were in good agreement with the values submitted in the [28–30]. Moreover, with the ultrasonic technique, longitudinal and transverse waves can be utilized for determining Young's modulus quantity [31,32]. The longitudinal and transverse waves of CaTiO₃ sample are shown in Figure 7. In this method, by measuring the waves velocity and density of specimen, the determination of Young's modulus quantity was carried out (Equation (6)).

$$E = \frac{4\rho\left(\frac{L}{t_s}\right)^2\left(3t_s^2 - 4t_l^2\right)}{t_s^2 - t_l^2} \tag{6}$$

where, t_s and t_l are differences between two echo in longitudinal and transverse waves, respectively [33,34]. According to the results shown in Figure 7, t_s and t_l values are calculated as 5.75 and 3.01 μs, respectively. In addition, the density of the specimen is recorded as 3857.30 $\frac{Kg}{m^3}$, and the length of the specimen after powder pressing reached 11.21 mm. After calculation, Young's modulus value of CaTiO$_3$ was 153.87 GPa. This value corresponds with the value reported by Ramajo et al. [35].

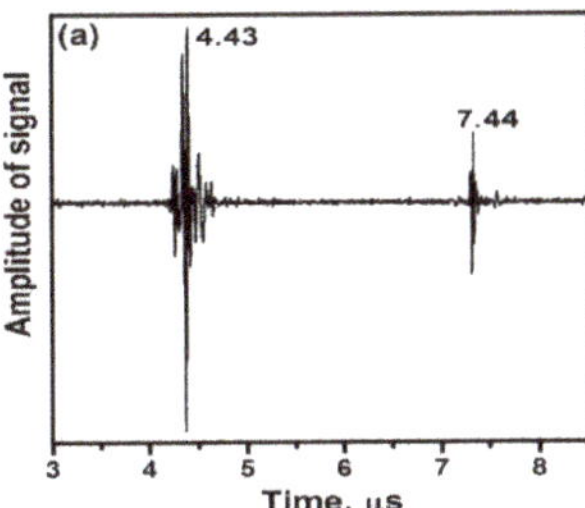

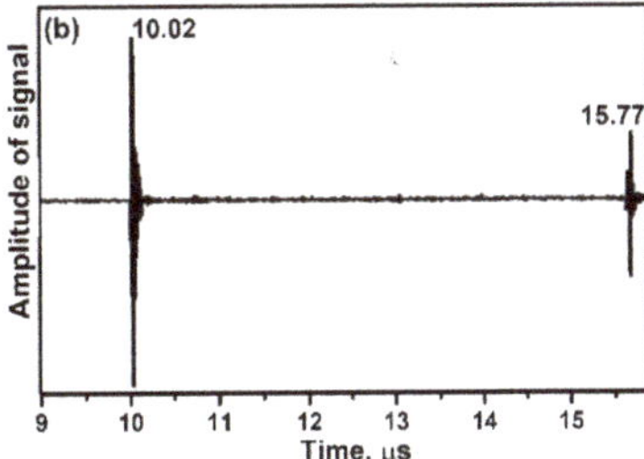

Figure 7. Recorded signals extracted via (**a**) longitudinal waves and (**b**) transverse waves of CaTiO$_3$ specimen.

3.3. Calculations: Relationship between Elastic Stiffness-Compliance Constants, Young's Modulus and Planar Density Extracted through the Unit Cell, Super Cells (2 × 2 × 2) and Symmetry Cells of CaTiO$_3$ Lattice

Three elastic constants of CaTiO$_3$ were calculated via the ultrasonic technique. For the cubic CaTiO$_3$ system, the relationship between stiffness (C_{ij}) and compliance constant (S_{ij}) are provided in Equations (7)–(9) [27,36]. The values resulted via Equations (7)–(9) are 0.0034, -0.0007 and 0.0105 GPa for S_{11}, S_{12} and S_{44}, respectively. Furthermore, Young's modulus of each diffracted plane of CaTiO$_3$ can be written as Equation (10) [37].

$$S_{11} = \frac{C_{11} + C_{12}}{(C_{11} - C_{12})(C_{11} + 2C_{12})} \tag{7}$$

$$S_{12} = \frac{-C_{12}}{(C_{11} - C_{12})(C_{11} + 2C_{12})} \tag{8}$$

$$S_{44} = \frac{1}{C_{44}} \tag{9}$$

$$\frac{1}{E_{hkl}} = S_{11} - 2\left[(S_{11} - S_{12}) - \frac{1}{2}S_{44}\right]\left[\frac{h^2k^2 + k^2l^2 + l^2h^2}{\left(h^2 + k^2 + l^2\right)}\right] \tag{10}$$

The planar density and Young's modulus values related to the each diffracted plane of the unit, super (2 × 2 × 2), symmetry and super (8 × 8 × 8) cells of CaTiO$_3$ lattice are tabulated in Table 2.

Table 2. Planar density and Young's modulus values of the unit cell, super cells (2 × 2 × 2) and symmetry cells of CaTiO$_3$.

Index	Planar Density of Unit Cell	Planar Density of Super Cell (2 × 2 × 2)	Planar Density of Symmetry Cells	Planar Density of Super Cell (8 × 8 × 8)	Young's Modulus (GPa)
(100)	0.93	0.93	0.93 in (2 × 2 × 2)	0.93	290.059
(110)	0.51	0.51	0.51 in (2 × 2 × 2)	0.51	221.652
(111)	0.04	0.04	0.04 in (2 × 2 × 2)	0.04	179.354
(200)	0.64	0.64	0.64 in (2 × 2 × 2)	0.64	290.059
(210)	0.41	0.41	0.41 in (2 × 2 × 2)	0.41	194.176
(211)	0.16	0.25	0.25 in (2 × 2 × 2)	0.25	150.612
(220)	0.6	0.6	0.6 in (2 × 2 × 2)	0.6	129.810
(221)	0.46	0.29	0.31 in (4 × 4 × 4)	0.31	109.622
(310)	0.24	0.24	0.23 in (4 × 4 × 4)	0.23	186.471
(311)	0.04	0.03	0.02 in (3 × 3 × 3)	0.02	140.386
(222)	0.99	0.88	0.88 in (3 × 3 × 3)	0.88	83.615

4. Discussion

According to Table 2 and Figures 3–6, the expanded cells have an optimum matrix, and in this case, achieving the optimum matrix is introduced as the symmetry cells. An optimum matrix is the minimum extension for a specific plane of the unit cell to a super cell from which the density plane of that plane does not change. For example, symmetry cell (optimum matrix) of (311) plane is ($3 \times 3 \times 3$), which means that after extending to a greater matrix such as ($4 \times 4 \times 4$) or ($8 \times 8 \times 8$), planar density values will be similar (Figure 5a–c). Real planar density values of each plane are obtained from the symmetry cell, because once the symmetry of each plane is reached, with the extension of that plane to infinity (real plane), the planar density does not change. In addition, to recognize the symmetry cell, knowing some parameters such as crystal lattice, locations of atoms in the planes and index of planes is essential. Therefore, to determine Young's modulus values based on the planar density of $CaTiO_3$, the symmetry cells should be found. It is very interesting that symmetrical or real values are related to the super cells of the ($8 \times 8 \times 8$) matrix, because in matrix ($8 \times 8 \times 8$), lattice correspondence in all directions is available; therefore, real planar density values should be calculated for the super cell of ($8 \times 8 \times 8$) matrix. To confirm this, calculations of real planar density and geometry of atoms and planes of (211), (221), (311) and (222) in super cells ($8 \times 8 \times 8$) are presented in Figures 3b, 4b, 5c and 6b, respectively. It is clear that finding the exact situation of planes and geometries is sophisticated, but with when they are known, the results obtained from Young's modulus values will have fewer errors. The basic supposition is that when the planar density is raised, the motion of atoms with the mechanism of dislocation movement needs high forces. Dislocations are regions in the lattice where an additional plane of atoms have been included abstracted from an ideal crystal (without imperfections). Dislocations are caused by losing acoustic energy, and this matter will affect the values of wavelength and time of ultrasonic waves [38].

The force (W), which is needed for the movement of atoms in each plane, is obtained from Equation (11) [39].

$$W = \frac{E}{2(1 + v)} \, b^2 l \tag{11}$$

In Equation (11), E is Young's modulus, b is Burgers vector, l is dislocation length and v is Poisson's ratio. The higher value of force is in accordance with the modulus of elasticity (Young's modulus), which would be higher.

To compare Young's modulus values of $CaTiO_3$ in a unit cell, super cells ($2 \times 2 \times 2$) and symmetry cells, the fitting of Young's modulus values extracted by each diffracted plane versus planar density values is presented in Figure 8. According to the results (shown in the Figure 8) and the straight fitting line, Young's modulus values of unit cell, super cells ($2 \times 2 \times 2$) and symmetry cells were calculated as 162.62 ± 0.4 GPa, 151.71 ± 0.4 GPa and 152.21 ± 0.4 GPa, respectively. As expected, the Young's modulus value of symmetry cells of $CaTiO_3$ (152.21 ± 0.4 GPa) is in good agreement with experimental Young's modulus value extracted via ultrasonic-echo technique (153.87 ± 0.2 GPa). Moreover, Young's modulus value of unit cell (162.62 ± 0.4 GPa) has a greater difference with experimental Young's modulus value, and as a result, the unit cell of $CaTiO_3$ cannot be represented as whole cells. This is because in a unit cell of $CaTiO_3$, crystalline defects are not considered and is especially controlling of deformation, and displacement of atoms in the planes is related to the dislocation networks [40]. Further, a unit cell of $CaTiO_3$ is not involved in imperfections (such as dislocations, Frenkel defect and Schottky defect) with respect to the super cell [41]; therefore, the slope line value of the unit cell is reported (37.23) to be less than the slope line value of super cells ($2 \times 2 \times 2$) (63.67) and symmetry cells (62.41). Consequently, the effect of imperfections in expanded cells (super cells) is very impressive, so the unit cell of $CaTiO_3$ is considered as the ideal lattice, while symmetry cells such as ($8 \times 8 \times 8$) of $CaTiO_3$ are real lattices [42]; this is consistent with the experimental Young's modulus. It is clear that each imperfection will be caused by a decreasing Young's modulus [43], and in Figure 8, this matter is confirmed when the Young's modulus value (intercept) in the unit cell of $CaTiO_3$ is higher than in super cells ($2 \times 2 \times 2$) and symmetry cells. Apparently, a unit cell of $CaTiO_3$ is represented by the volume of a real crystal, so the unit cell is useful to acquire theoretical density. Nevertheless, calculations of planar density based on the unit cell were obtained, but with errors.

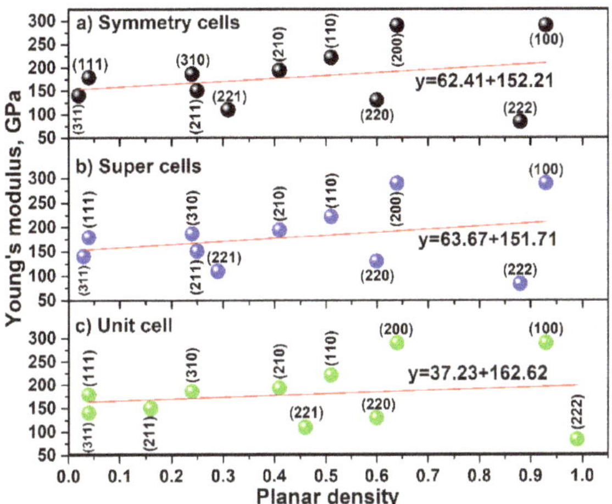

Figure 8. Young's modulus versus planar density values of each diffracted plane related to the (**a**) symmetry cells, (**b**) super cells (2 × 2 × 2) and (**c**) unit cell of $CaTiO_3$.

5. Conclusions

1. $CaTiO_3$ as a category of perovskite is successfully synthesized via solvothermal method.
2. Crystal size values of $CaTiO_3$ are calculated as ~59.10 and 63.02 through the Monshi-Scherrer method and BET analysis, and the crystal size values were confirmed by TEM image.
3. Planar density is responsible for modulus of elasticity of that plane; therefore, for the first time, comprehensive calculations of geometry, location and planar density values of $CaTiO_3$ were shown.
4. Elastic stiffness constants and Young's modulus values of $CaTiO_3$ were obtained by ultrasonic-echo method (C_{11} = 330.89, C_{12} = 93.03, C_{44} = 94.91 GPa and E = 153.87 ± 0.2 GPa).
5. Young's modulus values of $CaTiO_3$ extracted by planar density and least square method were calculated as 162.62 ± 0.4, 151.71 ± 0.4 and 152.21 ± 0.4 GPa for unit cell, super cells (2 × 2 × 2) and symmetry cells, respectively.
6. The Young's modulus value of $CaTiO_3$ reported by symmetry cells is in good agreement with Young's modulus value reported by ultrasonic-echo technique and the literatre.
7. A unit cell of $CaTiO_3$ is not representative of the distribution of atoms on the planes; therefore, to obtain the real value of planar density and find the symmetry of distribution of atoms on the planes, expanded cells and utilizing symmetry cells are suggested.
8. Obtaining the planar density values based on unit cell or each super cells except for (8 × 8 × 8) is an estimation.
9. The real value of Young's modulus of $CaTiO_3$ should be extracted by symmetry cells or super cells (8 × 8 × 8).
10. The value of Young's modulus of $CaTiO_3$ extracted with this method can be applied for industrial applications.

Supplementary Materials: The following are available online at https://www.mdpi.com/1996-1944/14/5/1258/s1, Figure S1: Synthesis route of CaTiO$_3$, Table S1: Crystallographic parameters of each individual XRD pattern related to the CaTiO$_3$, Figure S2: Linear plot of modified Scherrer equation related to the CaTiO$_3$, Figure S3: TEM image of CaTiO$_3$ powder, Figure S4: Geometry of planes and calculations of planar density of (a) (100), (b) (110), (c) (111), (d) (200), (e) (210), (f) (211), (g) (220), (h) (221), (i) (310), (j) (311) and (k) (222) related to the unit cell of CaTiO$_3$, Figure S5: Geometry of planes and calculations of planar density of (a) (100), (b) (110), (c) (111), (d) (200), (e) (210), (f) (211), (g) (220), (h) (221), (i) (310), (j) (311) and (k) (222) related to the super cells (2 × 2 × 2) of CaTiO$_3$, Figure S6: Geometry of planes and calculations of planar density of (a) (100), (b) (110), (c) (111), (d) (200), (e) (210), (f) (220), (g) (310) (4 × 4 × 4) and (h) (310) (8 × 8 × 8) related to the super cells (8 × 8 × 8) of CaTiO$_3$, Figure S7: Geometry of planes and calculations of planar density of (a) (211) super cell (4 × 4 × 4) and (b) (211) super cell (8 × 8 × 8), Figure S8: Geometry of planes and calculations of planar density of (a) (221) super cell (4 × 4 × 4) and (b) (221) super cell (8 × 8 × 8), Figure S9: Geometry of planes and calculations of planar density of (a) (311) super cell (3 × 3 × 3), (b) (311) super cell (4 × 4 × 4) and (c) (311) super cell (8 × 8 × 8), Figure S10: Geometry of planes and calculations of planar density of (a) (222) super cell (3 × 3 × 3), (b) (222) super cell (8 × 8 × 8).

Author Contributions: Conceptualization, M.R. and A.P.; methodology, M.R. and S.N.; investigation, A.V. and A.D.; data curation, A.D.; writing—original draft, M.R.; writing—review and editing, S.N. and G.J.; resources, G.J.; supervision and validation A.P. and G.J. All authors have read and agreed to the published version of the manuscript.

Funding: This research was funded by a grant No. S-MIP-19-43 from the Research Council of Lithuania.

Data Availability Statement: Data supporting the findings of this study are available from the corresponding author upon request.

Conflicts of Interest: The authors declare no conflict of interest.

References

1. Atta, N.F.; Galal, A.; El-Ads, E.H. *Perovskite Nanomaterials-Synthesis, Characterization, and Applications*; Intechopen: London, UK, 2016.
2. Reshmi Varma, P.C. Low-dimensional perovskites. In *Perovskite Photovoltaics: Basic to Advanced Concepts and Implementation*; Elsevier: Amsterdam, The Netherlands, 2018; pp. 197–229. ISBN 9780128129159.
3. Nakamura, T.; Sun, P.H.; Shan, Y.J.; Inaguma, Y.; Itoh, M.; Kim, I.N.S.; Sohn, J.H.O.; Ikeda, M.; Kitamura, T.; Konagaya, H. On the perovskite-related materials of high dielectric permittivity with small temperature dependence and low dielectric loss. *Ferroelectrics* **1997**, *196*, 205–209. [CrossRef]
4. Wang, X.; Xu, C.N.; Yamada, H.; Nishikubo, K.; Zheng, X.G. Electro-mechano-optical conversions in Pr^{3+}-doped BaTiO$_3$-CaTiO$_3$ ceramics. *Adv. Mater.* **2005**, *17*, 1254–1258. [CrossRef]
5. Ma, Y.Z.; Sobernheim, D.; Garzon, J.R. Glossary for Unconventional Oil and Gas Resource Evaluation and Development. In *Unconventional Oil and Gas Resources Handbook: Evaluation and Development*; Elsevier Inc.: Amsterdam, The Netherlands, 2016; pp. 513–526. ISBN 9780128025369.
6. Slaughter, W.S. *The Linearized Theory of Elasticity*; Birkhäuser Boston: New York, NY, USA, 2002.
7. Kuma, S.; Woldemariam, M.M. Structural, Electronic, Lattice Dynamic, and Elastic Properties of SnTiO$_3$ and PbTiO$_3$ Using Density Functional Theory. *Adv. Condens. Matter Phys.* **2019**, *2019*. [CrossRef]
8. Nakamura, M. Elastic constants of some transition- metal- disilicide single crystals. *Metall. Mater. Trans. A* **1994**, *25*, 331–340. [CrossRef]
9. Paszkiewicz, T.; Wolski, S. Elastic properties of cubic crystals: Every's versus Blackman's diagram. *J. Phys. Conf. Ser.* **2008**, *104*, 012038. [CrossRef]
10. Ching, W.Y.; Rulis, P.; Misra, A. Ab initio elastic properties and tensile strength of crystalline hydroxyapatite. *Acta Biomater.* **2009**, *5*, 3067–3075. [CrossRef]
11. Holec, D.; Friák, M.; Neugebauer, J.; Mayrhofer, P.H. Trends in the elastic response of binary early transition metal nitrides. *Phys. Rev. B-Condens. Matter Mater. Phys.* **2012**, *85*, 64101. [CrossRef]
12. Knowles, K.M.; Howie, P.R. The Directional Dependence of Elastic Stiffness and Compliance Shear Coefficients and Shear Moduli in Cubic Materials. *J. Elast.* **2015**, *120*, 87–108. [CrossRef]
13. Stair, K.; Liu, J.; Asta, M. Undefined Ultra-sonic measurement and computation of elastic constants. In Proceedings of the ASEE Annual Conference and Exposition, Kansas City, MO, USA, 13–15 September 2006.
14. Kelly, A.; Knowles, K.M. *Crystallography and Crystal Defects*; John Wiley & Sons, Ltd.: Chichester, UK, 2012; ISBN 9781119961468.
15. Epp, J. X-ray Diffraction (XRD) Techniques for Materials Characterization. In *Materials Characterization Using Nondestructive Evaluation (NDE) Methods*; Elsevier Inc.: Amsterdam, The Netherlands, 2016; pp. 81–124. ISBN 9780081000571.

16. Badawi, F.; Villain, P. Stress and elastic-constant analysis by X-ray diffraction in thin films. *J. Appl. Crystallogr.* **2003**, *36*, 869–879. [CrossRef]

17. Waseda, Y.; Matsubara, E.; Shinoda, K. *X-ray Diffraction Crystallography X-ray Diffraction Crystallography Introduction, Examples and Solved Problems*; Springer: Berlin/Heidelberg, Germany, 2011.

18. Dong, W.; Song, B.; Meng, W.; Zhao, G.; Han, G. A simple solvothermal process to synthesize $CaTiO_3$ microspheres and its photocatalytic properties. *Appl. Surf. Sci.* **2015**, *349*, 272–278. [CrossRef]

19. Wang, Y.; Niu, C.G.; Wang, L.; Wang, Y.; Zhang, X.G.; Zeng, G.M. Synthesis of fern-like $Ag/AgCl/CaTiO_3$ plasmonic photocatalysts and their enhanced visible-light photocatalytic properties. *RSC Adv.* **2016**, *6*, 47873–47882. [CrossRef]

20. *ASTM E797—95 Standard Practice for Measuring Thickness by Manual Ultrasonic Pulse-Echo Contact Method*; ASTM International: West Conshohocken, PA, USA, 2001.

21. *ASTM E214—05 Standard Practice for Immersed Ultrasonic Testing by the Reflection Method Using Pulsed Longitudinal Waves*; ASTM International: West Conshohocken, PA, USA, 2007.

22. Sahoo, S.; Parashar, S.K.S.; Ali, S.M. $CaTiO_3$ nano ceramic for NTCR thermistor based sensor application. *J. Adv. Ceram.* **2014**, *3*, 117–124. [CrossRef]

23. Cockayne, E.; Burton, B.P. Phonons and static dielectric constant in $CaTiO_3$ from first principles. *Phys. Rev. B-Condens. Matter Mater. Phys.* **2000**, *62*, 3735–3743. [CrossRef]

24. Janusas, T.; Urbaite, S.; Palevicius, A.; Nasiri, S.; Janusas, G. Biologically Compatible Lead-Free Piezoelectric Composite for Acoustophoresis Based Particle Manipulation Techniques. *Sensors* **2021**, *21*, 483. [CrossRef] [PubMed]

25. Rabiei, M.; Palevicius, A.; Monshi, A.; Nasiri, S.; Vilkauskas, A.; Janusas, G. Comparing Methods for Calculating Nano Crystal Size of Natural Hydroxyapatite Using X-ray Diffraction. *Nanomaterials* **2020**, *10*, 1627. [CrossRef]

26. Li, Y.; Thompson, R.B. Relations between elastic constants Cij and texture parameters for hexagonal materials. *J. Appl. Phys.* **1990**, *67*, 2663–2665. [CrossRef]

27. Huntington, H.B. The Elastic Constants of Crystals. *Solid State Phys.-Adv. Res. Appl.* **1958**, *7*, 213–351.

28. De Jong, M.; Chen, W.; Angsten, T.; Jain, A.; Notestine, R.; Gamst, A.; Sluiter, M.; Ande, C.K.; Van Der Zwaag, S.; Plata, J.J.; et al. Charting the complete elastic properties of inorganic crystalline compounds. *Sci. Data* **2015**, *2*, 150009. [CrossRef]

29. Tariq, S.; Ahmed, A.; Saad, S.; Tariq, S. Structural, electronic and elastic properties of the cubic $CaTiO_3$ under pressure: A DFT study. *AIP Adv.* **2015**, *5*, 077111. [CrossRef]

30. Sakhya, A.P.; Maibam, J.; Saha, S.; Chanda, S.; Dutta, A.; Sharma, B.I.; Thapa, R.; Sinha, T. Electronic structure and elastic properties of $ATiO_3$ (A = Ba, Sr, Ca) perovskites: A first principles study. *IJPAP* **2015**, *53*, 102–109.

31. Wang, H.; Prendiville, P.L.; McDonnell, P.J.; Chang, W.V. An ultrasonic technique for the measurement of the elastic moduli of human cornea. *J. Biomech.* **1996**, *29*, 1633–1636. [CrossRef]

32. Bray, D.E.; Stanley, R.K. *Nondestructive Evaluation: A Tool in Design, Manufacturing and Service*; CRC press: Boca Raton, FL, USA, 1996.

33. Figliola, R.S.; Beasley, D.E. *Theory and Design for Mechanical Measurements*, 7th ed.; John Wiley & Sons: Hoboken, NJ, USA, 2019.

34. Rabiei, M.; Palevicius, A.; Dashti, A.; Nasiri, S.; Monshi, A.; Vilkauskas, A.; Janusas, G. Measurement Modulus of Elasticity Related to the Atomic Density of Planes in Unit Cell of Crystal Lattices. *Materials* **2020**, *13*, 4380. [CrossRef]

35. Ramírez, M.A.; Parra, R.; Reboredo, M.M.; Varela, J.A.; Castro, M.S.; Ramajo, L. Elastic modulus and hardness of $CaTiO_3$, $CaCu_3Ti_4O_{12}$ and $CaTiO_3/CaCu_3Ti_4O_{12}$ mixture. *Mater. Lett.* **2010**, *64*, 1226–1228. [CrossRef]

36. Papaconstantopoulos, D.A.; Mehl, M.J. Tight-Binding Method in Electronic Structure. In *Encyclopedia of Condensed Matter Physics*; Elsevier Inc.: Amsterdam, The Netherlands, 2005; pp. 194–206. ISBN 9780123694010.

37. Zhang, J.M.; Zhang, Y.; Xu, K.W.; Ji, V. Young's modulus surface and Poisson's ratio curve for cubic metals. *J. Phys. Chem. Solids* **2007**, *68*, 503–510. [CrossRef]

38. Mason, W.P. Effect of Dislocations on Ultrasonic Wave Attenuation in Metals. *Bell Syst. Tech. J.* **1955**, *34*, 903–942. [CrossRef]

39. Reed-Hill, R.E.; Abbaschian, R. *PHYSICAL METALLURGY PRINCIPLES*; Boston PWS Publishing Company: Boston, MA, USA, 2009; ISBN 978-0-495-08254-5.

40. Berdichevsky, V. Energy of dislocation networks. *Int. J. Eng. Sci.* **2016**, *103*, 35–44. [CrossRef]

41. Hull, D.; Bacon, D.J. *Introduction to Dislocations*, 5th ed.; Butterworth-Heinemann: Burlington, MA, USA, 2011; ISBN 9780080966724.

42. Englert, U. Symmetry Relationships between Crystal Structures. Applications of Crystallographic Group Theory in Crystal Chemistry. By Ulrich Müller. *Angew. Chem. Int. Ed.* **2013**, *52*, 11973. [CrossRef]

43. Duparc, O.H.; Polytechnique, É.; Duparc, O.B.M.H. A review of some elements in the history of grain boundaries, centered on Georges Friedel, the coincident "site" lattice and the twin index Crystallography History View project ANR FluTi View project A review of some elements in the history of grain boundaries, centered on Georges Friedel, the coincident "site" lattice and the twin index. *Artic. J. Mater. Sci.* **2011**, *46*, 4116–4134.

Article

X-ray Diffraction Analysis and Williamson-Hall Method in USDM Model for Estimating More Accurate Values of Stress-Strain of Unit Cell and Super Cells (2 × 2 × 2) of Hydroxyapatite, Confirmed by Ultrasonic Pulse-Echo Test

Marzieh Rabiei [1,*], Arvydas Palevicius [1,*], Amir Dashti [2], Sohrab Nasiri [1], Ahmad Monshi [3], Akram Doustmohammadi [4], Andrius Vilkauskas [1] and Giedrius Janusas [1,*]

[1]   Faculty of Mechanical Engineering and Design, Kaunas University of Technology, LT-51424 Kaunas, Lithuania; sohrab.nasiri@ktu.edu (S.N.); andrius.vilkauskas@ktu.lt (A.V.)

[2]   Department of Materials Science and Engineering, Sharif University of Technology, Tehran 11365-9466, Iran; a.dashty@merc.ac.ir

[3]   Department of Materials Engineering, Isfahan University of Technology, Isfahan 84154, Iran; dr.ahmad.monshi@gmail.com

[4]   Materials and Energy Research Center (MERC), Meshkin-Dasht, Karaj 31787-316, Iran; hivadoostmohammadi91@gmail.com

*   Correspondence: marzieh.rabiei@ktu.edu (M.R.); arvydas.palevicius@ktu.lt (A.P); giedrius.janusas@ktu.lt (G.J.)

Citation: Rabiei, M.; Palevicius, A.; Dashti, A.; Nasiri, S.; Monshi, A.; Doustmohammadi, A.; Vilkauskas, A.; Janusas, G. X-ray Diffraction Analysis and Williamson-Hall Method in USDM Model for Estimating More Accurate Values of Stress-Strain of Unit Cell and Super Cells (2 × 2 × 2) of Hydroxyapatite, Confirmed by Ultrasonic Pulse-Echo Test. Materials 2021, 14, 2949. https://doi.org/10.3390/ma14112949

Academic Editor: Thomas Walter Cornelius

Received: 28 March 2021
Accepted: 28 May 2021
Published: 30 May 2021

Publisher's Note: MDPI stays neutral with regard to jurisdictional claims in published maps and institutional affiliations.

Abstract: Taking into account X-ray diffraction, one of the well-known methods for calculating the stress-strain of crystals is Williamson-Hall (W–H). The W-H method has three models, namely (1) Uniform deformation model (UDM); (2) Uniform stress deformation model (USDM); and (3) Uniform deformation energy density model (UDEDM). The USDM and UDEDM models are directly related to the modulus of elasticity (E). Young's modulus is a key parameter in engineering design and materials development. Young's modulus is considered in USDM and UDEDM models, but in all previous studies, researchers used the average values of Young's modulus or they calculated Young's modulus only for a sharp peak of an XRD pattern or they extracted Young's modulus from the literature. Therefore, these values are not representative of all peaks derived from X-ray diffraction; as a result, these values are not estimated with high accuracy. Nevertheless, in the current study, the W-H method is used considering the all diffracted planes of the unit cell and super cells (2 × 2 × 2) of Hydroxyapatite (HA), and a new method with the high accuracy of the W-H method in the USDM model is presented to calculate stress (σ) and strain (ε). The accounting for the planar density of atoms is the novelty of this work. Furthermore, the ultrasonic pulse-echo test is performed for the validation of the novelty assumptions.

Keywords: Williamson-Hall (W-H); uniform stress deformation model (USDM); Young's modulus; X-ray diffraction; hydroxyapatite; planar density; ultrasonic pulse-echo

1. Introduction

Young's modulus (E) plays an important role in the characterization of mechanical properties, and accurate knowledge of the engineering values of elastic modulus is vital for understanding materials' stiffness [1]. Recently, ceramic materials have been favored in industrial applications, because they exhibit good mechanical properties, such as high elastic moduli and high hardness [2]. One of the well-known bio ceramics is hydroxyapatite, which has bioactive properties very close to natural bone mineral and it has special biological significance [3]. The chemical formula of hydroxyapatite is $Ca_{10}(PO_4)_6(OH)_2$ and it differs very little from bone tissue [4,5]. Understanding the mechanical properties of hydroxyapatite during the crystallization and growth stages of the synthesis processes is important because the Young's modulus affects the growth of hydroxyapatite crystal in mechanically strained environments directly [6]. Therefore, paying attention to the mechanical properties and structural geometry of hydroxyapatite can be helpful for research and industrial applications. Paying attention to the details of the structural geometry of hydroxyapatite is essential for employing an easy, cost-effective and reliable method to determine the Young's modulus. In this research, we have developed a method based on the linear regression of the Young's modulus of each plane of the crystal lattice versus planar density to obtain a reliable total Young's modulus of materials. Hence, in this study, for the first time, we calculated the exact planar density derived from the diffracted planes of hydroxyapatite in unit cells and super cells ($2 \times 2 \times 2$). Then, we determined and investigated the total Young's modulus of samples. Furthermore, to determine the effect of cell size on the Young's modulus, an extensive, exact calculation of the planar density of super cells ($2 \times 2 \times 2$) of hydroxyapatite, and comparing it with the result obtained from unit cell calculations, was performed. The Williamson-Hall (W-H) method is a procedure to analyse stress and strain derived from X-ray diffraction. Ultrasonic pulse-echo is a scan representation commonly used for thickness measurement and sizing of the defect in an ultrasonic test in which the signal is reflected from a discontinuity in a test of material structure. This test is performed in this study for the confirmation of the validity of the novelty assumptions. The configuration of the ultrasonic pulse echo is called the acoustic sound energy and localizes the discontinuities or defect indication. In addition, ultrasonic pulse echo measurements can be used to determine the elastic constant and elastic compliance of compounds [7]. According to the W-H method, the basic calculation for the plot can be performed by using the XRD data. The big problem with utilizing the W-H method in the USDM model is attributed to the values of Young's modulus in the equation. Because in all studies and research, the values of Young's modulus have been reported for one sharp peak of an XRD pattern or the average values of Young's modulus; both of them have an error because the values are not representative of whole diffracted planes. For example, Ratan et al. calculated the stress (σ) and strain (ε) of cadmium selenide (CdSe) nanoparticles and they considered the average value of Young's modulus from the Williamson-Hall (W-H) method in the USDM model [8]. Furthermore, Khorsand et al. reported the Young's modulus value of ZnO nanoparticles, considering a sharp peak of X-ray diffraction of ZnO [9]. In another study, Rabiei et al. presented the USDM model of the Young's modulus value for hydroxyapatite and the Young's modulus value was considered an average of diffracted planes [5]. Moreover, Madhavi et al. utilized the USDM model of the Young's modulus value for VO_2 doped ZnS/CdS composite nanopowder and they calculated the Young's modulus value derived from average data. Rameshbabu et al. submitted the W-H method in the USDM model for calculating stress and strain values of hydroxyapatite and they also utilized the average value of Young's modulus [10]. In all research that has utilized the W-H method in the USDM model, the Young's modulus values of the investigated materials are reported as an average value or selected through the literature, and reported values are not inclusive of high accuracy. Interestingly, the values of the elastic compliance constant of studied materials were also derived from the literature, whereas each material has a special elastic compliances constant related to itself; therefore, with reference to the literature and utilizing the compliance constant values of studies, the accuracy of the report would be decreased. In addition, the W-H method is well suited to calculating and estimating the stress and strain of materials [10]. By applying our method and deploying the results of the W-H method, the stress-strain of a sample can be calculated with high accuracy. We have used both DFT calculations and ultrasonic measurements to compare and evaluate the validity of the proposed results (details are in the Supplementary materials). Overall, the evaluated results and the extracted values of this study were in good agreement with the theoretical, experimental and literature values.

2. Methods

The synthesis route of hydroxyapatite powder is explained completely in part 2 of the Support Information (preparation of hydroxyapatite powder).

2.1. Structural Analyses of Synthesized Hydroxyapatite

The XRD pattern of synthesized hydroxyapatite powder is shown in Figure 1. The XRD pattern exhibits several diffraction peaks, which can be indexed as the hexagonal hydroxyapatite. The XRD pattern was evaluated based on X'pert and the pattern was in agreement with the standard XRD peaks of hydroxyapatite (ICDD 9-432). Similar observations were reported in References [11,12]. In addition, crystallographic parameters and details of each diffracted plane of hydroxyapatite were evaluated by X'pert and the values are tabulated in Table 1 and Table S1.

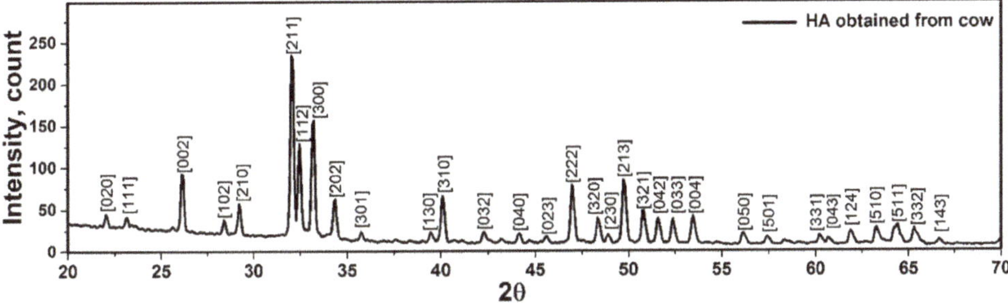

Figure 1. X-ray diffraction pattern of hydroxyapatite synthesized at 950 °C.

Table 1. Crystallographic parameters of hydroxyapatite structure.

Crystal System	a (Å)	c (Å)	Cell Volume (Å)^3	Crystal Density (g/cm^3)	Space Group
HCP	9.400	6.930	530.301	3.140	P6$_3$/m

According to Table S1, the values of the distance between planes are calculated by Equation (1). In this equation, h, k and l are indices of each plane, and a, c and d are lattice parameters and distance of planes, respectively [13].

$$\frac{1}{d^2} = \frac{4}{3}\left(\frac{h^2 + hk + k^2}{a^2}\right) + \frac{l^2}{c^2} \tag{1}$$

Hydroxyapatite has a hexagonal system with a P6$_3$/m space group and has little deviation from stoichiometry [14]. Figure 2 shows a sketch of a unit cell of hexagonal hydroxyapatite and a cif file of synthesized hydroxyapatite. There are two different situations of calcium ions and, in total, 18 ions are closely packed to create the hexagonal structure. At each hexagonal corner, a calcium ion is restricted by 12 calcium ions shared with 3 hexagons. Void spaces between two hexagons are filled with three phosphate tetrahedral per unit cell. Ions in hydroxyapatite can be interchangeably replaced with biologically useful ions due to the inherent versatility of this crystal structure and can also be referred to as doping. In addition, the substitution of calcium, phosphate and/or hydroxyl ions is possible [15]. Notably, the specific feature of hydroxyapatite is related to the OH$^-$ ions forming inner channels along the c axis. This property plays an important role in its mechanical and physical properties [16]. In addition, the Edax analysis of synthesized hydroxyapatite is presented in Figure S2. According to the EDX analysis, the value of the Ca/P ratio for hydroxyapatite obtained from bovine bone was recorded to be 1.60. In addition, thermal decomposition of hydroxyapatite into tricalcium phosphate and tetra calcium phosphate was not observed during the sintering, as in References [17,18], so the hydroxyapatite was successfully synthesized.

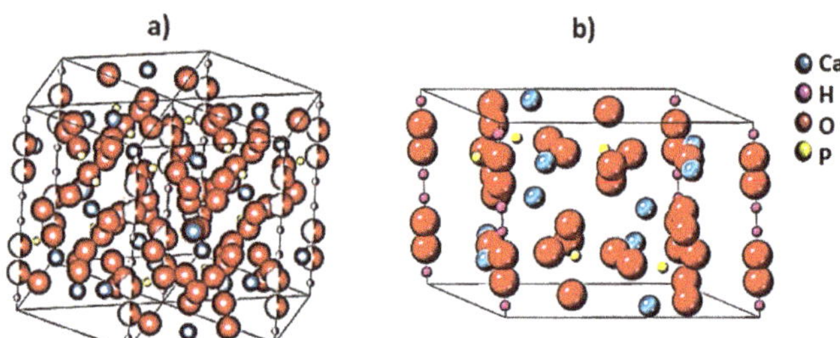

Figure 2. Schematic representation of (**a**) hydroxyapatite unit cell and (**b**) the hydroxyapatite structure extracted by cif file.

2.2. Planar Density of Unit Cell and Super Cells (2 × 2 × 2) of Hydroxyapatite

According to the resulting list of planes by X-ray diffraction (Figure 1, Table S1), the planar density values of each diffracted plane in unit cell and super cells (2 × 2 × 2) of hydroxyapatite are calculated in Figure 3, Figures S6 and S7. To calculate the planar density values, diffracted planes were selected from a low angle to a high angle in tandem. It is worth mentioning that the matrix of all super cell lattices was considered to be 2 × 2 × 2. According to the center of atoms, the planar density is calculated by the area of the atoms in the plane divided by the total area of that plane [19].

2.3. Elastic Stiffness Constant and Elastic Compliance of Hydroxyapatite

Hooke's law is shown in Equation (2); the stress corresponds to the strain for small displacements. It is the basic form, that this symmetry can be converted to the six items of stress (σ) and strain (ε) [20].

$$
\begin{pmatrix} \sigma_{xx} \\ \sigma_{yy} \\ \sigma_{zz} \\ \sigma_{yz} \\ \sigma_{zx} \\ \sigma_{xy} \end{pmatrix}
=
\begin{pmatrix}
C_{11} & C_{12} & C_{13} & C_{14} & C_{15} & C_{16} \\
C_{21} & C_{22} & C_{23} & C_{24} & C_{25} & C_{26} \\
C_{31} & C_{32} & C_{33} & C_{34} & C_{35} & C_{36} \\
C_{41} & C_{42} & C_{43} & C_{44} & C_{45} & C_{43} \\
C_{51} & C_{52} & C_{53} & C_{54} & C_{55} & C_{56} \\
C_{61} & C_{62} & C_{63} & C_{64} & C_{65} & C_{66}
\end{pmatrix}
=
\begin{pmatrix} \varepsilon_{xx} \\ \varepsilon_{yy} \\ \varepsilon_{zz} \\ \varepsilon_{yz} \\ \varepsilon_{zx} \\ \varepsilon_{xy} \end{pmatrix}
\tag{2}
$$

Additionally, Hooke's law can be written (Equation (3)):

$$
\begin{aligned}
\sigma_{xx} &= C_{11}\varepsilon_{xx}+C_{12}\varepsilon_{yy}+C_{13}\varepsilon_{zz}+C_{14}\varepsilon_{yz}+C_{15}\varepsilon_{zx}+C_{16}\varepsilon_{xy} \\
\sigma_{yy} &= C_{21}\varepsilon_{xx}+C_{22}\varepsilon_{yy}+C_{23}\varepsilon_{zz}+C_{24}\varepsilon_{yz}+C_{25}\varepsilon_{zx}+C_{26}\varepsilon_{xy} \\
\sigma_{zz} &= C_{31}\varepsilon_{xx}+C_{32}\varepsilon_{yy}+C_{33}\varepsilon_{zz}+C_{34}\varepsilon_{yz}+C_{35}\varepsilon_{zx}+C_{36}\varepsilon_{xy} \\
\sigma_{yz} &= C_{41}\varepsilon_{xx}+C_{42}\varepsilon_{yy}+C_{43}\varepsilon_{zz}+C_{44}\varepsilon_{yz}+C_{45}\varepsilon_{zx}+C_{46}\varepsilon_{xy} \\
\sigma_{zx} &= C_{51}\varepsilon_{xx}+C_{52}\varepsilon_{yy}+C_{53}\varepsilon_{zz}+C_{54}\varepsilon_{yz}+C_{55}\varepsilon_{zx}+C_{56}\varepsilon_{xy} \\
\sigma_{xy} &= C_{61}\varepsilon_{xx}+C_{62}\varepsilon_{yy}+C_{63}\varepsilon_{zz}+C_{64}\varepsilon_{yz}+C_{65}\varepsilon_{zx}+C_{66}\varepsilon_{xy}.
\end{aligned}
\tag{3}
$$

The elastic stiffness determines the response of crystal to an externally applied stress or strain and provides information about bonding characteristics and mechanical and structural stability [21]. The hydroxyapatite system has five elastic constants (Equation (4)). Therefore, the values of five independent C_{ij}, can be named $C_{11}, C_{12}, C_{13}, C_{33}, C_{44}$.

$$\text{Hydroxyapatite matrix} \begin{vmatrix} C_{11} & C_{12} & C_{13} & 0 & 0 & 0 \\ C_{12} & C_{11} & C_{13} & 0 & 0 & 0 \\ C_{13} & C_{13} & C_{33} & 0 & 0 & 0 \\ 0 & 0 & 0 & C_{44} & 0 & 0 \\ 0 & 0 & 0 & 0 & C_{44} & 0 \\ 0 & 0 & 0 & 0 & 0 & \frac{1}{2}(C_{11} - C_{12}) \end{vmatrix} \tag{4}$$

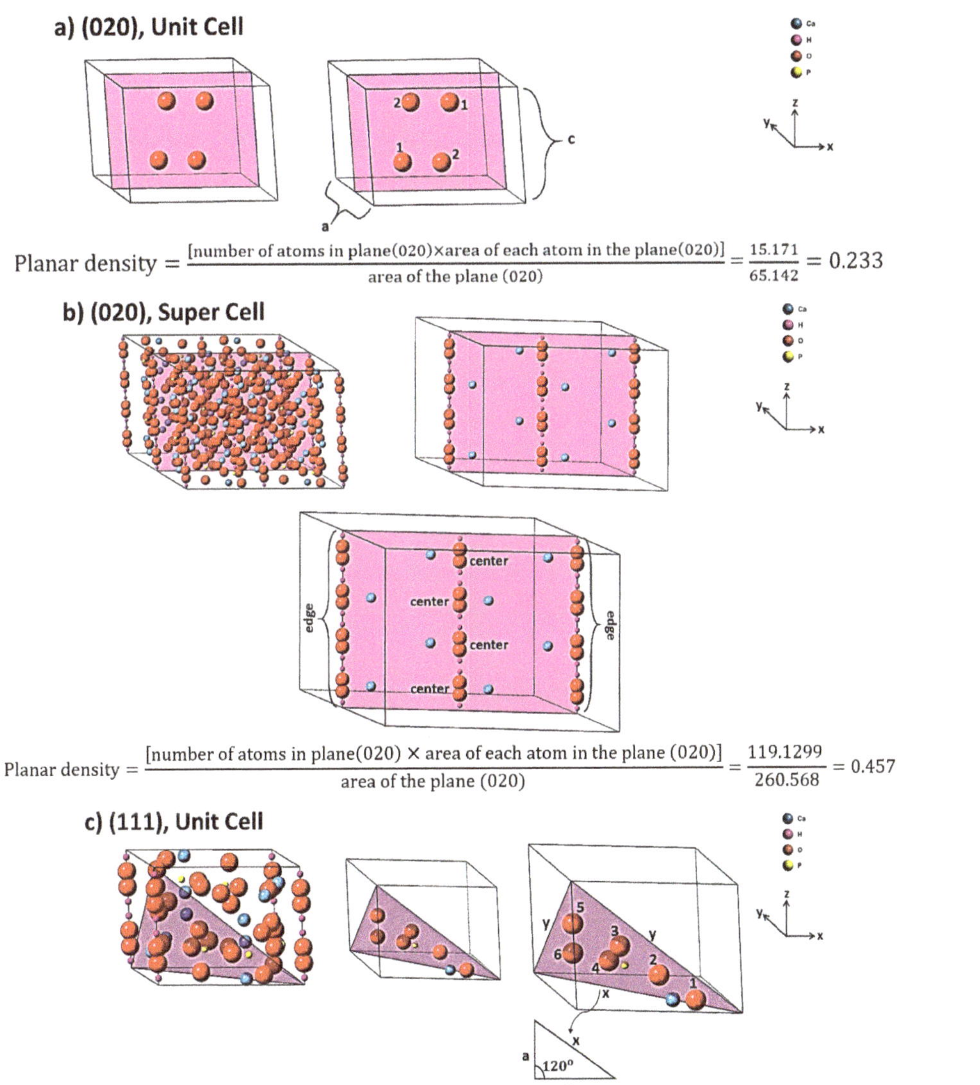

$$\text{Planar density} = \frac{[\text{number of atoms in plane(020)} \times \text{area of each atom in the plane(020)}]}{\text{area of the plane (020)}} = \frac{15.171}{65.142} = 0.233$$

$$\text{Planar density} = \frac{[\text{number of atoms in plane(020)} \times \text{area of each atom in the plane (020)}]}{\text{area of the plane (020)}} = \frac{119.1299}{260.568} = 0.457$$

Figure 3. *Cont.*

$$\text{Planar density} = \frac{[\text{number of atoms in plane}(111) \times \text{area of each atom in the plane}(111)]}{\text{area of the plane }(111)} = \frac{22.743}{68.171} = 0.334$$

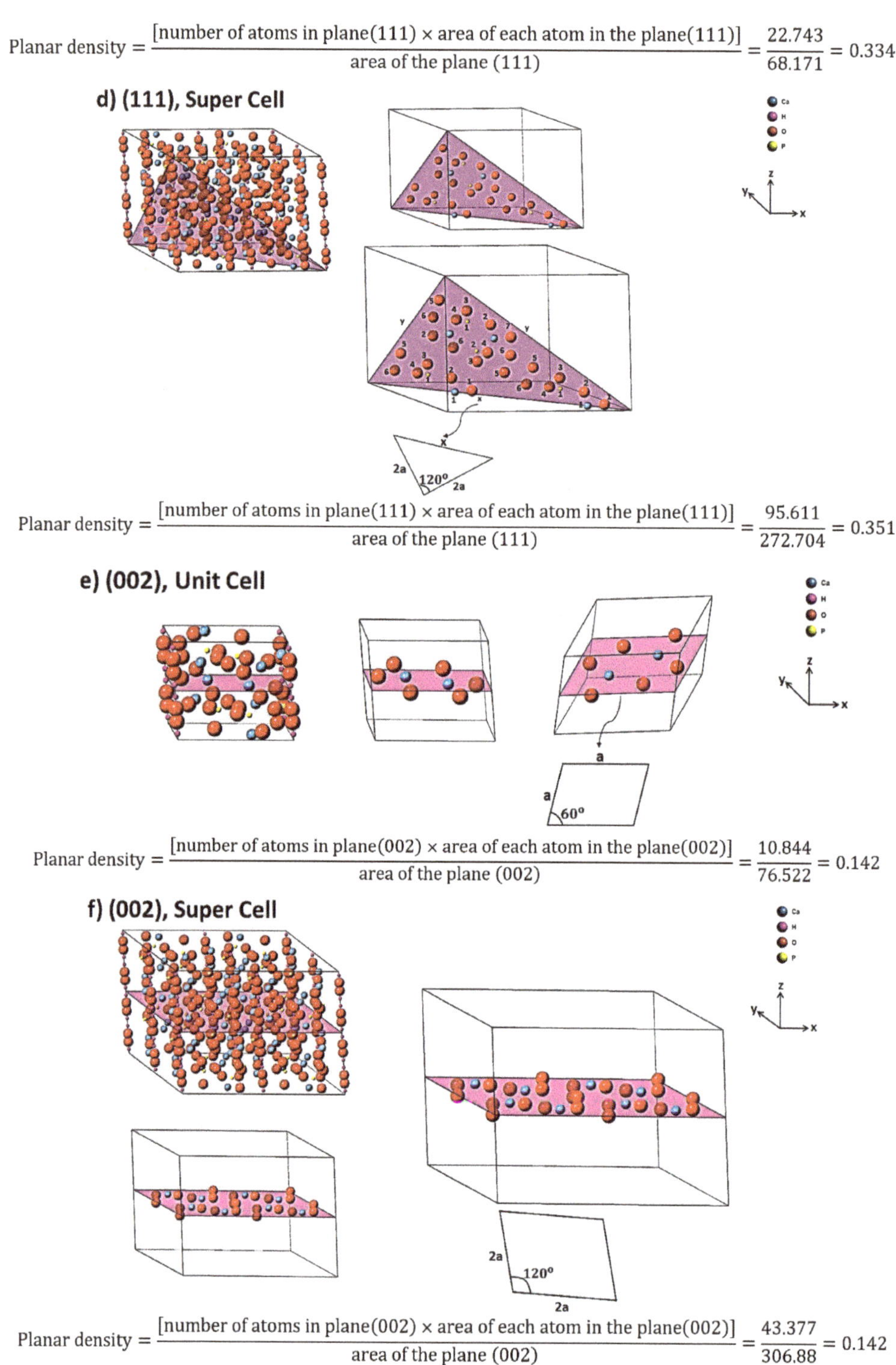

$$\text{Planar density} = \frac{[\text{number of atoms in plane}(111) \times \text{area of each atom in the plane}(111)]}{\text{area of the plane }(111)} = \frac{95.611}{272.704} = 0.351$$

$$\text{Planar density} = \frac{[\text{number of atoms in plane}(002) \times \text{area of each atom in the plane}(002)]}{\text{area of the plane }(002)} = \frac{10.844}{76.522} = 0.142$$

$$\text{Planar density} = \frac{[\text{number of atoms in plane}(002) \times \text{area of each atom in the plane}(002)]}{\text{area of the plane }(002)} = \frac{43.377}{306.88} = 0.142$$

Figure 3. *Cont.*

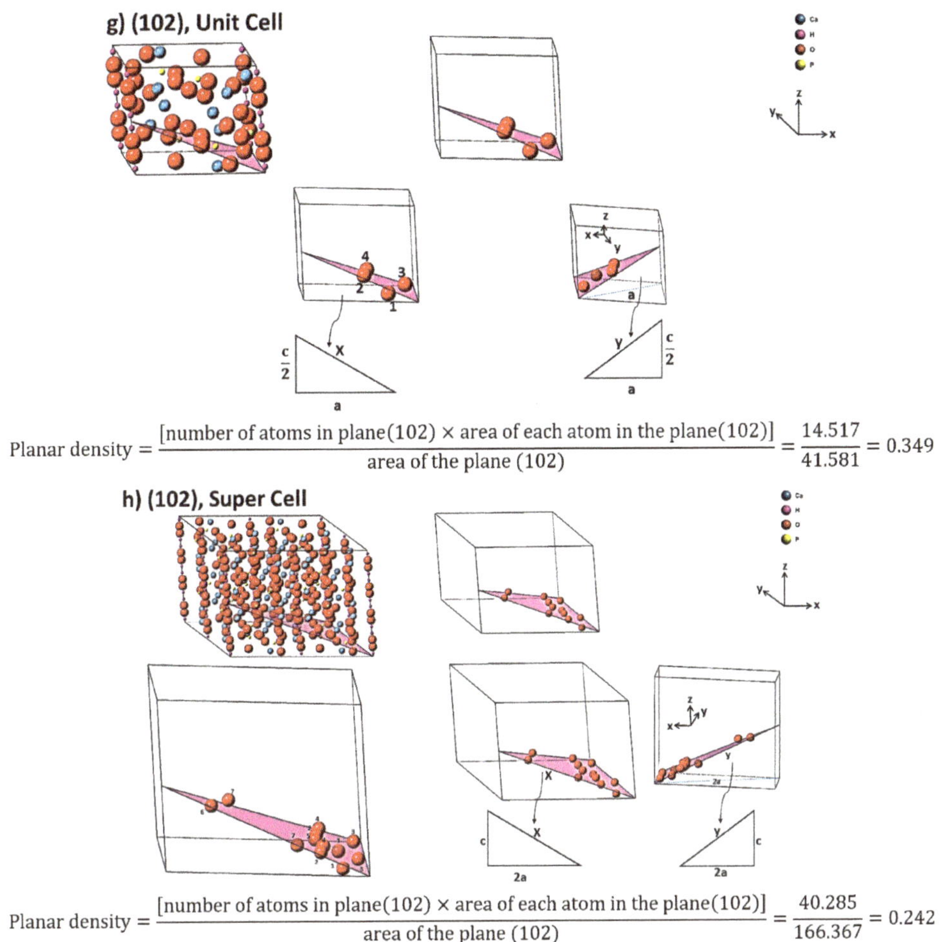

g) (102), Unit Cell

$$\text{Planar density} = \frac{[\text{number of atoms in plane}(102) \times \text{area of each atom in the plane}(102)]}{\text{area of the plane }(102)} = \frac{14.517}{41.581} = 0.349$$

h) (102), Super Cell

$$\text{Planar density} = \frac{[\text{number of atoms in plane}(102) \times \text{area of each atom in the plane}(102)]}{\text{area of the plane }(102)} = \frac{40.285}{166.367} = 0.242$$

Figure 3. Array and position of the involved atoms such as (**a**) (020) unit cell, (**b**) (020) super cell, (**c**) (111) unit cell, (**d**) (111) super cell, (**e**) (002) unit cell, (**f**) (002) super cell, (**g**) (102) unit cell and (**h**) (102) super cell (the calculations are in the Supplementary materials).

The crystallographic nature of the hexagonal structure is shown in Figure S3. Furthermore, for conventional hexagonal systems, such as hydroxyapatite, the relationship between C_{ij} and S_{ij} is introduced in Equations (5)–(9) [22,23].

$$S_{11} = \frac{1}{2}\left(\frac{C_{33}}{C_{33}(C_{11}+C_{12})-2(C_{13})^2} + \frac{1}{C_{11}-C_{12}}\right) \tag{5}$$

$$S_{12} = \frac{1}{2}\left(\frac{C_{33}}{C_{33}(C_{11}+C_{12})-2(C_{13})^2} - \frac{1}{C_{11}-C_{12}}\right) \tag{6}$$

$$S_{33} = \frac{C_{11}+C_{12}}{C_{33}(C_{11}+C_{12})-2(C_{13})^2} \tag{7}$$

$$S_{13} = -\frac{C_{13}}{C_{33}(C_{11}+C_{12})-2(C_{13})^2} \tag{8}$$

$$S_{44} = \frac{1}{C_{44}} \tag{9}$$

According to the Equations (5)–(9), to obtain S values C values are needed. We have used two approaches to obtain C values: First, theoretical calculations were performed via the CASTEP model of materials studio software version 6.0 in the GGA level of theory with a PBE basis set (Figure S4). The second approach is based on ultrasonic measurements (Table S2). The complete set of five elastic stiffness constant values (C_{11}, C_{12}, C_{13}, C_{33} and C_{44}) of the samples was found from ultrasonic measurements of the phase velocity anisotropy. The stiffness constant values were recorded by utilizing Equations (10)–(15). In these equations, ρ and V are density of sample and velocity, respectively [24–28].

$$C_{11} = \rho V_{1/1}^2, \; C_{22} = \rho V_{2/2}^2 \tag{10}$$

$$C_{66} = \rho V_{1/2}^2 = \rho V_{2/1}^2, \; C_{55} = \rho V_{1/3}^2 = \rho V_{3/1}^2 \tag{11}$$

$$C_{12} = \sqrt{\left(C_{11} + C_{66} - 2\rho V_{12/12}^2\right)\left(C_{22} + C_{66} - 2\rho V_{12/12}^2\right)} - C_{66} \tag{12}$$

$$C_{44} = \rho V_{2/3}^2 = \rho V_{3/2}^2 \tag{13}$$

$$C_{13} = \sqrt{\left(C_{11} + C_{55} - 2\rho V_{13/13}^2\right)\left(C_{33} + C_{55} - 2\rho V_{13/13}^2\right)} - C_{55} \tag{14}$$

$$C_{33} = \rho V_{3/3}^2 \tag{15}$$

For measuring the velocities, the standard ultrasonic pulse-echo ASTM E797/E797-M-15 was accomplished according to Reference [26]. In this study, the immersion procedure via water between probe and sample was utilized and the effect of pressure derived from a hand placed into the probe was decreased. In addition, the longitudinal frequency probe was 5.4 MHz, and to decrease the extension and depreciation of waves, especially the more energetic waves, a lens with a specific curvature was utilized according to the standard of Reference [28]. To measure the value of C66, a high amplitude of curve was carried out; therefore, changing the rotation angles was useful. To create the transverse waves, a probe of 2.3 MHz and a high viscos interface material (honey) were used and finally the pressure on the sample was adjusted by obtaining the better and smoother curve [29]. Taking into account each point of the samples, a three-dimensional axis, such as X_1, X_2 and X_3, can be performed. In this case, according to Figure S5, X_1 is the radial coordinate, X_2 is the circumferential coordinate and X_3 is the axial coordinate. $V_{i/j}$ is the denoted velocity of an ultrasound wave that can be propagated in the X_i direction with particle displacements in the Xj direction simultaneously. $V_{i/j}$, with the same i and j, is longitudinal and with $i \neq j$ being transverse waves. For measuring quasi-longitudinal or quasi-transverse velocity (Vij/ij), specimens should be cut (bezel) on the edges toward surfaces of perpendicular X directions (Figure S5). The obtained values of velocities are shown in Table S2. On the other hand, C_{11} is in good agreement with the longitudinal distortion and longitudinal compression/tension; thus, C_{11} can be introduced as the hardness. Moreover, the transverse distortion depends on the C_{12}, and C_{12} comes from the transverse expansion related to the Poisson's ratio. Additionally, C_{44} corresponds to the shear modulus, and C_{44} is in the settlement with C_{11} and C_{12} [22,30]. Accordingly, the shear modulus is proportional to the Burgers vector and the Young's modulus; in addition, dislocation density is in agreement with the Young's modulus [31,32].

In the ultrasonic method, longitudinal and transverse waves were utilized for measuring the Young's modulus value [33,34]. According to this method (Equation (16)), based on the velocity of ultrasound waves and density of sample, the Young's modulus value was determined.

$$E = \frac{\rho c_l^2 \left[3\left(\frac{c_l}{c_t}\right)^2 - 4\right]}{\left(\frac{c_l}{c_t}\right)^2 - 1} \tag{16}$$

In Equation (16), ρ, c_l and c_t are density, velocity of longitudinal and transverse ultrasound waves tandemly. Furthermore, according to Equation (17), the velocity of longitudinal and transverse waves can be registered by determining the length of specimen and the differences between two echoes ($t = t_2 - t_1$) in the signals [35].

$$c = \frac{2L}{t} \tag{17}$$

Here, L is the length of the sample and t is the difference between two echoes, and the density of the sample can be detected by measuring the mass and volume of the sample [36]. Additionally, with the substitution of Equations (16) and (17), the main equation for the calculation of Young's modulus is Equation (18).

$$E = \frac{4\rho \left(\frac{L}{t_s}\right)^2 \left(3t_s^2 - 4t_l^2\right)}{t_s^2 - t_l^2} \tag{18}$$

t_s and t_l are differences between two echoes in longitudinal and transverse waves separately [37]. The results of the theoretical calculation, and the experimental measurements of elastic stiffness constant values of hydroxyapatite (from the literature and this study), are shown in Table 2. In addition, taking into account Equations (10)–(15), the elastic compliance values of this study were calculated and are recorded in Table 3. As a result, the theoretical values were in good agreement with the theoretical values extracted by References [38,39]. The replicated values (five times) for measuring the time of transverse and longitudinal waves of the samples are presented in Table 4.

Table 2. Theoretical and experimental values of the Elastic constant of hydroxyapatite.

Stiffness Constant (C), (Gpa)	Theory. from Ref [40] (Ching et al.)	Theory. from Ref [41] (Leeuw et al.)	Experiment. from Ref [42] (Katz et al.)	Theory. This Study	Experiment. This Study
C_{11}	140.00	134.40	137.00	139.58	135.78
C_{12}	42.40	48.90	42.50	48.03	49.21
C_{13}	58.30	68.50	54.90	61.22	56.62
C_{33}	174.80	184.70	172.00	181.08	179.22
C_{44}	47.50	51.40	39.60	50.93	41.73

Table 3. Theoretical and experimental values of the Elastic compliance of hydroxyapatite.

Compliance Constant (S), (Gpa)	Theory. (Ching et al.)	Theory. (Leeuw et al.)	Experiment. (Katz et al.)	Theory. This Study	Experiment. This Study
S_{11}	0.008607002	0.009621806	0.008752330	0.008881114	0.009126549
S_{12}	−0.001638900	−0.002074100	−0.001829681	−0.002041879	−0.002424797
S_{13}	−0.002324030	−0.002799231	−0.002209613	−0.002312227	−0.002117248
S_{33}	0.007271063	0.007490496	0.007224509	0.007085868	0.006917516
S_{44}	0.021052632	0.019455253	0.025252525	0.019634793	0.023963575

Table 4. Resulted values of time for transverse and longitudinal waves of samples.

Number of Measurement	t_s (µs)	t_l (µs)	L (mm)
1	6.26	2.52	11.59
2	6.27	2.50	11.61
3	6.26	2.51	11.59
4	6.25	2.53	11.57
5	6.24	2.54	11.55

2.4. The Young's Modulus versus Planar Density of Unit Cell and Super Cells (2 × 2 × 2) of Hydroxyapatite

X-ray diffraction has provided data on diffracted planes and the location of atoms in each plane. In the previous section, the planar density of each diffracted plane was calculated and it could play an important role in the mechanical properties of each plane. The Young's modulus of each plane (E_{hkl}) of a hydroxyapatite lattice can be calculated as Equation (19) [5,43]. In this equation, h, k and l are the plane indices, a and c are the lattice parameters, C and S are the elastic stiffness constant and elastic compliance, respectively. The values of Young's modulus of 32 diffracted planes of hydroxyapatite in unit cells and super cells (2 × 2 × 2),$E_{(h_1k_1l_1)}$, $E_{(h_2k_2l_2)}$, $E_{(h_3k_3l_3)}$ ·············, $E_{(h_{32}k_{32}l_{32})}$, related to the literature and the present study, are reported in Table S3.

$$E_{hkl} = \frac{\left[h^2 + \frac{(h+2k)^2}{3} + \left(\frac{al}{c} \right)^2 \right]^2}{S_{11} \left(h^2 + \frac{(h+2k)^2}{3} \right)^2 + S_{33} \left(\frac{al}{c} \right)^4 + (2S_{13} + S_{44}) \left(h^2 + \frac{(h+2k)^2}{3} \right) \left(\frac{al}{c} \right)^2} \tag{19}$$

By using the least squares method between the Young's modulus and the planar density of diffracted planes (based on our proposed method), the Young's modulus value of hydroxyapatite was determined with high precision. Consequently, the calculation of C and S parameters for crystalline materials is essential for the application of this method. To show the feasibility and accuracy of our proposed method for determining the Young's modulus, the values of Young's modulus of each plane versus the planar density of the unit cell and supercells (2 × 2 × 2) are depicted in Figure 4. The Young's modulus values (intercept) of the unit cell and super cells (2 × 2 × 2) of hydroxyapatite, obtained from the least squares method, are tabulated in Table 5.

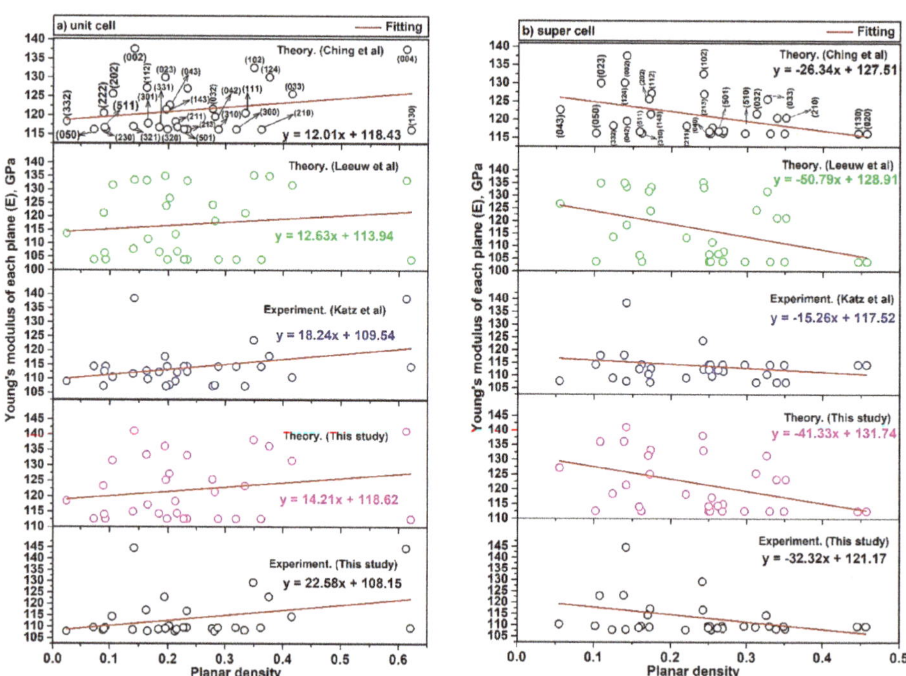

Figure 4. Young's modulus of each plane (**a**) unit-cell (**b**) super cells (2 × 2 × 2) of hydroxyapatite extracted by XRD patterns and planar density.

Table 5. Young's modulus values of unit cell and super cell lattices of hydroxyapatite.

Study	Young Modulus (E), (Gpa) in This Method (Intercept Value)	
	Unit Cell	Super Cells (2 × 2 × 2)
Theory. (Ching et al.)	118.43	127.51
Theory. (Leeuw et al.)	113.94	128.91
Experiment. (Katz et al.)	109.54	117.52
Theory. (This Study)	118.62	131.74
Experiment. (This Study)	108.15	121.17

According to the uncertainty measurement (the measurements were replicated five times (Table 4)) and Equation (18), the Young's modulus value gained 113.08 ± 0.14 GPa by ultrasonic measurement. This value is in good agreement with the reported values of our study in Table 5. In this study, the difference between theory and experiment values for both unit cell and super cells (2 × 2 × 2) are identical. This difference for unit cell and super cells (2 × 2 × 2) are 10.47 GPa and 10.57 GPa, respectively. This means that the theoretical calculation is valid and, by reducing it by about ~10 GPa, the experimental values can be obtained.

3. Result and Discussion

3.1. Positive and Negative Slope Values of Unit Cell and Super Cells (2 × 2 × 2)

The difference of Young's modulus values in the unit cell and super cells (2 × 2 × 2) is attributed to the locations of atoms. The calculated slope is a negative value in super cells (2 × 2 × 2). In the first aspect, it is clear that the slope is dependent on the planar density of diffracted planes, so the fitting based on the planar density of super cells with a matrix of eight unit cells could submit a better result of intercept [35]. It is because eight cells besides each other are completed and have more symmetry than two cells (Figure S8) [35]. In the second aspect, the reason for the positive slope in the unit cell and the negative slope in the super cells (2 × 2 × 2) is related to the defects (imperfections) including point defects (vacancies, substitutional and interstitials), line defects (screw and edge dislocation), surface defects (grain boundaries) and volume defects (lack of order of atoms due to amorphous region in a very tiny area). The effect of these imperfections is more effective in super cells (2 × 2 × 2), while the unit cell is more ideal and less affected by these imperfections. This means that when the density of atoms in planes is increased, lower forces for dislocation motion are required and the strength will be decreased and, consequently, in super cells (2 × 2 × 2) the slope is negative. In the case of super cells, when the number of atoms is increased, by the increase of planar density, the effect of dislocation motion is increased; the strength and Young's modulus will be decreased so the slope is negative. The intercept of the fitting line is a value of Young's modulus, which can show the Young's modulus of the plane with zero planar density as a plane without any specific atom. Therefore, a discussion of defects and imperfections cannot be considered for such a plane without an atom at the origin and so, in this case, the Young's modulus value of the unit cell is less than that of the super cells (2 × 2 × 2) and consequently the plane without an atom is more realistic in a smaller area of a unit cell than in a wider area of the super cells. This means that the intercept in a unit cell is closer to the real values of the Young's modulus. As mentioned above, the planar density depends on the array and position of atoms into the plane and the situation of the planes. For illustration, according to Figure 4b, the planar density of (210) > (510) > (102) > (023) > (043), because sequencing of the planes is not related to the diffraction (XRD), but it depends on the planar density values. With this method, it can be possible to consider two or more planes with similar planar density values; for example, in super cells (2 × 2 × 2), planar density values of (040) and (331) are equal to 0.250. The control of the displacement and deformation process of the atoms in the planes are associated with the dislocation networks. The work or energy (W) for the movement of the atoms in each plane corresponds to Equation (20) [44,45].

$$W = Gb^2 l \tag{20}$$

In this equation, G, b and l are shear modulus, Burgers vector and dislocation length, respectively. In addition, by merging Equations (20) and (21), Equation (22) can be registered. In Equations (21) and (22), E and ν are Young's modulus and Poisson's ratio tandemly.

$$G = \frac{E}{2(1+\nu)} \tag{21}$$

$$W = \frac{E}{2(1+\nu)} b^2 l \tag{22}$$

There are several studies on the properties of hydroxyapatite, especially with regard to biocompatibility and bioactivity, and these properties are dependent on the identification and recognition of hydroxyapatite structures [46,47] For example, when atoms are widely spaced, such as corner atoms (Ca) in (111) super cells (2 × 2 × 2), atoms require higher values of applied force for approaching, so knowledge on the planes of hydroxyapatite structure can be helpful for doping metals, polymers and other ceramics, for aims such as the controlled release of protein and also the fabrication of bioactive monolithic fragments in the biomaterials industry [48,49]. In addition, the hydroxyl ion is in the center of each unit cell. The hydroxyl group in the center of the hydroxyapatite lattice is surrounded by three calcium ions per hexagon, forming a ring (six calcium ions). A chord is formed by the structure, as these rings are responsible for many properties of hydroxyapatite, especially the biocompatibility and the position of the hydroxyl group in the planes have considerable importance for doping mechanisms [15,50].

3.2. Williamson-Hall Method in USDM Model

The W-H method has been used for determining different elastic properties. The best procedure is to mathematically reduce the errors and obtain the values of the Young's modulus by all the diffracted peaks, using the least squares method. The W-H method is a simplified integral expanse and, taking into account the peak width, strain-induced broadening is specified [51]. Taking into account the W-H method in the USDM model, Young's modulus values are examined. It is clear that in Equation (23) and Figure 5, the term of $\frac{4\sin\theta}{E_{hkl}}$ is along the X-axis and the term of $\beta_{hkl}.\cos\theta$ is along the Y-axis individually.

$$\beta_{hkl}. \cos \theta = \left(\frac{K\lambda}{L}\right) + 4\sigma \frac{\sin\theta}{E_{hkl}} \tag{23}$$

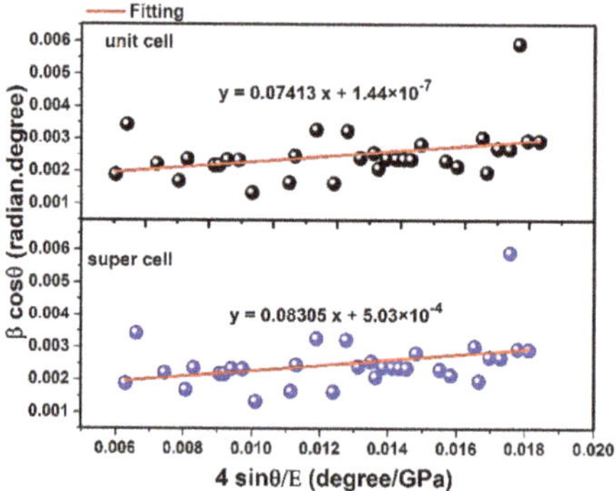

Figure 5. W-H in USDM model and plot of unit cell and super cells (2 × 2 × 2) of hydroxyapatite.

Here, β_{hkl} is the broadening peak from (hkl) plane and, in this study, the instrumental broadening is taken as 0.02 degree for each diffracted peak and is subtracted from β_{hkl} values before multiplying to $\frac{\pi}{180}$ to convert the degree to radian. In this model, the condition is performed to calculate the strain and the average Young's modulus. The average E value has been calculated in the research and studies on using the W-H method, but it is subject to errors, because if the average values of the Young's modulus are considered (through Equation (19)), the final value would be far from the standard value of Young's modulus in each peak extracted by X-ray diffraction. In addition, in some studies the Young's modulus value is considered a value that is listed in the literature but is not associated with the prepared materials. As an illustration, based on the study in Reference [10], which refers to the use of the W-H method to calculate the crystal size and Scherrer analysis of hydroxyapatite, the average value of Young's modulus has a larger deviation than the actual value of the Young's modulus of hydroxyapatite. The line broadening of the diffracted peak is extracted with Crystal Diffract 6.7.2.300 software. According to Figure 5, the slope values are associated with the stress (σ). The values obtained were positive, and the positive values of intrinsic strain and stress can prove the tensile stress and strain, and if the values were negative, they are associated with compressive stress and strain. Additionally, the resulting values of stress (σ) and strain (ε) by utilizing the W-H method in the USDM model are shown in Table 6. The value of σ is in good agreement with the values obtained from Reference [52].

Table 6. Mechanical properties values of unit cell and super cells ($2 \times 2 \times 2$) of hydroxyapatite.

Structure	Mechanical Properties		
	σ (GPa)	ε	E [a] (GPa)
Unit cell	0.07413	0.00068	108.15
Super cells ($2 \times 2 \times 2$)	0.08305	0.00068	121.17

a: Experiments of Young's modulus values (This Study).

4. Conclusions

In this study, crystal of hydroxyapatite was successfully synthesized with the thermal treatment process. Moreover, the position of atoms and extracted planar density values of each diffracted plane (32 planes) of hydroxyapatite were determined and calculated for the first time. In addition, a new method based on the relationship between the measurement of elastic modulus and atomic planar density of crystalline hydroxyapatite, consisting of a unit cell and super cells ($2 \times 2 \times 2$), was fully presented. According to the method of this study, the slope of modules of elasticity against planar density in super cells ($2 \times 2 \times 2$) was negative; the reason was related to imperfections, and this means that when the density of atoms in the planes is increased, lower forces are required for the dislocation motion. As a result, a plane without an atom is more realistic in a smaller area of the unit cell than a larger area of the super cells ($2 \times 2 \times 2$), and this means that the intercept of the unit cell is closer to the real values of the Young's modulus in the hydroxyapatite lattice. Moreover, a comparison of theoretical and experimental data of the Young's modulus of hydroxyapatite showed that there is a small difference between the values for both unit cell and super cells ($2 \times 2 \times 2$), namely 10.47 GPa and 10.57 GPa, respectively; this means that the theoretical calculation is valid and, by decreasing by about 10 GPa, the experimental value can be obtained. Furthermore, the Young's modulus values of hydroxyapatite in the unit cell and super cells ($2 \times 2 \times 2$) were achieved at 108.15 and 121.17 GPa tandemly. Finally, one of the applications of the presented method was carried out in this study and the Williamson-Hall method in the USDM model can be used to minimize the errors in the least squares method and to obtain the correct elastic modulus of hydroxyapatite, which is much more accurate than the average value.

Supplementary Materials: The following are available online at https://www.mdpi.com/article/10.3390/ma14112949/s1, Figure S1: Images of production route of hydroxyapatite obtained from cow bones, Table S1: Crystallographic parameters of XRD pattern related to the natural hydroxyapatite obtained from cow bones, Figure S2: EDX spectrum and stoichiometric composition of hydroxyapatite obtained from cow bones, Figure S3: (a) Raw hexagonal structure, (b) fundamental axis, (c) principal crystallographic directions and (d) between the orthogonal axes and crystallographic directions, Figure S4: Transfer of hydroxyapatite unite cell in P63/m to P63 space group, Figure S5: Schematic of a design of cutted samples, Table S2: The values of Longitudinal and Transvers velocity of samples, Figure S6: Geometry and the situation of involved atoms in diffracted planes related to the unit-cell of hydroxyapatite (a) 210, (b) 211, (c) 112, (d) 300, (e) 202, (f) (301), (g) (130), (h) (310), (i) (032), (j) (040), (k) (023), (l) (222), (m) (320), (n) (230), (o) (213), (p) (321), (q) (042), (r) (033), (s) (004), (t) (050), (u) (501), (v) (331), (w) (043), (x) (124), (y) (510), (z) (511), (a1) (332) and (b1) (143), Figure S7: Geometry and the situation of involved atoms in diffracted planes related to the super-cell of hydroxyapatite (a) 210, (b) 211, (c) 112, (d) 300, (e) 202, (f) (301), (g) (130), (h) (310), (i) (032), (j) (040), (k) (023), (l) (222), (m) (320), (n) (230), (o) (213), (p) (321), (q) (042), (r) (033), (s) (004), (t) (050), (u) (501), (v) (331), (w) (043), (x) (124), (y) (510), (z) (511), (a1) (332) and (b1) (143), Table S3: Young's modulus values of hydroxyapatite related to the literatures and present study, Figure S8: (a) The un-symmetry of two unit cells and (b) symmetry of eight unit cells of hydroxyapatite.

Author Contributions: M.R., A.P., G.J., A.V.: Investigation, Writing—original draft and Validation; S.N.: Writing—review and editing; A.D. (Amir Dashti), A.D. (Akram Doustmohammadi): Formal analysis and Software; A.M.: Methodology. All authors have read and agreed to the published version of the manuscript.

Funding: This research was funded by a grant No. S-MIP-19-43 from the Research Council of Lithuania.

Institutional Review Board Statement: Not applicable.

Informed Consent Statement: Not applicable.

Data Availability Statement: Data sharing is not applicable.

Conflicts of Interest: The authors declare no conflict of interest.

References

1. Sasmita, F.; Wibisono, G.; Judawisastra, H.; Priambodo, T.A. Determination of elastic modulus of ceramics using ultrasonic testing. *AIP Conf. Proc.* **2018**, *1945*, 020017. [CrossRef]
2. Shimada, M.; Matsushita, K.; Kuratani, S.; Okamoto, T.; Koizumi, M.; Tsukuma, K.; Tsukidate, T. Temperature dependence of young's modulus and internal friction in alumina, silicon nitride, and partially stabilized zirconia ceramics. *J. Am. Ceram. Soc.* **1984**, *67*, C23–C24. [CrossRef]
3. Vallet-Regí, M. Revisiting ceramics for medical applications. *Dalton Trans.* **2006**, 5211–5220. [CrossRef] [PubMed]
4. Marzieh, R.; Sohrab, N.; Arvydas, P.; Giedrius, J. Preparation and investigation of bioactive organic-inorganic nano-composite derived from PVB-co-VA-co-VAc/HA. In Proceedings of the 15th International Conference Mechatronic Systems and Materials, MSM 2020, Bialystok, Poland, 1–3 July 2020; Institute of Electrical and Electronics Engineers Inc.: Piscataway, NJ, USA, 2020.
5. Rabiei, M.; Palevicius, A.; Monshi, A.; Nasiri, S.; Vilkauskas, A.; Janusas, G. Comparing methods for calculating nano crystal size of natural hydroxyapatite using X-ray diffraction. *Nanomaterials* **2020**, *10*, 1627. [CrossRef] [PubMed]
6. Shih, W.J.; Wang, M.C.; Hon, M.H. Morphology and crystallinity of the nanosized hydroxyapatite synthesized by hydrolysis using cetyltrimethylammonium bromide (CTAB) as a surfactant. *J. Cryst. Growth* **2005**, *275*, e2339–e2344. [CrossRef]
7. Loganathan, T.M.; Sultan, M.T.H.; Gobalakrishnan, M.K. Ultrasonic inspection of natural fiber-reinforced composites. In *Sustainable Composites for Aerospace Applications*; Elsevier: Amsterdam, The Netherlands, 2018; pp. 227–251, ISBN 9780081021316.
8. Nath, D.; Singh, F.; Das, R. X-ray diffraction analysis by Williamson-Hall, Halder-Wagner and size-strain plot methods of CdSe nanoparticles—A comparative study. *Mater. Chem. Phys.* **2020**, *239*, 122021. [CrossRef]
9. Khorsand Zak, A.; Abd Majid, W.H.; Abrishami, M.E.; Yousefi, R. X-ray analysis of ZnO nanoparticles by Williamson-Hall and size-strain plot methods. *Solid State Sci.* **2011**, *13*, 251–256. [CrossRef]
10. Venkateswarlu, K.; Chandra Bose, A.; Rameshbabu, N. X-ray peak broadening studies of nanocrystalline hydroxyapatite by WilliamsonHall analysis. *Phys. B Condens. Matter* **2010**, *405*, 4256–4261. [CrossRef]
11. Bahrololoom, M.; Javidi, M.; Javadpour, S. Characterisation of natural hydroxyapatite extracted from bovine cortical bone ash. *J. Ceram. Process. Res.* **2009**, *10*, 129–138.
12. Shahabi, S.; Najafi, F.; Majdabadi, A.; Hooshmand, T.; Haghbin Nazarpak, M.; Karimi, B.; Fatemi, S.M. Effect of gamma irradiation on structural and biological properties of a PLGA-PEG-hydroxyapatite composite. *Sci. World J.* **2014**, *2014*. [CrossRef]

13. Pasero, M.; Kampf, A.R.; Ferraris, C.; Pekov, I.V.; Rakovan, J.; White, T.J. Nomenclature of the apatite supergroup minerals. *Eur. J. Mineral.* **2010**, *22*, 163–179. [CrossRef]
14. Hench, L.L. *An Introduction to Bioceramics*, 2nd ed.; Imperial College Press: London, UK, 2013; ISBN 9781908977168.
15. Lin, K.; Chang, J. Structure and properties of hydroxyapatite for biomedical applications. In *Hydroxyapatite (Hap) for Biomedical Applications*; Elsevier: Amsterdam, The Netherlands, 2015; pp. 3–19.
16. Bystrov, V.S.; Coutinho, J.; Bystrova, A.V.; Dekhtyar, Y.D.; Pullar, R.C.; Poronin, A.; Palcevskis, E.; Dindune, A.; Alkan, B.; Durucan, C.; et al. Computational study of hydroxyapatite structures, properties and defects. *J. Phys. D Appl. Phys.* **2015**, *48*, 195302. [CrossRef]
17. Rajkumar, M.; Sundaram, N.; Rajendran, V. Preparation of size controlled, stoichiometric and bioresorbable hydroxyapatite nanorod by varying initial pH, Ca/P ratio and sintering temperature. *Digest J. Nanomater. Biostruct.* **2011**, *6*, 169–179.
18. Landi, E.; Riccobelli, S.; Sangiorgi, N.; Sanson, A.; Doghieri, F.; Miccio, F. Porous apatites as novel high temperature sorbents for carbon dioxide. *Chem. Eng. J.* **2014**, *254*, 586–596. [CrossRef]
19. Rabiei, M.; Palevicius, A.; Dashti, A.; Nasiri, S.; Monshi, A.; Vilkauskas, A.; Janusas, G. Measurement Modulus of elasticity related to the atomic density of planes in unit cell of crystal lattices. *Materials* **2020**, *13*, 4380. [CrossRef] [PubMed]
20. Rajabi, K.; Hosseini-Hashemi, S. Application of the generalized Hooke's law for viscoelastic materials (GHVMs) in nanoscale mass sensing applications of viscoelastic nanoplates: A theoretical study. *Eur. J. Mech. A Solids* **2018**, *67*, 71–83. [CrossRef]
21. Kanoun, M.B.; Goumri-Said, S.; Abdullah, K. Theoretical study of physical properties and oxygen incorporation effect in nanolaminated ternary carbides 211-MAX phases. In *Advances in Science and Technology of Mn+1AXn Phases*; Elsevier: Amsterdam, The Netherlands, 2012; pp. 177–196.
22. Li, Y.; Thompson, R.B. Relations between elastic constants Cij and texture parameters for hexagonal materials. *J. Appl. Phys.* **1990**, *67*, 2663–2665. [CrossRef]
23. Huntington, H.B. The elastic constants of crystals. *Solid State Phys. Adv. Res. Appl.* **1958**, *7*, 213–351. [CrossRef]
24. Mah, M.; Schmitt, D.R. Determination of the complete elastic stiffnesses from ultrasonic phase velocity measurements. *J. Geophys. Res. Solid Earth* **2003**, *108*. [CrossRef]
25. Neighbours, J.R.; Schacher, G.E. Determination of elastic constants from sound-velocity measurements in crystals of general symmetry. *J. Appl. Phys.* **1967**, *38*, 5366–5375. [CrossRef]
26. ASTM E797/E797M-15. Standard Practice for Measuring Thickness by Manual Ultrasonic Pulse-Echo Contact Method. Available online: https://www.astm.org/Standards/E797.htm (accessed on 19 November 2020).
27. Van Buskirk, W.C.; Cowin, S.C.; Ward, R.N. Ultrasonic measurement of orthotropic elastic constants of bovine femoral bone. *J. Biomech. Eng.* **1981**, *103*, 67–72. [CrossRef]
28. ASTM E214-05. Standard Practice for Immersed Ultrasonic Testing by the Reflection Method Using Pulsed Longitudinal Waves (Withdrawn 2007). Available online: https://www.astm.org/Standards/E214.htm (accessed on 19 November 2020).
29. ASTM E1001-16. Standard Practice for Detection and Evaluation of Discontinuities by the Immersed Pulse-Echo Ultrasonic Method Using Longitudinal Waves. Available online: https://www.astm.org/Standards/E1001.htm (accessed on 19 November 2020).
30. Elliot, S. *The Physics and Chemistry of Solids*; Wiley: Hoboken, NJ, USA, 1998. Available online: https://www.wiley.com/en-us/The+Physics+and+Chemistry+of+Solids-p-9780471981954 (accessed on 19 November 2020).
31. McHugh, P.E. Introduction to crystal plasticity theory. In *Mechanics of Microstructured Materials*; Springer: Cham, Switzerland, 2004; pp. 125–171.
32. Pandech, N.; Sarasamak, K.; Limpijumnong, S. Elastic properties of perovskite A TiO$_3$ (A = Be, Mg, Ca, Sr, and Ba) and PbBO$_3$ (B = Ti, Zr, and Hf): First principles calculations. *J. Appl. Phys.* **2015**, *117*, 174108. [CrossRef]
33. Wang, H.; Prendiville, P.L.; McDonnell, P.J.; Chang, W.V. An ultrasonic technique for the measurement of the elastic moduli of human cornea. *J. Biomech.* **1996**, *29*, 1633–1636. [CrossRef]
34. Bray, D.E.; Stanley, R.K. *Nondestructive Evaluation: A Tool in Design, Manufacturing and Service*; CRC Press: Boca Raton, FL, USA, 1996. Available online: https://books.google.com.et/books?id=5WtmjwEACAAJ&printsec=copyright#v=onepage&q&f=false (accessed on 19 November 2020).
35. Rabiei, M.; Palevicius, A.; Nasiri, S.; Dashti, A.; Vilkauskas, A.; Janusas, G. Relationship between young's modulus and planar density of unit cell, super cells (2 × 2 × 2), symmetry cells of perovskite (CaTiO$_3$) LATTICE. *Materials* **2021**, *14*, 1258. [CrossRef]
36. Bodke, M.; Gawai, U.; Patil, A.; Dole, B. Estimation of accurate size, lattice strain using Williamson-Hall models, SSP and TEM of Al doped ZnO nanocrystals. *Mater. Tech.* **2018**, *106*. [CrossRef]
37. Figliola, R.S.; Beasley, D.E. *Theory and Design for Mechanical Measurements*, 7th ed.; Wiley: Hoboken, NJ, USA, 2014. Available online: https://www.wiley.com/en-us/Theory+and+Design+for+Mechanical+Measurements%2C+7th+Edition-p-97811194 75651 (accessed on 19 November 2020).
38. Moradi, K.; Sabbagh Alvani, A.A. First-Principles study on Sr-doped hydroxyapatite as a biocompatible filler for photo-cured dental composites. *J. Aust. Ceram. Soc.* **2020**, *56*, 591–598. [CrossRef]
39. Bhat, S.S.; Waghmare, U.V.; Ramamurty, U. First-Principles study of structure, vibrational, and elastic properties of stoichiometric and calcium-deficient hydroxyapatite. *Cryst. Growth Des.* **2014**, *14*, 3131–3141. [CrossRef]
40. Ching, W.Y.; Rulis, P.; Misra, A. Ab initio elastic properties and tensile strength of crystalline hydroxyapatite. *Acta Biomater.* **2009**, *5*, 3067–3075. [CrossRef]

41. De Leeuw, N.H.; Bowe, J.R.; Rabone, J.A.L. A computational investigation of stoichiometric and calcium-deficient oxy- and hydroxy-apatites. *Faraday Discuss.* **2007**, *134*, 195–214. [CrossRef]
42. Katz, J.L.; Ukraincik, K. On the anisotropic elastic properties of hydroxyapatite. *J. Biomech.* **1971**, *4*, 221–227. [CrossRef]
43. Zhang, J.M.; Zhang, Y.; Xu, K.W.; Ji, V. Anisotropic elasticity in hexagonal crystals. *Thin Solid Films* **2007**, *515*, 7020–7024. [CrossRef]
44. Berdichevsky, V. Energy of dislocation networks. *Int. J. Eng. Sci.* **2016**, *103*, 35–44. [CrossRef]
45. Reed-Hill, R.E.; Abbaschian, R. *Physical Metallurgy Principles*; Cengage: Boston, MA, USA, 1992.
46. Shi, D.; Jiang, G.; Bauer, J. The effect of structural characteristics on the in vitro bioactivity of hydroxyapatite. *J. Biomed. Mater. Res.* **2002**, *63*, 71–78. [CrossRef] [PubMed]
47. Ragel, C.V.; Vallet-Regí, M.; Rodríguez-Lorenzo, L.M. Preparation and in vitro bioactivity of hydroxyapatite/solgel glass biphasic material. *Biomaterials* **2002**, *23*, 1865–1872. [CrossRef]
48. Dasgupta, S.; Banerjee, S.S.; Bandyopadhyay, A.; Bose, S. Zn- and Mg-doped hydroxyapatite nanoparticles for controlled release of protein. *Langmuir* **2010**, *26*, 4958–4964. [CrossRef] [PubMed]
49. Uysal, I.; Severcan, F.; Tezcaner, A.; Evis, Z. Co-Doping of hydroxyapatite with zinc and fluoride improves mechanical and biological properties of hydroxyapatite. *Prog. Nat. Sci. Mater. Int.* **2014**, *24*, 340–349. [CrossRef]
50. Ptáček, P. *Apatites and Their Synthetic Analogues—Synthesis, Structure, Properties and Applications*; InTech: London, UK, 2016.
51. Suryanarayana, C.; Norton, M.G.; Suryanarayana, C.; Norton, M.G. Practical aspects of X-ray diffraction. In *X-ray Diffraction*; Springer: New York, NY, USA, 1998; pp. 63–94.
52. Itatani, K.; Tsuchiya, K.; Sakka, Y.; Davies, I.J.; Koda, S. Superplastic deformation of hydroxyapatite ceramics with B_2O_3 or Na_2O addition fabricated by pulse current pressure sintering. *J. Eur. Ceram. Soc.* **2011**, *31*, 2641–2648. [CrossRef]

Communication

Bicontinuous Gyroid Phase of a Water-Swollen Wedge-Shaped Amphiphile: Studies with In-Situ Grazing-Incidence X-ray Scattering and Atomic Force Microscopy

Kseniia N. Grafskaia [1,2], Azaliia F. Akhkiamova [2,3], Dmitry V. Vashurkin [2,4], Denis S. Kotlyarskiy [2,4], Diego Pontoni [3], Denis V. Anokhin [1,2], Xiaomin Zhu [5,6] and Dimitri A. Ivanov [1,2,4,7,*]

[1] Moscow Institute of Physics and Technology, Instituskiy per. 9, 141700 Dolgoprudny, Russia; grafskayaxeniya@gmail.com (K.N.G.); deniano@yahoo.com (D.V.A.)
[2] Institute of Problems of Chemical Physics, Russian Academy of Sciences, Chernogolovka, 142432 Moscow, Russia; azigy@mail.ru (A.F.A.); dmvav21@gmail.com (D.V.V.); 9920860@gmail.com (D.S.K.)
[3] Partnership for Soft Condensed Matter (PSCM), ESRF–The European Synchrotron, 71 Avenue des Martyrs, 38043 Grenoble, France; diego.pontoni@esrf.fr
[4] Faculty of Chemistry, Lomonosov Moscow State University (MSU), GSP-1, 1-3 Leninskiye Gory, 119991 Moscow, Russia
[5] DWI–Leibniz-Institute for Interactive Materials e.V., D-52056 Aachen, Germany; zhu@dwi.rwth-aachen.de
[6] Institute for Technical and Macromolecular Chemistry of RWTH Aachen University, Forkenbeckstr. 50, D-52056 Aachen, Germany
[7] Institut de Sciences des Matériaux de Mulhouse-IS2M, CNRS UMR 7361, Jean Starcky 15, F-68057 Mulhouse, France
[*] Correspondence: dimitri.ivanov@uha.fr

Citation: Grafskaia, K.N.; Akhkiamova, A.F.; Vashurkin, D.V.; Kotlyarskiy, D.S.; Pontoni, D.; Anokhin, D.V.; Zhu, X.; Ivanov, D.A. Bicontinuous Gyroid Phase of a Water-Swollen Wedge-Shaped Amphiphile: Studies with In-Situ Grazing-Incidence X-ray Scattering and Atomic Force Microscopy. *Materials* **2021**, *14*, 2892. https://doi.org/10.3390/ma14112892

Academic Editors: Thomas Walter Cornelius and Souren Grigorian

Received: 10 May 2021
Accepted: 24 May 2021
Published: 28 May 2021

Abstract: We report on formation of a bicontinuous double gyroid phase by a wedge-shaped amphiphilic mesogen, pyridinium 4'-[3″,4″,5″-tris-(octyloxy)benzoyloxy]azobenzene-4-sulfonate. It is found that this compound can self-organize in zeolite-like structures adaptive to environmental conditions (e.g., temperature, humidity, solvent vapors). Depending on the type of the phase, the structure contains 1D, 2D, or 3D networks of nanometer-sized ion channels. Of particular interest are bicontinuous phases, such as the double gyroid phase, as they hold promise for applications in separation and energy. Specially designed environmental cells compatible with grazing-incidence X-ray scattering and atomic force microscopy enable simultaneous measurements of structural parameters/morphology during vapor-annealing treatment at different temperatures. Such in-situ approach allows finding the environmental conditions at which the double gyroid phase can be formed and provide insights on the supramolecular structure of thin films at different spatial levels.

Keywords: wedge-shaped amphiphile; double gyroid phase; grazing-incidence X-ray scattering; environmental atomic force microscopy; vapor annealing

1. Introduction

Molecular self-assembly is a spontaneous process during which the constitutive elements self-organize under action of non-covalent bonding forces [1–3]. Among self-assembling species, amphiphilic low-molecular-weight mesogens have been extensively studied because of their fascinating phase behavior [4,5]. In particular, wedge-shaped amphiphiles can self-assemble into a remarkable range of lyotropic liquid crystalline (LLC) and thermotropic liquid crystalline (TLC) mesophases with intricate one-, two-, and three-dimensional (1D, 2D, and 3D) periodic nanostructures [6,7]. Due to the unique structures of non-lamellar LC mesophases, they have attracted significant interest over the past few decades particularly for their applications in such areas as drug delivery [8,9], membrane protein crystallization [10,11], energy conversion and storage, gas storage, chemical sensing, and others [12,13].

Nanoporous membranes are already widely used in various applications, like chemical separation and purification, fuel conversion, and ecology [14]. They are perspective not only for classical chemical production but also for design of new sensors, electronics, medicine, and, especially, for fuel cells [15].

The most desirable structures for nanoporous materials are columnar or cubic phases. The use of columnar phases is more developed for conducting proposes but the control of channels orientation in such phases can be an issue. Thus, for optimizing the ion transport across the membrane in the columnar phase, one has to induce vertical or, the so-called homeotropic, orientation of the columns [16]. This is a nontrivial task because, in many instances, such orientation is not thermodynamically stable. The bicontinuous cubic phases do not require macroscopic orientation because the channels in such structures are running along all three directions in space. The cubic phases can be described in terms of triply periodic minimal surfaces (TPMSs). The TPMSs represent nonintersecting surfaces with three-dimensional (3D) periodicity and vanishing mean curvature H, where $H = \frac{1}{2}(\kappa_1 + \kappa_2)$, and κ_1 and κ_2 stand for the principal curvatures at the point [17]. Such continuous surface separates the space into two interwoven nets of channels known as bicontinuous structure. The most commonly known TPMSs are the Schwarz primitive (P), Schwarz diamond (D), and Schoen gyroid (G) surface structures.

The TPMSs are intriguing due to their ability to exhibit unique physical properties. For example, such surfaces are used to model different crystalline structures in both natural and synthetic systems. Examples of these structures include LLCs and TLCs [18,19], block copolymer (BCP) self-assemblies [2,20–22] and organic zeolites [23]. It is noteworthy that one important structural difference between the LC and BCP systems consists in the values of the unit cell parameter, which ranges between ca. 10 and 100 nm for BCP-based systems [24], while it can rarely exceed several nanometers for LC systems [25]. In some instances, it is interesting to make the characteristic distances of both systems closer to each other, i.e., by increasing the sizes of the LC systems and/or by bringing down the dimensions of phases generated by the BCPs in order to make them interesting for the developing areas of nanotechnologies and nanopatterning [26].

The wedge-shaped molecules can be considered as building units for design of zeolite-like materials that can be excellent candidates for fabrication of nanoporous membranes [27–30]. It is already established that these compounds are able to self-organize in complex morphologies pertinent to the cubic bicontinuous phases, such as diamond, gyroid, and primitive cubic phases. The family of bicontinuous cubic phases can be considered as an organic equivalent of zeolites due to their developed network of well-defined nanochannels. Among these structures, gyroid bicontinuous phases are considered as being more accessible for practical applications because of self-supporting frameworks with better mechanical strength, ensuring open-cell character, high and uniform porosity, and large and predictable specific surface area. However, the gyroid phases formation mechanism and corresponding functionality are still not sufficiently studied. This is in part accounted for by experimental difficulties in studying the structure formation process in different environments with the necessity to perform in-situ structural characterizations. Therefore, several scientific and technical challenges have still to be taken to get a full understanding of the corresponding structure-property relationships, as well as details of the gyroid phase formation process.

For several years, our group has been involved in studies of self-assembly processes in various wedge-shaped compounds based on sulfonic acid. We have previously shown that, during heating or UV-irradiation (for the systems with light-sensitive groups), these systems form a variety of LC morphologies, such as smectic, columnar, and cubic (double diamond and gyroid). Swelling of thin films in solvent vapors in some cases leads to development of ion channel networks, with the lattice parameter becoming comparable to the one in the BCP systems [31–35]. Earlier, we have reported on formation of well-organized cubic phase in the films of wedge-shaped salts with linear alkyl chains during swelling in methanol. Surprisingly, this mesophase was found to be rather stable in a wide temperature range, and, upon preparation, the methanol solvent can be replaced by water in ion channels, which makes these systems interesting for future applications [33–35]. Synchrotron Grazing-Incidence X-ray scattering (GISAXS) experiments in combination with computer simulation reveal development of water channels of ca. 2 nm in diameter (cf. Figure 1), which makes them a promising candidate for fabrication of mechanically stable membranes with excellent proton conductivity.

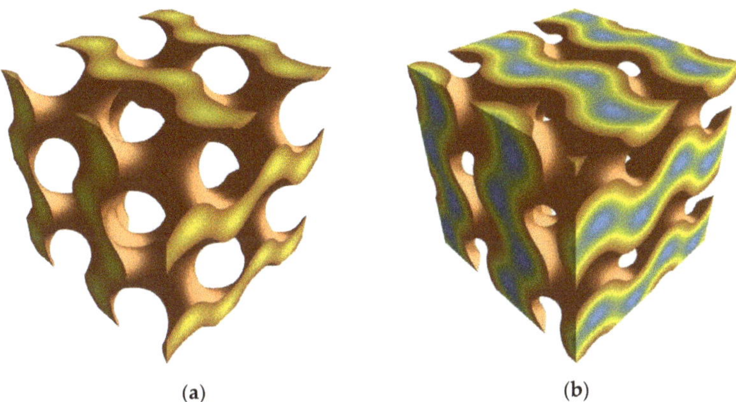

(a)   (b)

Figure 1. Three-dimensional model of a bicontinuous double gyroid phase of a wedge-shaped amphiphile in a dry (**a**) and swollen (**b**) state. Yellow chocolate color indicates the hydrophilic part of amphiphilic molecules, while the blue color denotes the water inside the molecular channels. The structure in (**b**) contains water fraction 33 wt.%.

In the present study, we address the development of ion-channel network during swelling in humid atmosphere by in-situ monitoring the structural evolution and topography of thin films of a wedge-shaped amphiphile.

2. Materials and Methods

The synthesis of pyridinium salt of 4′-[3″,4″,5″-tris-(octyloxy)benzoyloxy]azobenzene-4-sulfonic acid (C8AzoPyr) was described in previous reports [31,36]. The chemical structure of C8AzoPyr is given in Scheme 1.

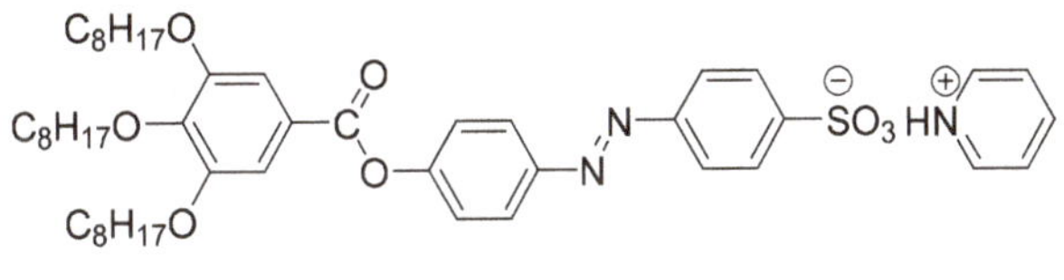

Scheme 1. The chemical structure of the studied wedge-shaped sulfonate molecule C8AzoPyr.

Differential Scanning Calorimetry (DSC) measurements were carried out using a Netzsch DSC 204 unit. Samples (typical weight of 5 mg) were enclosed in standard Netzsch 25 μL aluminum crucibles. Heating and cooling rate was 10 °C/min.

Grazing-Incidence Small- and Wide-Angle X-ray scattering (GISAXS-GIWAXS) measurements were performed at the BM26 and ID10 beamlines of the European Synchrotron Radiation Facility (ESRF) in Grenoble (France) using a custom-designed environmental chamber [34,35].

The energy of X-ray photons was 12 keV. The s-axis ($|s| = 2\sin\theta/\lambda$, where θ is the Bragg angle, λ is the wavelength, and $|s|$ is the norm of the s-vector) was calibrated using several diffraction orders of silver behenate. X-ray patterns were recorded using a 2D

Pilatus 1M camera. The X-ray data analysis, including background subtraction and radial integrations of the 2D patterns were accomplished using home-built routines designed within the IgorPro software package (Version 6.37, Wavemetrics Ltd., Portland, OR, USA). For the GISAXS experiments, thin films of C8AzoPyr were prepared from a chloroform solution (20 mg/mL) by spin-coating (500 rpm/min) on a silicon wafer substrate. The phase composition of the thin films at different temperatures was addressed by in-situ heating and cooling of the samples in temperature range from –50° to 100 °C. The change in the phase structure during the water uptake process in thin films was also monitored using in-situ control of relative humidity (RH) and temperature.

A specially designed environmental cell was used for in-situ studies of the structural evolution in thin films (Figure 2). The compactness of experimental cell makes it possible to use it on the synchrotron beamlines, i.e., in the sample chambers of a diffractometer or spectrophotometer, and can be also combined with an optical microscope. The experimental cell allows one to study thin films by the GIWAXS/GISAXS methods with simultaneous control of external factors, such as UV radiation, temperature, and atmosphere of solvent vapors. The experimental cell includes the following components: a sealed chamber, consisting of a heating element, a lid to create the desired atmosphere over the sample, a frame for mounting, and a temperature and humidity control system.

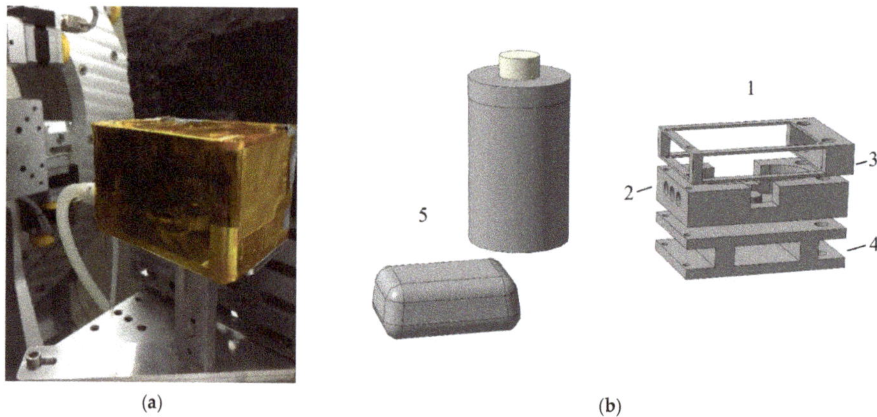

(a)   (b)

Figure 2. (**a**): Environmental cell for in-situ control of the atmosphere and temperature installed at BM26 beamline of ESRF; (**b**): schematics of the environmental cell components: (1)-sealed chamber, composed of a heating element (2), lid (3) to create controlled atmosphere around the sample, frame (4) for mounting, temperature and humidity control system. The latter consists of a dewar with liquid nitrogen (depicted as a cylinder) and a control panel/computer (depicted as a box) (5).

AFM imaging was performed in Tapping Mode using a Cypher S Asylum Research Atomic Force Microscope. In the experiments, Oxford Instruments AC240TS medium soft silicon cantilevers (Abingdon, UK) with a resonance frequency 67 kHz and spring constant 1.82 N/m were used. The spring constant of the cantilevers was measured by the thermal noise method. To conduct measurements at variable humidity, a special portable air humidity control system was employed. The system shown on Scheme 2a consists of a computer connected to a controller and aeraulic circuit of humidity regulator (Scheme 2b). The main controller regulates the flow of the compressed air through the system. In order to create humidity-controlled air, dry and humid air are mixed in the desired proportions with the help of Mass Flow Controllers (MFC) for different flow ranges connected to the main controller. This air is then redirected through the Atomic Force Microscope Cypher S Asylum Research equipped with humidity sensor that allows to maintain the required humidity value during the experiment. The AFM images were analyzed using the open source software Gwyddion (Version 2.58).

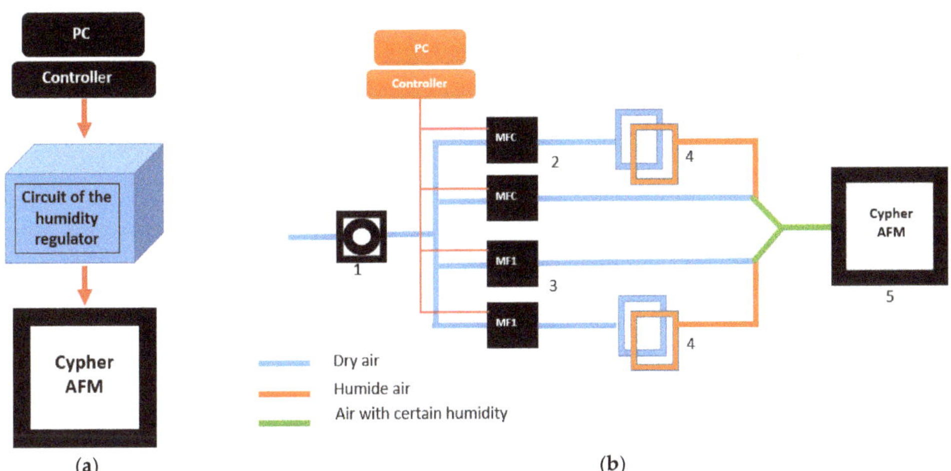

Scheme 2. (**a**): Scheme of portable air humidity control system; (**b**): aeraulic circuit of the humidity regulator connected to controller and PC: 1–manometer, 2–Mass Flow Controller with flow range of 10,000 sccm, 3–Mass Flow Controller MF-1 with flow range of 100 sccm, 4–laboratory flask with water, 5–Atomic Force Microscope Cypher S Asylum Research.

3. Results

The thermal behavior of C8AzoPyr was assessed by DSC. The DSC trace corresponding to the first heating exhibits two endothermic peaks (Figure 3, black line). The first peak with onset at 72 °C, and enthalpy 0.7 J/g is associated with a solid-solid state transition; the second intense endothermic peak with onset at 96 °C, and enthalpy 26 J/g can be attributed to melting of a crystalline structure. Interestingly, the second heating curve does not exhibit endothermic peaks anymore (Figure 3, red line). In our earlier report, crystallization of C8AzoPyr from a LC phase during annealing at room temperature was explained by a process of local ordering of linear octyl chains [35]. Figure S1a shows FTIR spectrum of C8AzoPyr after a long storage at room temperature. In the magnified region from 3000 to 2800 cm^{-1}, one can see that conformation-sensitive anti-symmetric and symmetric stretching vibration modes of CH_2 group are positioned at 2921 and 2851 cm^{-1}, respectively (cf. Figure S1b). These values are close to those characteristic of all-trans conformation of n-alkanes–2920 and 2850 cm^{-1}, respectively [37]. Consequently, in the crystalline state, C8AzoPyr possesses ordered octyl chains in a quasi-extended conformation. Above the melting point, the structure reorganizes to a columnar hexagonal LC phase with disordered alkyl chains (Figure S2a,b). Since the second heating performed after 3 min at room temperature does not reveal any transitions, the LC-to-crystal phase transition at room temperature is likely to be a slow process (Figure S2c,d). This gives one the time to manipulate the structure while in the mobile LC state by varying the external factors. In contrast, in the crystalline state, a rigid framework of side chains prevents any structural evolution in the atmosphere of water vapors or in vapors of organic solvents [35].

Figure 4a displays GISAXS pattern after long storage at room temperature. The ordered phase was indexed to a columnar monoclinic unit cell (Col$_{mon}$), which has the following parameters: a = 58.9 Å, b = 50.0 Å, γ = 61°. However, in the LC state with liquid-like side chains which is generated upon cooling from 100 °C, the structure of C8AzoPyr is more sensitive to vapors of polar solvents, such as water or alcohols, due to their more effective diffusion through the amorphous alkyl periphery of the molecular wedges. Thus, swelling in methanol atmosphere for 2 h results in formation of a layer-like lamellar structure with parameter a$_{lam}$ = 51.2 Å (Figure 4b). The corresponding indexed 1D-reduced diffractograms are presented in Figure S3.

Earlier, we demonstrated that during slow cooling from 100 °C in a methanol atmosphere the high-temperature columnar hexagonal structure transforms to a stable cubic double gyroid structure [35]. It is noteworthy that the normal to the film surface becomes parallel to the 211 reciprocal space direction, as was previously noted for another wedge-shaped mesogen [31]. After several months annealing at ambient conditions, the gyroid structure not only persists but even shows a definite improvement of its organization. POM images of the dry film (not shown here) confirm the absence of birefringence which is typical of cubic phases. The diffraction pattern of C8AzoPyr thin film demonstrates a set of peaks with d-spacings ratio of $\sqrt{6}:\sqrt{8}$, which is characteristic of the Cub_{gyr} phase (symmetry $Ia\bar{3}d$). The corresponding lattice parameter a_{gyr} is 117 Å (Figure 4c). The summary of the extinction rules for the different bicontinuous phases can be found in ref. [38].

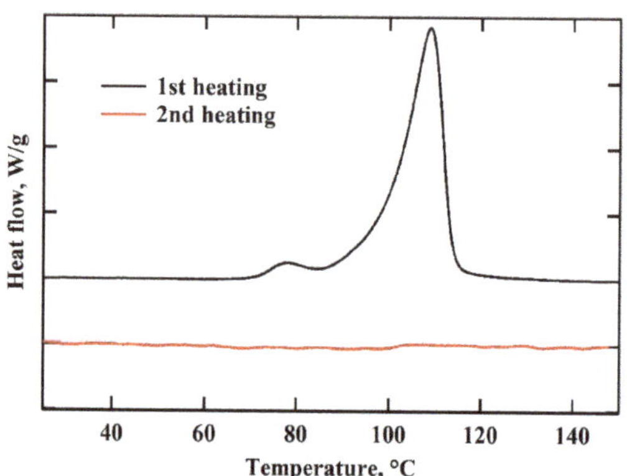

Figure 3. DSC curves: first (**black**) and second (**red**) heating traces of bulk C8AzoPyr after long annealing at room temperature.

The presence of bicontinuous network of opened hydrophilic channels makes the film capable of efficiently swelling in a humid atmosphere. A 2D GISAXS pattern of the thin film with the Cub_{gyr} phase corresponding to the sample kept for two days in saturated water vapors reveals a significant increase of the lattice parameter a_{gyr} to 138 Å. This is an indication of formation of bicontinuous networks of water nanochannels (Figure 4d). The average radii of water channels in gyroid phase in the as-prepared and swollen film is found to be 8 ± 1 and 10 ± 2 Å, respectively [35]. Consequently, formation of Cub_{gyr} phase stabilized by ordered alkyl chains demonstrates the outstanding ability of the system to absorb water without destruction of supramolecular organization with long-range order. The swelling process of C8AzoPyr was further studied by in-situ AFM.

Figure 5a shows a topographic image of a C8AzoPyr thin film exposed to relative humidity of 42%. The film clearly exhibits a terrace-like structure that can be quantitatively analyzed using height cross-sections, as shown in Figure 5b. In this case, the cross-section traced along the black solid line in Figure 5a reveals terraces with an average height of 47.5 Å. Since the vertical direction of the film corresponds to the 211 reciprocal space vector of the double gyroid phase according to GISAXS, it is logical to assume that the corresponding step height should give the reticular distance of the (211) planes of $\frac{a}{\sqrt{6}}$. This allows computing the lattice parameter a_{gyr} of 116 Å.

However, if one wants to analyze the gyroid structure with more scrutiny, it would be necessary to image the film surface with higher resolution. An example of such measurement is given in Figure 6, where AFM phase images of C8Pyr thin films at different RH values are shown. Despite some noise present in the images (the images after filtering are shown in the insets), one can recognize the unique pattern of the 211 plane of the double gyroid phase (see, e.g., Reference [39]). The observed morphological features of the 211 plane can be viewed relative to the $1\bar{1}1$ and $01\bar{1}$ reciprocal space vectors that direct along and perpendicular to the characteristic "knitting" features of the plane, respectively. The analysis in the Fourier space given in the right column of Figure 6 allows visualizing the characteristic in-plane periodicities. Thus, the image taken in the dry (RH = 20%) state of the film reveals the fundamental periodicity along the $01\bar{1}$ direction of 0.057 nm^{-1}, which is equivalent to the distance of 17.5 nm in direct space. The periodicity in the vertical direction of 0.09 nm^{-1} provides the corresponding direct-space repeat of 11.1 nm. Taking into account the expressions for both distances given in ref. [39], one obtains the lattice parameter of the gyroid phase to be 12.4 and 12.8 nm, respectively. These values agree relatively well with the corresponding SAXS unit cell parameter, which validates such AFM-based analysis. A small difference (of ca. 3%) in the values computed for the horizontal and vertical directions can be accounted for by a small drift of the piezo along the slow scan direction.

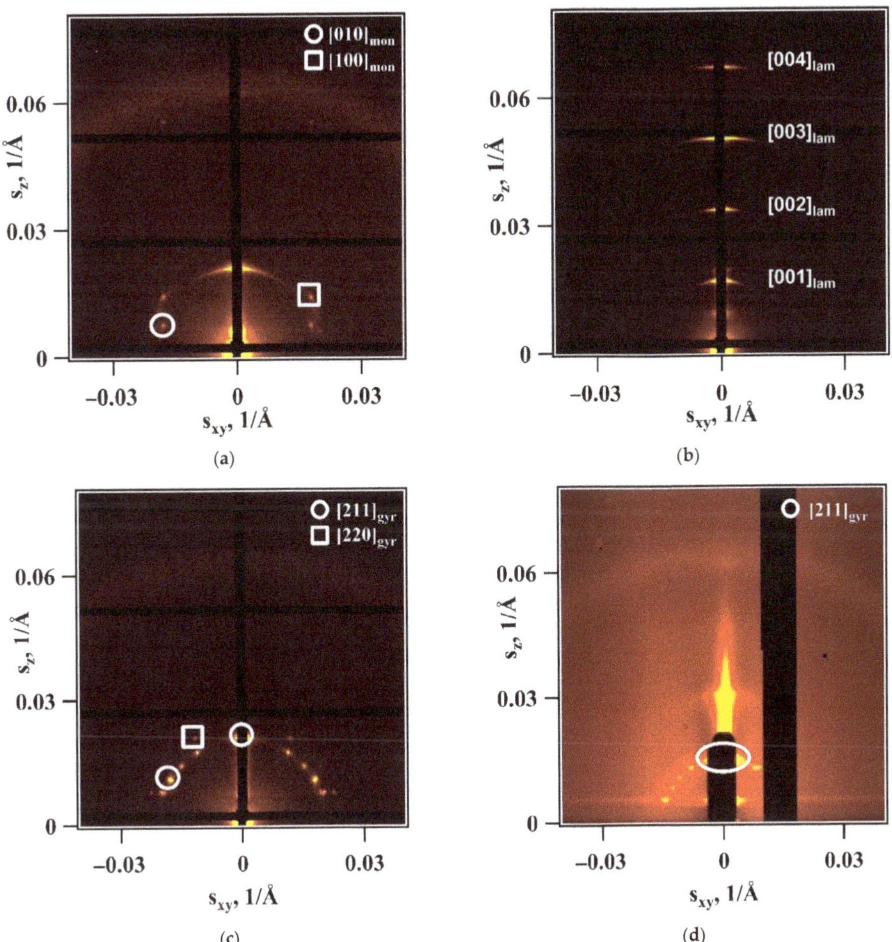

Figure 4. Two-dimensional GISAXS pattern of a dry C8AzoPyr thin film. (**a**) Col$_{mon}$ phase formed after long annealing at room temperature; (**b**) lamellar phase formed upon swelling in methanol vapors for two hours at room temperature; (**c**) Cub$_{gyr}$ phase generated after long-term annealing at ambient conditions; (**d**) Cub$_{gyr}$ phase swollen in water vapors for two days at room temperature.

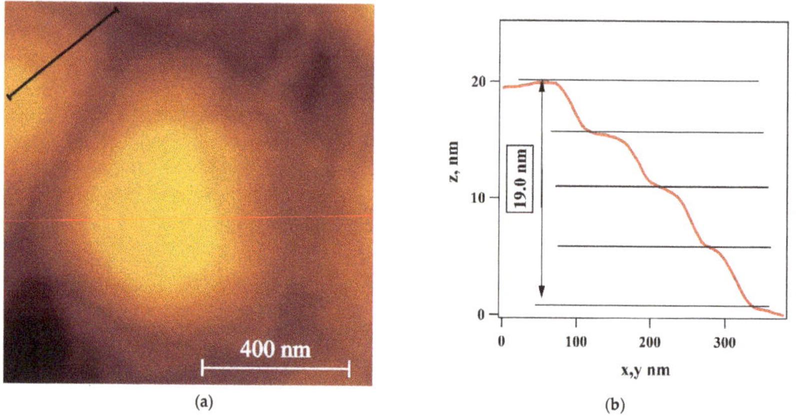

(a)   (b)

Figure 5. AFM image of a C8AzoPyr thin film: (**a**) height trace of the film surface at RH = 42%; (**b**) height cross-section traced along the black solid line in (**a**) showing quantized terraces.

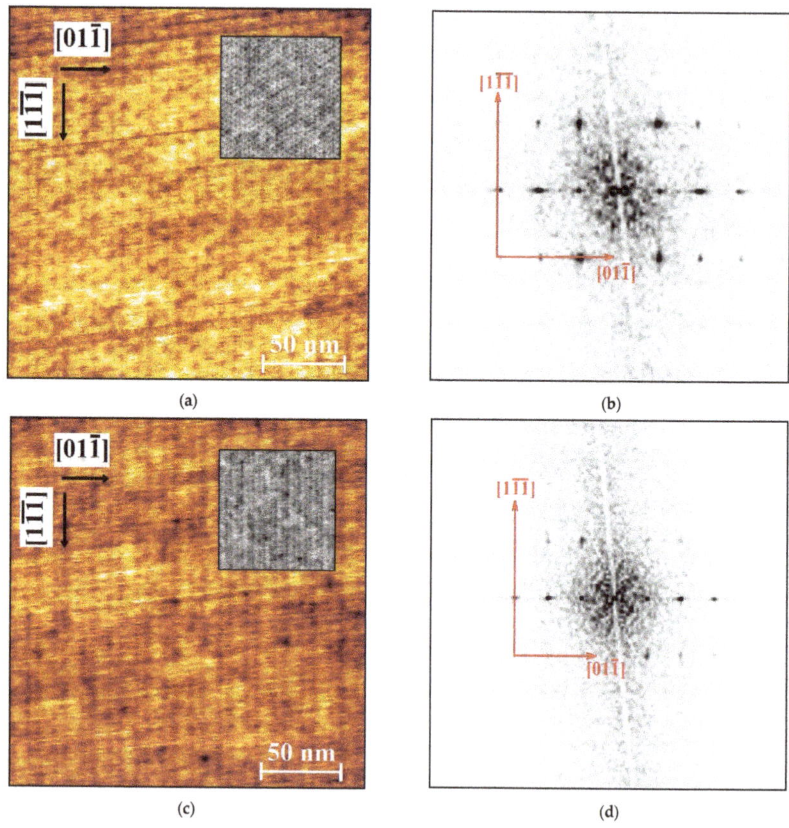

(a)   (b)
(c)   (d)

Figure 6. AFM images of the C8AzoPyr sample: (**a**) phase image and (**b**) FT pattern of the dry film (RH = 20%); (**c**) phase image and (**d**) FT pattern of the swollen film (RH = 100%). The insets show the corresponding AFM images after filtering in the Fourier space.

The described reciprocal-space analysis makes it possible to quantitatively compare the AFM phase images of the gyroid phase taken at different values of RH. Indeed, the two images in Figure 6 can be confronted based on the positions of the characteristic peaks in the Fourier space. The computation of the unit cell parameter for the image taken at 100% RH gives the value of 13.5 nm, which is noticeably larger than the corresponding value in the dry state and is also close to the SAXS unit cell parameter measured for the long-term annealing of the sample in the saturated water vapors. The presented AFM results confirm the unusually high stability of the Cub_{gyr} phase in swelling-drying cycles.

4. Discussion

The present work provides an example of a wedge-shaped C8AzoPyr amphiphile bearing promise for applications in ion-selective membranes for separation, catalysis, and energy conversion. To optimize the ion transport, one has to prepare the C8AzoPyr membranes in a bicontinuous cubic phase, such as the double gyroid $Ia\bar{3}d$. The experiments reveal a very rich polymorphic behavior of C8AzoPyr both in LLC and TLC phases, which makes any search for the desired cubic phase exclusively with ex-situ techniques inefficient. Here, we show that such search employing a combination of environmental in-situ GISAXS-GIWAXS and AFM experiments is much more productive as one can easily operate the external factors, such as solvent vapors and temperature, while continuously monitoring the film structure.

An interesting particularity of the studied system is that the gyroid phase can be locked in the film and persist even in its fully dry state. This is completely unusual for the LLC phases for which the gyroid phase exists only in a certain range of solvent fractions. Such locking the gyroid phase is accounted for by ordering of the octyl side chains that keep the gyroid phase stable in the course of repeated swelling-drying cycles. Such metastable state can be erased only if one melts the side-chains' locks at high temperature. However, as long as the side chains keep their local ordering, the formed gyroid phase will persist, and the material will preserve its functionality. Therefore, it is shown that, by combining in-situ structure monitoring with reciprocal- and direct-space experimental techniques and optimized molecular architecture, one can design the materials with enhanced stability of the bicontinuous structure.

5. Conclusions

Using environmental in-situ grazing-incidence X-ray diffraction and atomic force microscopy, the phase behavior of thin films of a wedge-shaped mesogen in dry and swollen state was investigated. The studied samples show the presence of unusually stable bicontinuous cubic double gyroid phase, whereas, in the swollen thin films, they reveal formation of swollen water channels. The obtained results can help understanding the structure formation in supramolecular systems at different hierarchical levels and, consequently, contribute to the development of new approaches in fabrication of self-organized films with the desired morphology.

Supplementary Materials: The following are available online at https://www.mdpi.com/article/10.3390/ma14112892/s1, Figure S1: (a) FTIR spectrum of C8AzoPyr powder after long-term storage at room temperature; (b) magnified region of (a) with anti-symmetric and symmetric stretching vibration modes of CH2 groups, Figure S2: 2D GISAXS patterns of a dry C8AzoPyr thin film recorded during first heating from 25 °C at 80 °C (a) and at 100 °C (b), and during second heating from 25 °C at 80 °C (c) and at 100 °C(d), Figure S3: 1D-reduced GISAXS pattern of a dry C8AzoPyr thin film. (a) Columnar monoclinic (Colmon) phase formed after long-term annealing at room temperature; (b) lamellar phase (lam) formed upon swelling in methanol vapors for two hours at room temperature; (c) double gyroid phase (Cubgyr) generated after long-term annealing at ambient conditions; (d) Cubgyr phase swollen in water vapors for two days at room temperature.

Author Contributions: Conceptualization, K.N.G.; Data curation, D.S.K.; Formal analysis, D.V.A.; Investigation, A.F.A.; Methodology, D.V.V., A.F.A., X.Z., D.P.; Supervision, D.A.I.; writing—original draft preparation, K.N.G.; Writing–original draft, D.V.A., D.A.I.; Writing—review and editing, D.A.I. All authors have read and agreed to the published version of the manuscript.

Funding: This research was funded by the Ministry of Science and High Education of the Russian Federation (contract No 05.605.21.0188 from 3 December 2019 (RFMEFI60519X0188)).

Institutional Review Board Statement: Not applicable.

Informed Consent Statement: Not applicable.

Data Availability Statement: Data available on request due to restrictions e.g., privacy or ethical. The data presented in this study are available on request from the corresponding author. The data are not publicly available due to organization policy.

Acknowledgments: We thank ESRF for beamtime at BM26, ESRF's Partnership for Soft Condensed Matter (PSCM) for support and A.Panzarella and M.Capron for help in AFM experiments and A.Rychkov for the sketch of the GISAXS environmental cell.

Conflicts of Interest: The authors declare no conflict of interest. The funders had no role in the design of the study; in the collection, analyses, or interpretation of data; in the writing of the manuscript, or in the decision to publish the results.

References

1. Petsko, G.A.; Ringe, D. *Protein Structure and Function*; New Science Press, Ltd.: London, UK, 2004.
2. Mai, Y.; Eisenberg, A. Self-assembly of block copolymers. *Chem. Soc. Rev.* **2012**, *41*, 5969–5985. [CrossRef]
3. Yashima, E.; Ousaka, N.; Taura, D.; Shimomura, K.; Ikai, T.; Maeda, K. Supramolecular helical systems: Helical assemblies of small molecules, foldamers, and polymers with chiral amplification and their functions. *Chem. Rev.* **2016**, *116*, 13752–13990. [CrossRef] [PubMed]
4. Borisch, K.; Diele, S.; Göring, P.; Kresse, H.; Tschierske, C. Tailoring thermotropic cubic mesophases: Amphiphilic polyhydroxy derivatives. *J. Mater. Chem.* **1998**, *8*, 529–543. [CrossRef]
5. Borisch, K.; Tschierske, C.; Göring, P.; Diele, S. Molecular design of thermotropic liquid crystalline polyhydroxy amphiphiles forming type 1 columnar and cubic mesophases. *Langmuir* **2000**, *16*, 6701–6708. [CrossRef]
6. Fong, C.; Le, T.; Drummond, C.J. Lyotropic liquid crystal engineering–ordered nanostructured small molecule amphiphile self-assembly materials by design. *Chem. Soc. Rev.* **2012**, *41*, 1297–1322. [CrossRef]
7. Mezzenga, R.; Seddon, J.M.; Drummond, C.J.; Boyd, B.J.; Schröder-Turk, G.E.; Sagalowicz, L. Nature-Inspired Design and Application of Lipidic Lyotropic Liquid Crystals. *Adv. Mater.* **2019**, *31*, 1900818. [CrossRef] [PubMed]
8. Zhai, J.; Fong, C.; Tran, N.; Drummond, C.J. Non-Lamellar Lyotropic Liquid Crystalline Lipid Nanoparticles for the Next Generation of Nanomedicine. *ACS Nano* **2019**, *13*, 6178–6206. [CrossRef]
9. Barriga, H.M.G.; Holme, M.N.; Stevens, M.M. Cubosomes: The Next Generation of Smart Lipid Nanoparticles? *Angew. Chem. Int. Ed.* **2019**, *58*, 2958–2978. [CrossRef]
10. Caffrey, M. A comprehensive review of the lipid cubic phase or in meso method for crystallizing membrane and soluble proteins and complexes. *Struct. Biol. Commun.* **2015**, *71*, 3–18.
11. Zabara, A.; Meikle, T.G.; Newman, J.; Peat, T.S.; Conn, C.E.; Drummond, C.J. The nanoscience behind the art of in-meso crystallization of membrane proteins. *Nanoscale* **2017**, *9*, 754–763. [CrossRef]
12. Reppe, T.; Poppe, S.; Tschierske, C. Controlling Mirror Symmetry Breaking and Network Formation in Liquid Crystalline Cubic, Isotropic Liquid and Crystalline Phases of Benzil-Based Polycatenars. *Chem. Eur. J.* **2020**, *26*, 16066–16079. [CrossRef] [PubMed]
13. Zhang, L.; Jaroniec, M. Strategies for development of nanoporous materials with 2D building units. *Chem. Soc. Rev.* **2020**, *49*, 6039–6055. [CrossRef] [PubMed]
14. Wang, Z.; Yu, J.; Xu, R. Needs and trends in rational synthesis of zeolitic materials. *Chem. Soc. Rev.* **2012**, *41*, 1729–1741. [CrossRef] [PubMed]
15. Xiang, Z.; Cao, D.; Dai, L. Well-defined two dimensional covalent organic polymers: Rational design, controlled syntheses, and potential applications. *Polym. Chem.* **2015**, *6*, 1896–1911. [CrossRef]
16. Gearba, R.I.; Anokhin, D.V.; Bondar, A.I.; Bras, W.; Lehmann, M.; Ivanov, D.A. Homeotropic Alignment of Columnar Liquid Crystals in Open Films by Means of Surface Nano-Patterning. *Adv. Mater.* **2007**, *19*, 815–820. [CrossRef]
17. Hyde, S.; Blum, Z.; Landh, T.; Lidin, S.; Ninham, B.W.; Andersson, S.; Larsson, K. *The Language of Shape: The Role of Curvature in Condensed Matter: Physics, Chemistry and Biology*; Elsevier: Amsterdam, The Netherlands, 1996.
18. Fontell, K. Cubic phases in surfactant and surfactant-like lipid systems. *Colloid Polym. Sci.* **1990**, *268*, 264–285. [CrossRef]
19. Seddon, J.M. An inverse face-centered cubic phase formed by diacylglycerol-phosphatidylcholine mixtures. *Biochemistry* **1990**, *29*, 7997–8002. [CrossRef]

20. Orilall, M.C.; Wiesner, U. Block copolymer based composition and morphology control in nanostructured hybrid materials for energy conversion and storage: Solar cells, batteries, and fuel cells. *Chem. Soc. Rev.* **2011**, *40*, 520–535. [CrossRef]

21. Cochran, E.W.; Garcia-Cervera, C.J.; Fredrickson, G.H. Stability of the gyroid phase in diblock copolymers at strong segregation. *Macromolecules* **2006**, *39*, 2449–2451. [CrossRef]

22. Stefik, M.; Guldin, S.; Vignolini, S.; Wiesner, U.; Steiner, U. Block copolymer self-assembly for nanophotonics. *Chem. Soc. Rev.* **2015**, *44*, 5076–5091. [CrossRef]

23. Sun, J.; Bonneau, C.; Cantín, A.; Corma, A.; Díaz-Cabañas, M.J.; Moliner, M.; Zhang, D.; Li, M.; Zou, X. The ITQ-37 mesoporous chiral zeolite. *Nature* **2009**, *458*, 1154–1157. [CrossRef] [PubMed]

24. Meuler, A.J.; Hillmyer, M.A.; Bates, F.S. Ordered network mesostructures in block polymer materials. *Macromolecules* **2009**, *42*, 7221–7250. [CrossRef]

25. Beginn, U. Thermotropic columnar mesophases from N-H···O, and N···H-O hydrogen bond supramolecular mesogenes. *Prog. Polym. Sci.* **2003**, *28*, 1049–1105. [CrossRef]

26. Miskaki, C.; Moutsios, I.; Manesi, G.-M.; Artopoiadis, K.; Chang, C.-Y.; Bersenev, E.A.; Moschovas, D.; Ivanov, D.A.; Ho, R.-M.; Avgeropoulos, A. Self-Assembly of Low-Molecular-Weight Asymmetric Linear Triblock Terpolymers: How Low Can We Go? *Molecules* **2020**, *25*, 5527. [CrossRef] [PubMed]

27. Zhu, X.; Beginn, U.; Möller, M.; Gearba, R.I.; Anokhin, D.V.; Ivanov, D.A. Self-Organization of polybases neutralized with mesogenic wedge-shaped sulfonic acid molecules: An approach toward supramolecular cylinders. *Am. Chem. Soc.* **2006**, *128*, 16928–16937. [CrossRef]

28. Hernandez Rueda, J.J.; Zhang, H.; Rosenthal, M.; Möller, M.; Zhu, X.; Ivanov, D.A. Polymerizable wedge-shaped ionic liquid crystals for fabrication of ion-conducting membranes: Impact of the counterion on the phase structure and conductivity. *Eur. Polym. J.* **2016**, *81*, 674–685. [CrossRef]

29. Chen, Y.; Lingwood, M.D.; Goswami, M.; Kidd, B.E.; Hernandez, J.J.; Rosenthal, M.; Ivanov, D.A.; Perlich, J.; Zhang, H.; Zhu, X.; et al. Humidity-Modulated Phase Control and Nanoscopic Transport in Supramolecular Assemblies. *J. Phys. Chem. B* **2014**, *118*, 3207–3217. [CrossRef] [PubMed]

30. Hernandez, J.J.; Zhang, H.; Chen, Y.; Rosenthal, M.; Lingwood, M.D.; Goswami, M.; Zhu, X.; Moeller, M.; Madsen, L.A.; Ivanov, D.A. Bottom-Up Fabrication of Nanostructured Bicontinuous and Hexagonal Ion-Conducting Polymer Membranes. *Macromolecules* **2017**, *50*, 5392–5401. [CrossRef]

31. Zhang, H.; Li, L.; Moller, M.; Zhu, X.; Hernandez Rueda, J.J.; Rosenthal, M.; Ivanov, D.A. From channel-forming ionic liquid crystals exhibiting humidity-induced phase transitions to nanostructured ion-conducting polymer membranes. *Adv. Mater.* **2013**, *25*, 3543–3548. [CrossRef]

32. Grafskaia, K.N.; Hernandz Rueda, J.J.; Zhu, X.; Nekipelov, V.M.; Anokhin, D.V.; Moeller, M.; Ivanov, D.A. Designing the topology of ion nano-channels in the mesophases of amphiphilic wedge-shaped molecules. *Phys. Chem. Chem. Phys.* **2015**, *17*, 30240–30247. [CrossRef]

33. Grafskaia, K.N.; Anokhin, D.V.; Hernandez Rueda, J.J.; Ivanov, D.A. In situ studies of molecular self-assembling during the formation of ion-conducting membranes for fuel cells. *Appl. Mech. Mater.* **2015**, *792*, 623–628. [CrossRef]

34. Grafskaia, K.; Zimka, B.; Zhu, X.; Anokhin, D.; Ivanov, D. Engineering of ion channels topology in self-assembled wedge-shaped amphiphiles by combination of temperature and solvent vapor treatment. *AIP Conf. Proc.* **2016**, *1748*, 040009.

35. Grafskaia, K.N.; Anokhin, D.V.; Zimka, B.I.; Izdelieva, I.A.; Zhu, X.; Ivanov, D.A. An "on–off" switchable cubic phase with exceptional thermal stability and water sorption capacity. *Chem. Commun.* **2017**, *53*, 13217–13220. [CrossRef]

36. Zhu, X.; Tartsch, B.; Beginn, U.; Möller, M. Wedge-Shaped Molecules with a Sulfonate Group at the Tip—A New Class of Self-Assembling Amphiphiles. *Chem. Eur. J.* **2004**, *10*, 3871–3878. [CrossRef]

37. Snyder, R.G.; Strauss, H.L.; Elliger, C.A. Carbon-hydrogen Stretching Modes and the Structure of n-Alkyl Chains. 1. Long Disordered Chains. *J. Phys. Chem.* **1982**, *86*, 5145–5150. [CrossRef]

38. Moschovas, D.; Manesi, G.-M.; Karydis-Messinis, A.; Zapsas, G.; Ntetsikas, K.; Zafeiropoulos, N.E.; Piryazev, A.A.; Thomas, E.L.; Hadjichristidis, N.; Ivanov, D.A.; et al. Alternating Gyroid Network Structure in an ABC Miktoarm Terpolymer Comprised of Polystyrene and Two Polydienes. *Nanomaterials* **2020**, *10*, 1497. [CrossRef] [PubMed]

39. Feng, X.; Zhuo, M.; Guo, H.; Thomas, E.L. Visualizing the double-gyroid twin. *Proc. Natl. Acad. Sci. USA* **2021**, *118*, e2018977118. [CrossRef] [PubMed]

 *materials*

Article

Multiblock Thermoplastic Polyurethanes: In Situ Studies of Structural and Morphological Evolution under Strain

Denis V. Anokhin [1,2,*], Marina A. Gorbunova [1], Ainur F. Abukaev [1,3] and Dimitri A. Ivanov [1,2,4]

[1] Institute for Problems of Chemical Physics Russian Academy of Sciences, Semenov Prospect 1, 142432 Chernogolovka, Russia; mflute2008@yandex.ru (M.A.G.); abukaev.af@phystech.edu (A.F.A.); dimitri.ivanov.2014@gmail.com (D.A.I.)

[2] Faculty of Fundamental Physical and Chemical Engineering, Lomonosov Moscow State University, 1 Leninskie Gory, 119991 Moscow, Russia

[3] Moscow Institute of Physics and Technology, Institutskiy per. 9, 141700 Dolgoprudny, Russia

[4] Institut de Sciences des Matériaux de Mulhouse-IS2M, CNRS UMR 7361, Jean Starcky, 15, F-68057 Mulhouse, France

* Correspondence: deniano@yahoo.com; Tel.: +7-905-509-35-21

Citation: Anokhin, D.V.; Gorbunova, M.A.; Abukaev, A.F.; Ivanov, D.A. Multiblock Thermoplastic Polyurethanes: In Situ Studies of Structural and Morphological Evolution under Strain. *Materials* **2021**, *14*, 3009. https://doi.org/10.3390/ma14113009

Academic Editor: Dario Pasini

Received: 6 May 2021
Accepted: 29 May 2021
Published: 1 June 2021

Publisher's Note: MDPI stays neutral with regard to jurisdictional claims in published maps and institutional affiliations.

Abstract: The structural evolution of multiblock thermoplastic polyurethane ureas based on two polydiols, poly(1,4-butylene adipate (PBA) and poly-ε-caprolactone (PCL), as soft blocks and two diisocyanites, 2,4-toluylene diisocyanate (TDI) and 1,6-hexamethylene diisocyanate (HMDI), as hard blocks is monitored during in situ deformation by small- and wide-angle X-ray scattering. It was shown that the urethane environment determines the crystal structure of the soft block. Consequently, two populations of crystalline domains of polydiols are formed. Aromatic TDI forms rigid domains and imposes constrains on the crystallization of bounded polydiol. During stretching, the TDI–polydiol domains reveal limited elastic deformation without reorganization of the crystalline phase. The constrained lamellae of polydiol form an additional physical network that contributes to the elastic modulus and strength of the material. In contrast, polydiols connected to the linear semi-flexible HMDI have a higher crystallization rate and exhibit a more regular lamellar morphology. During deformation, the HMDI-PBA domains show a typical thermoplastic behavior with plastic flow and necking because of the high degree of crystallinity of PBA at room temperature. Materials with HMDI-PCL bonding exhibit elastic deformation due to the low degree of crystallinity of the PCL block in the isotropic state. At higher strain, hardening of the material is observed due to the stress-induced crystallization of PCL.

Keywords: thermoplastic polyurethane ureas; shape memory materials; synchrotron SAXS/WAXS; polymer deformation; lamellar morphology; poly-ε-caprolactone; poly(1,4-butylene adipate)

1. Introduction

Recently, a new class of materials adaptive to external factors has started to rapidly develop. They can be exemplified by thermoplastic elastomers based on semi-crystalline polyurethanes (TPU) and polyurethane-ureas (TPUU). Such materials combining elasticity and strength have been already actively used in medicine for the engineering of tissues, plasters, sutures, implants, as well as in other areas [1–3]. TPU and TPUU are linear block copolymers consisting of thermodynamically incompatible soft (SB) and hard (HB) blocks that undergo a microphase separation. The soft segment is formed by polyester or polyether diols such as poly-ε-caprolactone (PCL) [4,5], poly(1,4-butylene adipate) (PBA), and their mixtures [6,7], poly-L-lactide [8], poly(ethylene glycol) [9], etc. The hard blocks are diol-urethane or urea fragments synthesized from aromatic diisocyanates (2,4-toluene diisocyanate), cycloaliphatic (isophoron diisocyanate), and aliphatic (1,6-hexamethylenediisocyanate) compounds and chain extenders (1,4-butanediol, diamine, and 2-aminoethanol).

89

Apart from chemical composition, the mechanical, thermal, and relaxation properties of TPU and TPUU depend on the thermal history [10–12]. Thus, semi-crystalline polyurethanes are able to fix their temporary shape during deformation and further restore the original shape after heating above a switching temperature, i.e., they show the shape memory effect (SME) [13,14]. In a previous publication, it was shown that SME is determined by transition of the TPUU between three morphological states: (1) semi-crystalline lamellar morphology of isotropic film in a permanent shape, (2) fibrillar morphology with extended-chain crystals of the soft block after film stretching to a temporary shape, and (3) phase-separated state of the block copolymer in the amorphous state after shape recovery above the switching temperature [15]. At present, the attention of researchers is directed to studies of thermomechanical properties of adaptive TPUU [16–19]. However, the relationship between the chemical composition, nanoscale morphology, and thermomechanical characteristics is far from being fully understood.

Synchrotron small- and wide-angle X-ray scattering (SAXS/WAXS) is a powerful technique for analysis of polymer structure. The high brilliance of the X-ray source allows monitoring structural and morphological evolution with adequate time resolution. In particular, a combination of SAXS/WAXS with DSC [20–25], FTIR [26], and relaxometry [27] is capable of providing important information about structural evolution under external factors.

Structural investigations of TPUs are focused on a combination of two main processes: phase separation and ordering of HB [22,28]. The size and packing quality depend on the type of diisocyanate, its content, and preparation conditions. Particularly, the growth of HB content and annealing at elevated temperature stimulate an association process with an increase of HB domain size and relaxation of internal stress [21,22].

The most efficient approach combines SAXS/WAXS with uniaxial strain at different temperatures. Classical models of deformation of semi-crystalline polymers and block copolymers consider the continuous increase of a long period during stretching [29]. As a result of the specific chemical structure, polyurethanes possess fundamentally different structural evolution during deformation [30–33]. Particularly, TPU diffractograms often show a retardation of SAXS peak shift during film deformation, which indicates that the morphology is not related to the sequence of hard and soft domains [34,35]. A detailed study of TPU deformation was performed using in situ SAXS/WAXS experiments during deformation [36–38].

In studies of TPU based on polytetrahydrofurane (PTHF1000), methylene diphenyl diisocyanate (MDI), and 1,4-butanediol, Stribeck et al. explained a considerable retardation of the SAXS peak with respect to deformation by "paracrystalline" morphology of the HB. The authors suggested that a considerable fraction of the macroscopic strain must be realized in "poorly arranged entities" that do not contribute to the SAXS maximum because of absence of quasiperiodicity. In contrast, "well-arranged ensembles" are hardly deformed during strain and give a main contribution to the SAXS peak. At high deformation, tie polymer chains start to be pulled out from "well-arranged ensembles", resulting in a release of stress and shrinking of the long period (sacrifice-and-relief mechanism) [39,40].

It should be mentioned that despite numerous publications on structural evolution of TPUs under strain, only a few of them discuss the crystallization of soft block [41,42]. This can be explained by the low soft block content and small molecular weight of polydiols, which are used in most of the studies. Thus, the impact of soft block crystallinity, distribution of crystalline domains, and the size and kinetics of formation on macroscopic properties has been not well studied yet.

In the present work, the structural evolution of multiblock thermoplastic polyurethane urea based on two polydiols (poly(1,4-butylene adipate diol) and poly-ε-caprolactone diol) and two diisocyanites (diisocyanates (2,4-toluylene diisocyanate and 1,6-hexamethylene diisocyanate) is monitored as a function of temperature and strain during in-situ SAXS/WAXS experiments.

2. Materials and Methods

2.1. Materials

Poly(1,4-butylene adipate) diol (PBA) (Huakai Resin Co., Ltd., Shandong, China, Mn = 2000 Da) and poly-ε-caprolactone diol (PCL) (Merck KGaA, Darmstadt, Germany, Mn = 2000 Da) were dried in vacuum for 4 h at 80 °C. The OH groups content determined by chemical method [43] was 1.7 w/w%. Diisocyanates (2,4-toluylene diisocyanate (TDI) and 1,6-hexamethylene diisocyanate (HMDI)) (Merck KGaA, Darmstadt, Germany) were distilled in vacuum at 50–55 °C/12 mm Hg and stored in sealed ampoules. Chain elongation agents (2-amino-1-ethanol (EA) and 1,4-butanediol (BD)) were distilled over freshly powdered calcium hydride under a reduced pressure. Dibutyltin dilaurate (DBTDL) catalyst purchased from Aldrich was used as received. The disappearance of OH and NCO groups and the appearance of urethane groups were monitored by IR spectroscopy.

2.2. Synthesis of Multiblock Thermoplastic Polyurethane Urea

The synthesis of multiblock copolymers was carried out by the three-reactor method developed earlier by us [15]. At the first stage (reactor 1) of TPUU synthesis, a macrodiol from polydiol (PBA or PCL) and TDI in the presence of two chain extenders (BD and EA) was prepared. In the second stage (reactor 2), a macrodiol from polydiol (PBA or PCL), HMDI, and chain extender (BD) were synthesized. Then, the reaction products from the two reactors were mixed and linked by HMDI added in a stoichiometric ratio [NCO]/[OH] = 1. In the result, four polymers have been synthesized, where PCL and PBA are linked in different combinations with bulky aromatic TDI and more flexible and elongated HMDI (Table 1). Upon reaching the degree of conversion of NCO groups ≈98%, the reaction mass was poured into a flat Teflon container and dried at 40 °C during the day until constant weight. The control of the full course of the reaction was carried out by infrared spectra on the complete disappearance of absorption bands of isocyanate (νNCO = 2271 cm^{-1}) and hydroxyl groups (νOH = 3620 cm^{-1}). Molecular characteristics of the TPUUs were determined by gel permeation chromatography (GPC) in tetrahydrofuran using a Waters GPCV 2000 chromatograph equipped with refractometric and viscosimetric detectors. The number and weight average molecular weight of all four polymers were the following: Mn = 40,000 Da, Mw = 80,000 Da, Mw/Mn = 2. The scheme of synthesis of TPUM is presented in Scheme 1.

The molar ratio of the reaction was determined from which HS content was estimated. The hard segment (HS) was defined as:

$$HS = \frac{m(TDI + HMDI) + m(AE + BD)}{m(TDI + HMDI) + m(AE + BD) + mPolyols}.$$

Table 1 shows the composition and ratio of components in the synthesized TPUUs.

Table 1. Chemical composition of studied TPUUs.

Polymer	Polymer Composition		Mass Fraction of Reagents, %			
	P1	P2	Polyol	Diisocyanate	Chain Extender	SS/HS
TPUU-AA	PBA	PBA	69	23	8	2.2
TPUU-BB	PCL	PCL	69	23	8	2.2
TPUU-AB	PBA	PCL	69	23	8	2.2
TPUU-BA	PCL	PBA	68	23	8	2.2

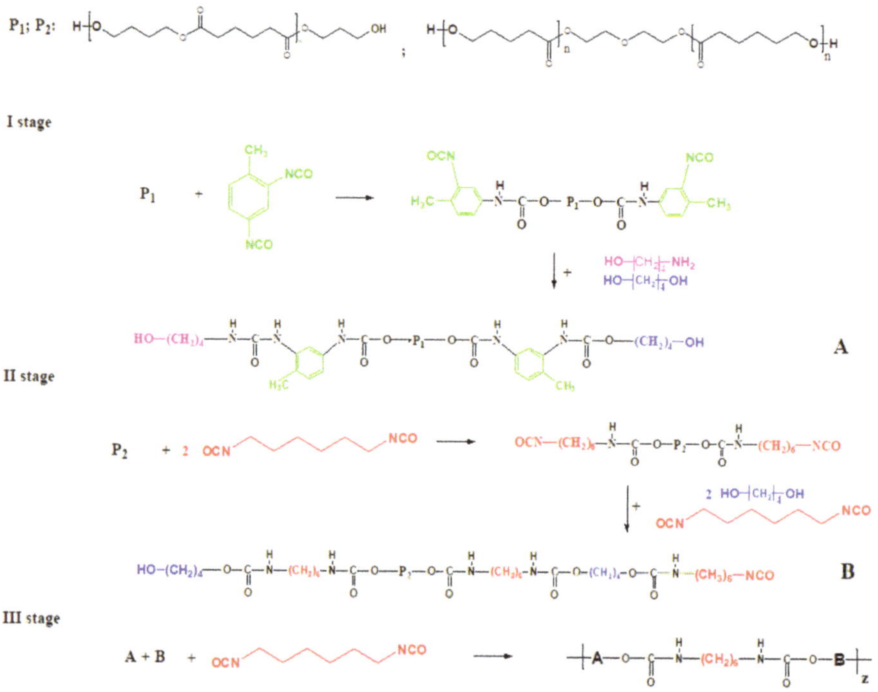

Scheme 1. General scheme of synthesis of macrodiol A (I stage), macroisocyanate B (II stage) and studied TPUUs (III stage).

2.3. Experimental Techniques

The structural investigation was performed at the BM26 beamline of the European Synchrotron Radiation Facility (ESRF) in Grenoble, France. The beamline is equipped with Pilatus 1M (SAXS, s-range 0.002–0.04 Å^{-1}) and Pilatus 300k (WAXS, s-range 0.08–0.5 Å^{-1}) detectors. The experiments used X-ray photons with a wavelength of 1.04 Å, and the spot size of the beam on the sample was 0.65 × 0.65 mm. The central part of the dogbone-like test bars for analysis had dimensions of 22 × 6 × 0.7 mm. The in situ strain experiments were carried out in a Linkam TST350 tensile stage with a maximum force of 20N and strain rate of 1 mm/min. The strain rate was decreased compared to tensile tests to minimize the variation of deformation during SAXS/WAXS exposure (12 s). Shape recovery was measured during heating of stretched film at a heating rate 5 °C/min to a temperature above the melting point of the soft block e.g., 80 °C, and rapid cooling to room temperature. The two-dimensional diffraction patterns were analyzed using a program package developed in Igor Pro Program package (Wavemetrics Ltd., Portland, OR, USA) [44–46].

The ex situ tensile tests were performed on a Zwick TC-FR010TH Material Testing Machine at 50 mm/min stretching rate using the same samples as for SAXS/SAXS strain experiments. This test was performed according to the ASTM-D638 Type IV.

3. Results and Discussions

The stress–strain curves of TPUU depending on the composition of the soft block are shown in Figure 1. It can be seen that TPUU-BB with PCL as a soft block demonstrates the stress–strain curve of a typical elastomer with low Young's modulus (E) and the highest elongation at break (ε_r) (cf. black curve in Figure 1). The elasticity in TPUU-BB is provided by the connection of PCL with a more flexible linear urethane fragment. The mechanical characteristics of TPUUs are summarized in Table 1. TPUU-AA and TPUU-BA containing the PBA block linked with HMDI demonstrate the mechanical behavior typical of stiff thermoplastics: high Young's modulus and plastic flow with the formation of a neck in the strain range of 50–400% (green and blue curves in Figure 1). The stiffness of the materials stems from a limited mobility of PBA chains bounded with bulky rigid TDI-based urethane (Figure 1). This clearly indicates that the type of polydiol is important in the design of SME polymers. The mechanical characteristics are summarized in Table 2.

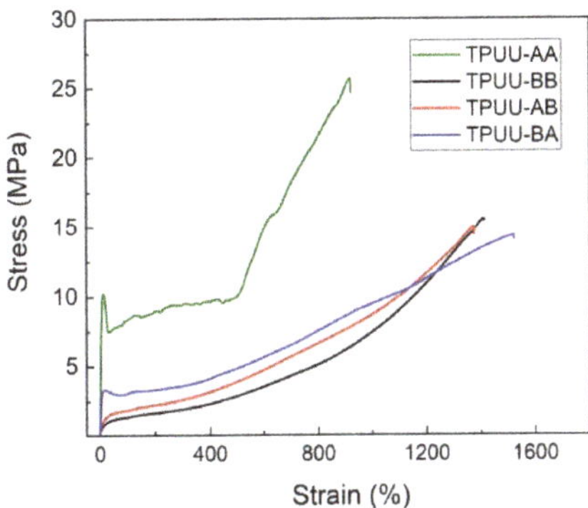

Figure 1. Stress–strain curves of the studied TPUUs.

Table 2. Mechanical characteristics of TPUUs.

Polymer	ε_r, %	σ_r, MPa	E, MPa
TPUU-AA	807 ± 78	22 ± 3	199 ± 4
TPUU-BB	1445 ± 27	15 ± 1	7 ± 1
TPUU-AB	1445 ± 52	16 ± 1	11 ± 1
TPUU-BA	1340 ± 84	13 ± 2	59 ± 2

The variation of mechanical properties is related to the specific morphology of the films with different bonding of urethane of polydiol. In this paper, the reorganization of the morphology was studied in situ by SAXS/WAXS analysis using a brilliant synchrotron source.

Figure 2 shows changes in the film organization during stretching. Isotropic sample TPUU-AA reveals a ring characteristic of PBA crystalline lamellar ordering with a long period of L_{SB} = 14.7 nm (Figure 2A). In addition, on the 1D-reduced curve, a broad peak with a maximum at 9.7 nm can be identified (Figure 3A, black curve). This peak cannot be attributed to the second order of the long period, and its origin can be understood using the in situ film deformation experiment. Under small reversible strains, the isotropic ring transforms into a four-spot pattern, indicating reorientation of the PBA lamellae along the drawing direction [47]. An increase of half-width of the crystalline peaks in the WAXS region indicates the fragmentation of large PBA lamellae under stress (Figure 4A, blue curve). The d-spacing of the second peak increases to 10.4 nm, and it organizes in broad "clouds" positioned along the direction of stretching (Figure 3A, blue curve). At further strains, the four-spot lamellar peak disappears due to the lack of phase contrast during plastic flowing of the material in the neck. Meantime, the equatorial orientation of the second peak becomes more pronounced, and its position gradually increases to 12.9 nm for ε = 300% (Figure 3A, dark blue curve). After release of stress, the film of TPUU-AA shows high residual deformation—200% in the neck. The intensity of the second peak increases because of enhanced phase contrast and the spacing decreases to L_{HB} = 11.6 nm (Figure 3A, green curve). Interestingly, this peak is preserved even after heating to above the melting point of the PBA crystal phase (Figures 3A and 4A, yellow curves). Consequently, the peak can be assigned to the phase-separated morphology of soft and hard segments rather than to the stacking of PBA crystals [15]. We expect that this spacing reflects the distance between the rigid TDI domains that surround the PBA block (see Figure 5A, left). The PBA chains bonded to TDI crystallize under hard geometric constraints [48,49]. The presence of stressed tie chains in such domains prevents the constrained PBA lamellae from unfolding during stretching with small elastic deformation of the amorphous phase (Figure 5A, right). In contrast, the PBA lamellae surrounded by relatively flexible linear urethane fragments are formed in soft confinement. During application of strain, these regions behave as typical semi-crystalline flexible-chain polymers with plastic flow and the formation of fibrillar crystals. The difference in crystallization conditions of PBA connected to aromatic and linear urethanes can illustrate the appearance of SAXS maximum at 20.4 nm during the deformation of amorphous TPUU-AA after heating to 80 °C (Table 3), indicating crystallization of the HMDI-surrounded soft blocks before the TDI-surrounded ones. Two populations of the PBA crystals generate two melting peaks in the DSC curve of the fresh TPUU-AA sample [14].

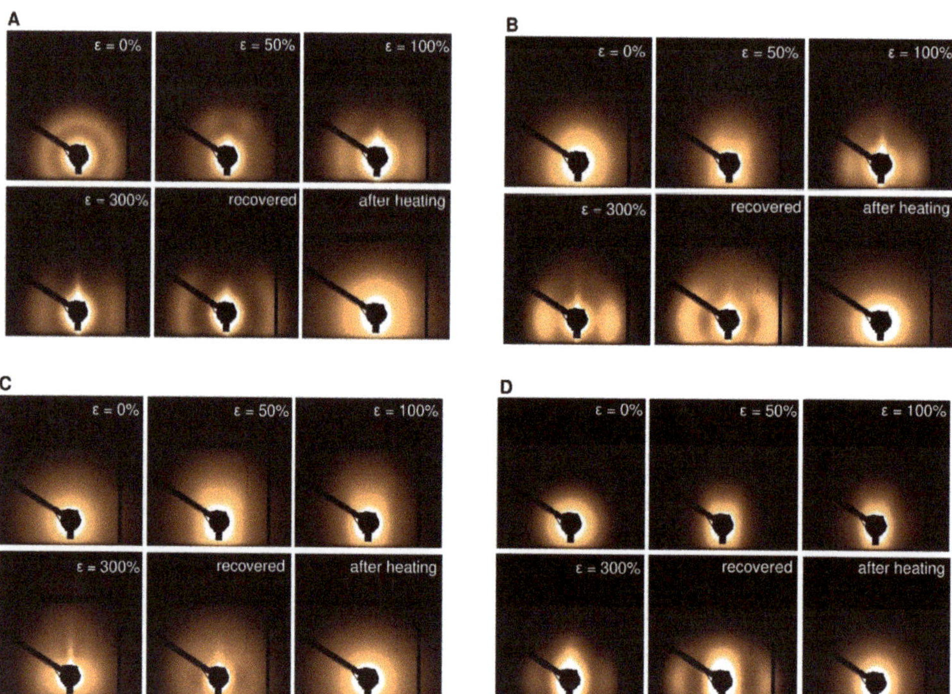

Figure 2. Two-dimensional (2D) SAXS patterns of the studied films: (**A**) TPUU-AA; (**B**) TPUU-BB; (**C**) TPUU-AB; (**D**) TPUU-BA, measured during strain, recovered after release of stress, and after heating to 60 °C. Drawing direction is horizontal.

The replacement of well-crystallizable PBA soft block by PCL with a moderate crystallization rate at room temperature changes the morphology and corresponding mechanical characteristics of TPUU (Figure 2B). In the isotropic film, the lamellar SB peak is broader and shifted to smaller angles (Figures 2B and 3B, red curve). During deformations up to 50%, one can see an increase of both SB and HB periodicities (Figure 3B, black curve, Table 3). The intensity of the SB peak is concentrated in the direction normal to drawing (Figure 2B). On further stretching, the SB maximum from folded lamellae disappears without plastic flow of PCL chains (Figures 2B and 3B, blue curve). At even higher deformations, WAXS reveals growing crystalline peaks corresponding to the stress-induced formation of fibrillar PCL crystals (Figure 4B, blue curve). For ε = 300%, one observes a certain decrease of phase-separated L_{HB} distance probably caused by "sacrifice-and-relief" mechanism described by Stribeck for MDI-based polyurethanes [39].

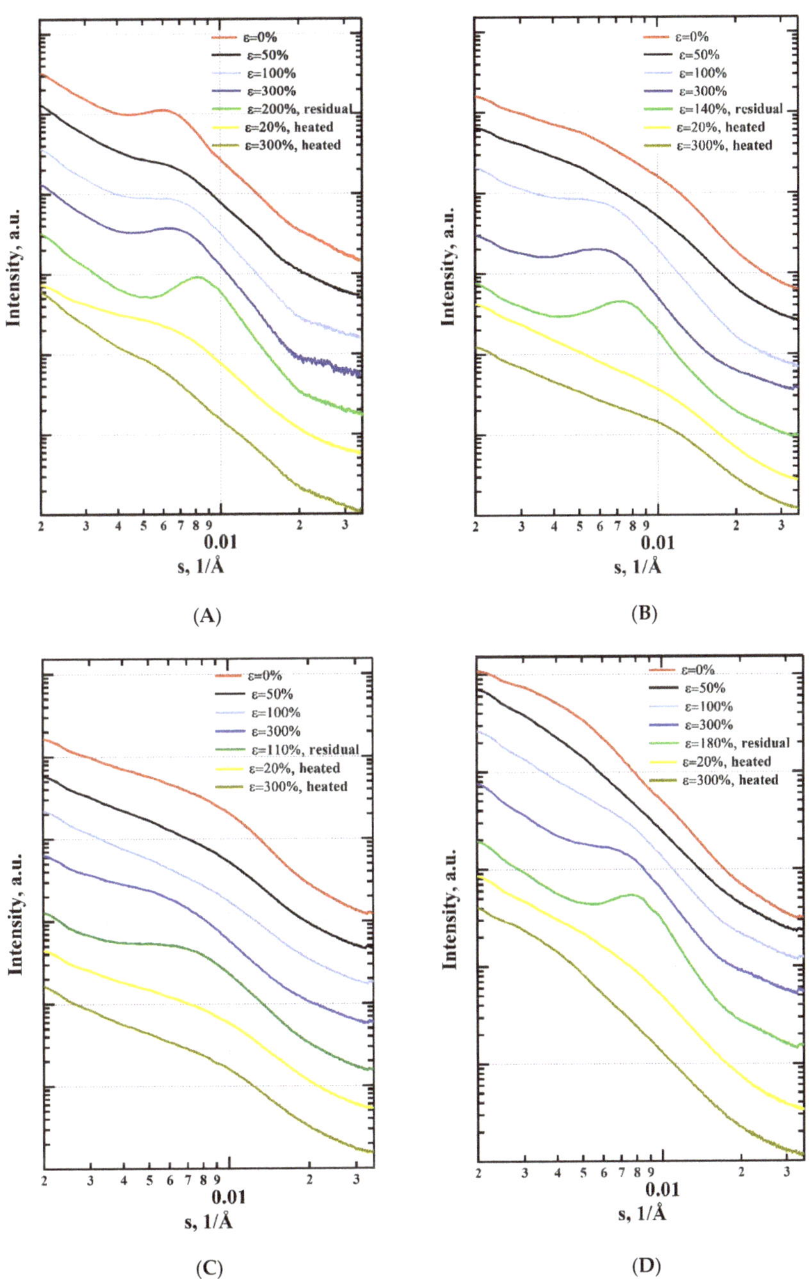

Figure 3. One-dimensional (1D)-reduced SAXS curves of the studied films: (**A**) TPUU-AA; (**B**) TPUU-BB; (**C**) TPUU-AB; (**D**) TPUU-BA, measured during deformation, recovered after release of stress, and after heating to 60 °C.

The same effect of decrease of L_{HB} was detected during stretching of the amorphous film preliminarily heated above the melting point of PCL (Table 3). The peak due to phase separation is less pronounced probably because of the better miscibility of PCL with linear urethanes (Figure 3B, yellow curve). Interestingly, stretching of the initially amorphous film does not show any stress-induced crystallization (Figure 3B, brown curve). Generally, TPUU-BB with a PCL soft block possesses less regular lamellar organization. Rubber-like stress–strain curves and a higher increase of L_{SB} and L_{HB} indicate that both TDI-bounded and HMDI-bounded PCL chains do not form regularly folded lamellae.

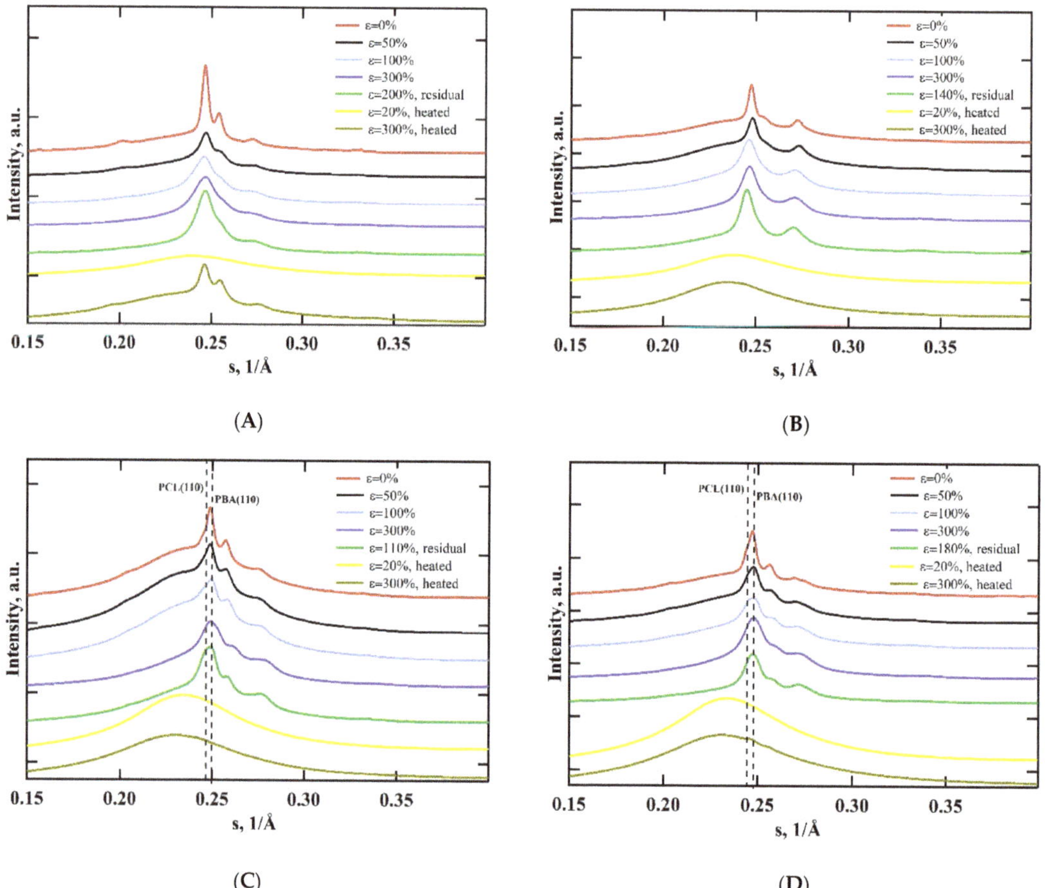

Figure 4. One-dimensional (1D)-reduced WAXS curves of the studied films: (**A**) TPUU-AA; (**B**) TPUU-BB; (**C**) TPUU-AB; (**D**) TPUU-BA, measured during deformation, recovered after release of stress, and after heating to 60 °C. Vertical dashed lines point on positions of the (110) peaks of PBA and PCL.

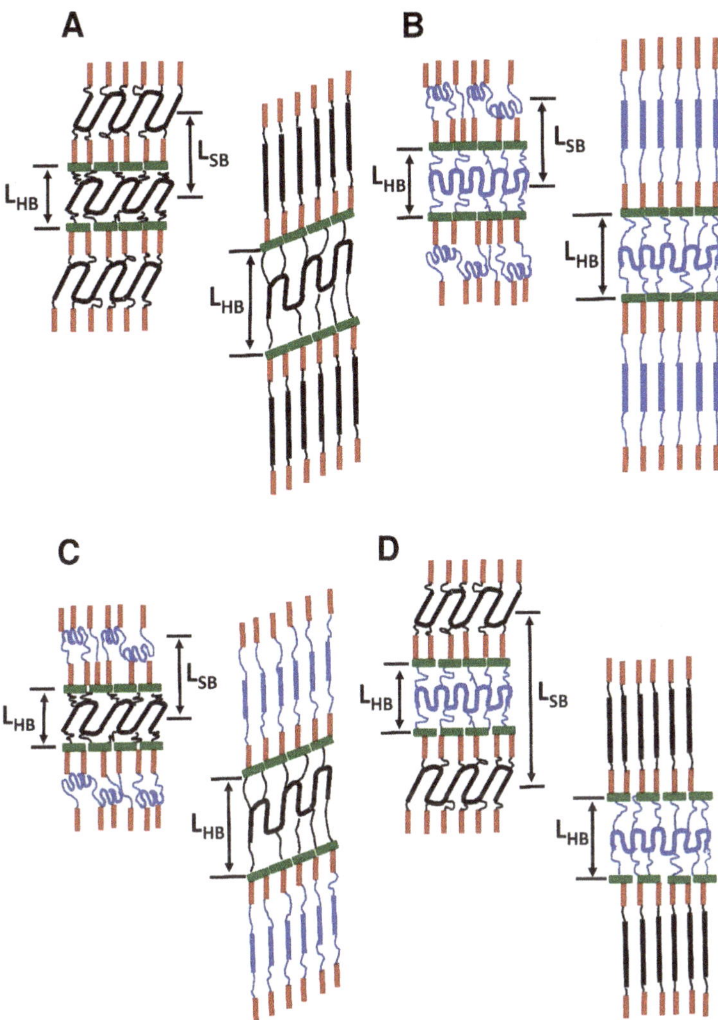

Figure 5. Schematics of morphology of the TDI (green), HMDI (red), PBA (black), and PCL (blue) regions in isotropic (left) and stretched to ε = 300% (right) films: (**A**) TPUU-AA; (**B**) TPUU-BB; (**C**) TPUU-AB; (**D**) TPUU-BA. Stretching direction is vertical.

In the case of TPUUs with both PBA and PCL soft blocks, an important role is played by the position of a particular polydiol in the multiblock chain. Isotropic sample TPUU-AB reveals the presence of a mainly crystalline phase of TDI-bounded PBA (Figure 4C, red curve). We suppose that PCL blocks surrounded by semi-flexible linear urethanes still have a low crystallization rate, resulting in the absence of correlation between PCL and PBA lamellae (Table 3). During deformation, the crystal fraction of PCL increases due to stress-induced crystallization and the weak SAXS maximum at L_{SB} = 18.4 nm is visible up to ε = 100% (Figures 3C and 4C blue curves). PBA crystals surrounded by hard domains of TDI show a relatively small increase of L_{HB} on deformation and absence of the peak orientation in the drawing direction (Figures 2C and 3C). After the release of stress, the orientation of the phase-separated morphology almost completely disappears due to the high flexibility of the PCL blocks (Figure 3C, green curve). This sample demonstrates mechanical behavior that is intermediate between rubber-like and thermoplastic without necking but with orientational hardening (Figure 1, red curve and Figure 5C).

Table 3. Structural parameters of samples at different strain.

Polymer	ε, %	L_{HB}, nm	L_{SB}, nm
TPUU-AA	0	9.7	14.6
	50	10.8	14.0
	100	12.8	-
	300	12.9	-
	200, restored	11.6	-
	20, after heating	9.7	-
	300, after heating	9.4	20.4
TPUU-BB	0	9.6	15.6
	50	10.3	17.8
	100	14.7	-
	300	14.3	-
	140, restored	12.5	-
	20, after heating	9.1	-
	300, after heating	8.6	-
TPUU-AB	0	10.8	-
	50	11.2	16.3
	100	12.4	18.4
	300	14.8	-
	110, restored	12.3	-
	20, after heating	10.1	-
	300, after heating	9.9	-
PTUU-BA	0	10.0	20.6
	50	11.7	24.8
	100	13.6	-
	300	13.0	-
	180, restored	11.4	-
	20, after heating	11.1	-
	300, after heating	13.8	24.6

In contrast, WAXS patterns of TPUU-BA with HMDI-bounded PBA chains reveal a high degree of crystallinity in the isotropic state with an approximately equal content of PBA and PCL crystals (Figure 4D, red curve). Thus, PCL chains constrained by a neighbor TDI hard domain show a better tendency to crystallize at room temperature than in the more mobile HMDI domains (Figure 5D). However, regular lamellae of HMDI-surrounded PBA blocks give high phase contrast, resulting in the enhanced SAXS maximum with increased distance L_{SB} = 20.6 nm (Figure 3D, red curve). This peak transforms to a four-spot pattern at ε = 50% but further disappears due to the lack of phase contrast between HB and extended-chain PBA crystals (Figures 2D and 3D, blue curve). Interestingly, PCL chains in the TDI domain do not reveal significant deformation and stress-induced crystallization (Figure 4D, dark blue curve). In contrast to TPUU-AA, the HB peak of TPUU-BA shows very good azimuthal orientation in the drawing direction at high strain (Figure 2A,D). This can be attributed to bigger chain tilt in constrained PBA lamellae compared to the lamellae of PCL (Figure 5A,D). After heating above the melting point of PBA and PCL crystals, HMDI-surrounded PBA chains rapidly crystallize with the reappearance of maximum at 24.6 nm during the stretching of amorphous film (Figure 3D, yellow curve). This confirms that the crystallization of HMDI-surrounded PBA chains occurs faster than that of TDI-surrounded PBA with the formation of more regularly stacked lamellae. The presence of HMDI-surrounded PBA in TPUU provides relatively high residual deformation after the release of stress and a slow relaxation of strain that is important for the design of shape memory materials.

4. Conclusions

In conclusion, the variation of chemical nature of polydiol and adjusted diisocyanates allows tuning the mechanical properties of resulting TPUUs from soft elastomers to rigid thermoplastics. Synchrotron SAXS/WAXS studies of a series of TPUUs reveal a complex change in morphology during deformation related to the superposition of phase separation of soft and hard blocks and crystallization of the polydiols. It was shown that TDI-bounded polydiols are constrained in rigid domains which, on the one hand, decrease the crystallization rate and regularity of lamellae but, on the other hand, preserve crystals from plastic flow during strain. In this case, constrained crystals of soft block play the role of an additional physical network imparting higher stiffness to the material. In the meantime, HMDI-bounded polydiols reveal higher crystallinity and faster crystallization from the melt. During deformation, these crystalline domains behave as typical semi-crystalline flexible-chain polymers. The TPUUs with HMDI-bounded PBA show plastic flow with the formation of extended-chain crystals, while HMDI-bounded PCL fragments reveal stress-induced crystallization. Thus, HMDI-rich domains are responsible for the elastic characteristics of the material: elongation at break, residual deformation, etc. Consequently, rigid aromatic TDI or semi-rigid linear HMDI form a different interface for two populations of crystallites of polydiols. A significant difference in crystallization kinetics at room temperature between PBA and PCL provides an additional tool for fine-tuning the thermoplastic properties and shape memory behavior.

In situ investigation of the structure and morphology of TPUU at different external stimuli allows finding the relationship between the structure and deformational properties that helps optimizing the composition of the soft block for desired mechanical characteristics.

Author Contributions: Conceptualization, D.V.A.; Data curation, A.F.A.; Formal analysis, D.V.A.; Investigation, M.A.G.; Methodology, M.A.G.; Supervision, D.V.A.; writing—original draft preparation, D.V.A.; Writing—review and editing, D.A.I. All authors have read and agreed to the published version of the manuscript.

Funding: This research was funded by Russian Foundation for Basic Research (Project No. 19-53-15016) and carried out within the framework of the State Assignment (Topic No. 0089-2019-0012, State Registration No. AAA-A19-119032690060-9).

Institutional Review Board Statement: Not applicable.

Informed Consent Statement: Not applicable.

Data Availability Statement: The data presented in this study are available in article.

References

1. Wang, W.; Wang, C. *Polyurethane for Biomedical Applications: A Review of Recent Developments*; Elsevier Masson SAS: Cambridge, UK, 2012. [CrossRef]
2. Liu, Y.F.; Wu, J.L.; Zhang, J.X.; Peng, W. Feasible Evaluation of the Thermo-Mechanical Properties of Shape Memory Polyurethane for Orthodontic Archwire. *J. Med. Biol. Eng.* **2017**, *37*, 666–674. [CrossRef]
3. Huang, W. Thermo-Moisture Responsive Polyurethane Shape Memory Polymer for Biomedical Devices. *Open Med. Devices J.* **2010**, *2*, 11–19. [CrossRef]
4. Haryńska, A.; Kucinska-Lipka, J.; Sulowska, A.; Gubanska, I.; Kostrzewa, M.; Janik, H. Medical-Grade PCL Based Polyurethane System for FDM 3D Printing—Characterization and Fabrication. *Materials* **2019**, *12*, 887. [CrossRef]
5. Kuang, W.; Mather, P.T. A Latent Crosslinkable PCL-Based Polyurethane: Synthesis, Shape Memory, and Enzymatic Degradation. *J. Mater. Res.* **2018**, *33*, 2463–2476. [CrossRef]
6. Gorbunova, M.A.; Shukhardin, D.M.; Lesnichaya, V.A.; Badamshina, E.R.; Anokhin, D.V. New Polyurethane Urea Thermoplastic Elastomers with Controlled Mechanical and Thermal Properties for Medical Applications. *Key Eng. Mater.* **2019**, *816*, 187–191. [CrossRef]

7. Gorbunova, M.; Komratova, V.; Grishchuk, A.; Badamshina, E.; Anokhin, D. The Effect of Addition of Low-Layer Graphene Nanoparticles on Structure and Mechanical Properties of Polyurethane-Based Block Copolymers. *Polym. Bull.* **2019**, *76*, 5813–5829. [CrossRef]
8. Shi, L.; Zhang, R.-Y.; Ying, W.-B.; Hu, H.; Wang, Y.-B.; Guo, Y.-Q.; Wang, W.-Q.; Tang, Z.-B.; Zhu, J. Polyether-Polyester and HMDI Based Polyurethanes: Effect of PLLA Content on Structure and Property. *Chin. J. Polym. Sci.* **2019**, *37*, 1152–1161. [CrossRef]
9. Callegari, D.; Colombi, S.; Nitti, A.; Simari, C.; Nicotera, I.; Ferrara, C.; Mustarelli, P.; Pasini, D.; Quartarone, E. Autonomous Self-Healing Strategy for Stable Sodium-Ion Battery: A Case Study of Black Phosphorus Anodes. *ACS Appl. Mater. Interfaces* **2021**, *13*, 13170–13182. [CrossRef] [PubMed]
10. Li, X.; Stribeck, A.; Schulz, I.; Pöselt, E.; Eling, B.; Hoell, A. Nanostructure of Thermally Aged Thermoplastic Polyurethane and Its Evolution under Strain. *Eur. Polym. J.* **2016**, *81*, 569–581. [CrossRef]
11. He, Y.; Xie, D.; Zhang, X. The Structure, Microphase-Separated Morphology, and Property of Polyurethanes and Polyureas. *J. Mater. Sci.* **2014**, *49*, 7339–7352. [CrossRef]
12. Yilg, E.; Yilg, I.; Yilgör, I.; Yilgör, E.; Wilkes, G.L. Critical Parameters in Designing Segmented Polyurethanes and Their Effect on Morphology and Properties: A Comprehensive Review. *Polymer* **2015**, *58*, A1–A36. [CrossRef]
13. Gorbunova, M.A.; Anokhin, D.V.; Badamshina, E.R. Recent Advances in the Synthesis and Application of Thermoplastic Semicrystalline Shape Memory Polyurethanes. *Polym. Sci. Ser. B* **2020**, *62*, 427–450. [CrossRef]
14. Yan, W.; Fang, L.; Noechel, U.; Kratz, K.; Lendlein, A. Influence of Deformation Temperature on Structural Variation and Shape-Memory Effect of a Thermoplastic Semi-Crystalline Multiblock Copolymer. *Express Polym. Lett.* **2015**, *9*, 624–635. [CrossRef]
15. Anokhin, D.V.; Gorbunova, M.A.; Estrin, Y.I.; Komratova, V.V.; Badamshina, E.R. The Role of Fast and Slow Processes in the Formation of Structure and Properties of Thermoplastic Polyurethanes. *Phys. Chem. Chem. Phys.* **2016**, *18*, 31769–31776. [CrossRef] [PubMed]
16. Li, M.; Zhang, R.; Li, X.; Wu, Q.; Chen, T.; Sun, P. High-Performance Recyclable Cross-Linked Polyurethane with Orthogonal Dynamic Bonds: The Molecular Design, Microstructures, and Macroscopic Properties. *Polymer* **2018**, *148*, 127–137. [CrossRef]
17. Zhang, H.; Zhao, H.; Zhuo, K.; Hua, Y.; Chen, J.; He, X.; Weng, W.; Xia, H. "Carbolong" Polymers With Near Infrared Triggered, Spatially Resolved and Rapid Self-Healing Properties. *Polym. Chem.* **2019**, *10*, 386–394. [CrossRef]
18. Liu, S.; Yuan, Y.; Li, J.; Sun, S.; Chen, Y. An Optomechanical Study of Mechanoluminescent Elastomeric Polyurethanes with Different Hard Segments. *Polym. Chem.* **2020**, *11*, 1877–1884. [CrossRef]
19. Hermida-Merino, D.; O'Driscoll, B.; Hart, L.R.; Harris, P.J.; Colquhoun, H.M.; Slark, A.T.; Prisacariu, C.; Hamley, I.W.; Hayes, W. Enhancement of Microphase Ordering and Mechanical Properties of Supramolecular Hydrogen-Bonded Polyurethane Networks. *Polym. Chem.* **2018**, *9*, 3406–3414. [CrossRef]
20. Koberstein, J.T.; Russell, T.P. Simultaneous SAXS-DSC Study of Multiple Endothermic Behavior in Polyether-Based Polyurethane Block Copolymers. *Macromolecules* **1986**, *19*, 714–720. [CrossRef]
21. Stribeck, A.; Dabbous, R.; Eling, B.; Pöselt, E.; Malfois, M.; Schander, E. Scattering of X-Rays during Melting and Solidification of Thermoplastic Polyurethane. Graphite as Nucleating Agent and Stabilizer of the Colloidal Melt. *Polymer* **2018**, *153*, 565–573. [CrossRef]
22. Stribeck, A.; Eling, B.; Pöselt, E.; Malfois, M.; Schander, E. Melting, Solidification, and Crystallization of a Thermoplastic Polyurethane as a Function of Hard Segment Content. *Macromol. Chem. Phys.* **2019**, *220*, 1900074. [CrossRef]
23. Tian, Q.; Yan, G.; Bai, L.; Li, X.; Zou, L.; Rosta, L.; Wacha, A.; Li, Q.; Krakovský, I.; Yan, M.; et al. Phase Mixing and Separation in Polyester Polyurethane Studied by Small-Angle Scattering: A Polydisperse Hard Sphere Model Analysis. *Polymer* **2018**, *147*, 1–7. [CrossRef]
24. Kong, Z.; Tian, Q.; Zhang, R.; Yin, J.; Shi, L.; Ying, W.B.; Hu, H.; Yao, C.; Wang, K.; Zhu, J. Reexamination of the Microphase Separation in MDI and PTMG Based Polyurethane: Fast and Continuous Association/Dissociation Processes of Hydrogen Bonding. *Polymer* **2019**, *185*, 121943. [CrossRef]
25. Sui, T.; Salvati, E.; Zhang, H.; Dolbnya, I.P.; Korsunsky, A.M. Multiscale Synchrotron Scattering Studies of the Temperature-Dependent Changes in the Structure and Deformation Response of a Thermoplastic Polyurethane Elastomer. *Mater. Today Adv.* **2019**, *4*, 100024. [CrossRef]
26. Wang, Z.; Li, X.; Pöselt, E.; Eling, B.; Wang, Z. Melting Behavior of Polymorphic MDI/BD-Block TPU Investigated by Using in-Situ SAXS/WAXS and FTIR Techniques. Hydrogen Bonding Formation Causing the Inhomogeneous Melt. *Polym. Test.* **2021**, *96*, 107065. [CrossRef]
27. Stribeck, A.; Schneider, K.; Eling, B.; Poselt, E. Short-Term Morphology Relaxation of Thermoplastic Polyurethane Elastomers after Fast Strain Steps. *Macromol. Mater. Eng.* **2020**, *2000386*. [CrossRef]
28. Li, X.; Wang, H.; Xiong, B.; Pöselt, E.; Eling, B.; Men, Y. Destruction and Reorganization of Physically Cross-Linked Network of Thermoplastic Polyurethane Depending on Its Glass Transition Temperature. *ACS Appl. Polym. Mater.* **2019**, *1*, 3074–3083. [CrossRef]
29. Stribeck, N. *X-Ray Scattering of Soft Matter*; Springer: Berlin/Heidelberg, Germany, 2007. [CrossRef]
30. Briber, R.M.; Thomas, E.L. Physics Investigation of Two Crystal Forms in MDI/BDO-Based Polyurethanes. *J. Macromol. Sci. Part B* **2015**, 37–41. [CrossRef]
31. Blackwell, J.; Lee, C.D. Hard-Segment Domain Sizes in MDI/Diol Polyurethane Elastomers. *J. Polym. Sci. Polym. Phys. Ed.* **1983**, *21*, 2169–2180. [CrossRef]

32. Fu, B.X.; Hsiao, B.S.; Pagola, S.; Stephens, P.; White, H.; Rafailovich, M.; Sokolov, J.; Mather, P.T.; Jeon, H.G.; Phillips, S.; et al. Structural Development during Deformation of Polyurethane Containing Polyhedral Oligomeric Silsesquioxanes (POSS) Molecules. *Polymer* **2001**, *42*, 599–611. [CrossRef]

33. Korley, L.T.J.; Pate, B.D.; Thomas, E.L.; Hammond, P.T. Effect of the Degree of Soft and Hard Segment Ordering on the Morphology and Mechanical Behavior of Semicrystalline Segmented Polyurethanes. *Polymer* **2006**, *47*, 3073–3082. [CrossRef]

34. Bonart, R. X-Ray Investigations Concerning the Physical Structure of Cross-Linking in Segmented Urethane Elastomers. *J. Macromol. Sci. Part B* **1968**, *2*, 115–138. [CrossRef]

35. Bonart, R.; Boetzl, F.; Schmid, J. Cross Interferences in the Small-Angle X-ray Patterns of Strained Segmented PU Elastomers. *J. Makromol. Chem.* **1987**, *188*, 907–919. [CrossRef]

36. Martin, C.; Eeckhaut, G.; Mahendrasingam, A.; Blundell, D.J.; Fuller, W.; Oldman, R.J.; Bingham, S.J.; Dieing, T.; Riekel, C. Micro-SAXS and Force/Strain Measurements during the Tensile Deformation of Single Struts of an Elastomeric Polyurethane Foam. *J. Synchrotron Radiat.* **2000**, *7*, 245–250. [CrossRef] [PubMed]

37. Blundell, D.J.; Eeckhaut, G.; Fuller, W.; Mahendrasingam, A.; Martin, C. Time-Resolved SAXS/Stress-Strain Studies of Thermoplastic Polyurethanes During Mechanical Cycling at Large Strains. *J. Macromol. Sci. Phys.* **2004**, *43*, 125–142. [CrossRef]

38. Waletzko, R.S.; James Korley, L.S.T.; Pate, B.D.; Thomas, E.L.; Hammond, P.T. Role of Increased Crystallinity in Deformation-Induced Structure of Segmented Thermoplastic Polyurethane Elastomers with PEO and PEO-PPO-PEO Soft Segments and HDI Hard Segments. *Macromolecules* **2009**, *42*, 2041–2053. [CrossRef]

39. Stribeck, N.; Zeinolebadi, A.; Harpen, F.; Luinstra, G.; Eling, B.; Botta, S. Thermoplastic Polyurethane Cross-Linked by Functionalized Silica. Nanostructure Evolution under Mechanical Load. *Macromolecules* **2013**, *46*, 4041–4052. [CrossRef]

40. Stribeck, A.; Li, X.; Zeinolebadi, A.; Pöselt, E.; Eling, B.; Funari, S. Morphological Changes under Strain for Different Thermoplastic Polyurethanes Monitored by SAXS Related to Strain at Break. *Macromol. Chem. Phys.* **2015**, *216*, 2318–2330. [CrossRef]

41. Lee, B.S.; Chun, B.C.; Chung, Y.; Sul, K.I.; Cho, J.W. Structure and Thermomechanical Properties of Polyurethane Block Copolymers with Shape Memory Effect. *Macromolecules* **2001**, *34*, 6431–6437. [CrossRef]

42. Pereira, I.M.; Oréfice, R.L. The Morphology and Phase Mixing Studies on Poly(Ester–Urethane) during Shape Memory Cycle. *J. Mater. Sci.* **2010**, *45*, 511–522. [CrossRef]

43. Tarasov, A.E.; Lodygina, V.P.; Komratova, V.V.; Gorbunova, M.A.; Badamshina, E.R. New IR-Spectroscopic Methods for Determining the Hydroxyl Content in Oligomers. *J. Appl. Spectrosc.* **2017**, *84*, 211–216. [CrossRef]

44. May, J.R.; Gentilini, C.; Clarke, D.E.; Odarchenko, Y.I.; Anokhin, D.V.; Ivanov, D.A.; Feldman, K.; Smith, P.; Stevens, M.M. Tailoring of Mechanical Properties of Derivatized Natural Polyamino Acids through Esterification and Tensile Deformation. *RSC Adv.* **2014**, *4*, 2096–2102. [CrossRef]

45. Odarchenko, Y.I.; Sijbrandi, N.J.; Rosenthal, M.; Kimenai, A.J.; Mes, E.P.C.; Broos, R.; Bar, G.; Dijkstra, P.J.; Feijen, J.; Ivanov, D.A. Structure formation and hydrogen bonding in all-aliphatic segmented copolymers with uniform hard segments. *Acta Biomater.* **2013**, *9*, 6143–6149. [CrossRef] [PubMed]

46. Sijbrandi, N.J.; Kimenai, A.J.; Mes, E.P.C.; Broos, R.; Bar, G.; Rosenthal, M.; Odarchenko, Y.; Ivanov, D.A.; Dijkstra, P.J.; Feijen, J. Synthesis, morphology and properties of segmented poly(ether amide)s with uniform oxalamide based hard segments. *Macromolecules* **2012**, *45*, 3948–3961. [CrossRef]

47. Yeh, F.; Hsiao, B.S.; Sauer, B.B.; Michel, S.; Siesler, H.W. In-Situ Studies of Structure Development during Deformation of a Segmented Poly(Urethane–urea) Elastomer. *Macromolecules* **2003**, *36*, 1940–1954. [CrossRef]

48. Zhu, L.; Mimnaugh, B.R.; Ge, Q.; Quirk, R.P.; Cheng, S.Z.D.; Thomas, E.L.; Lotz, B.; Hsiao, B.S.; Yeh, F.; Liu, L. Hard and Soft Confinement Effects on Polymer Crystallization in Microphase Separated Cylinder-Forming PEO-b-PS/PS Blends. *Polymer* **2001**, *42*, 9121–9131. [CrossRef]

49. Zhou, Y.; Ahn, S.K.; Lakhman, R.K.; Gopinadhan, M.; Osuji, C.O.; Kasi, R.M. Tailoring Crystallization Behavior of PEO-Based Liquid Crystalline Block Copolymers through Variation in Liquid Crystalline Content. *Macromolecules* **2011**, *44*, 3924–3934. [CrossRef]

Article

Direct Observations of the Structural Properties of Semiconducting Polymer: Fullerene Blends under Tensile Stretching

Mouaad Yassine Aliouat [1,2], Dmitriy Ksenzov [3], Stephanie Escoubas [1,*], Jörg Ackermann [4], Dominique Thiaudière [5], Cristian Mocuta [5], Mohamed Cherif Benoudia [2], David Duche [1], Olivier Thomas [1] and Souren Grigorian [3,*]

[1] Aix Marseille Univ, U. Toulon, CNRS, IM2NP (Institut Matériaux Microélectronique et Nanosciences de Provence), Campus St-Jérôme, 13397 Marseille CEDEX 20, France; mouaad-yassine.aliouat@im2np.fr (M.Y.A.); david.duche@univ-amu.fr (D.D.); olivier.thomas@im2np.fr (O.T.)

[2] Ecole Nationale Supérieure des Mines et de la Métallurgie, L3M, Annaba, Sidi Amar W129, Algeria; mohamed-cherif.benoudia@ensmm-annaba.dz

[3] Institute of Physics, University of Siegen, D-57068 Siegen, Germany; ksenzov@physik.uni-siegen.de

[4] Aix-Marseille University, CNRS, CINAM, 13007 Marseille, France; jorg.ackermann@univ-amu.fr

[5] Synchrotron SOLEIL, L'Orme des Merisiers Saint-Aubin, CEDEX BP 48, 91192 Gif-sur-Yvette CEDEX, France; dominique.thiaudiere@synchrotron-soleil.fr (D.T.); cristian.mocuta@synchrotron-soleil.fr (C.M.)

* Correspondence: stephanie.escoubas@univ-amu.fr (S.E.); grigorian@physik.uni-siegen.de (S.G.)

Received: 15 June 2020; Accepted: 6 July 2020; Published: 10 July 2020

Abstract: We describe the impact of tensile strains on the structural properties of thin films composed of PffBT4T-2OD π-conjugated polymer and $PC_{71}BM$ fullerenes coated on a stretchable substrate, based on a novel approach using in situ studies of flexible organic thin films. In situ grazing incidence X-ray diffraction (GIXD) measurements were carried out to probe the ordering of polymers and to measure the strain of the polymer chains under uniaxial tensile tests. A maximum 10% tensile stretching was applied (i.e., beyond the relaxation threshold). Interestingly we found different behaviors upon stretching the polymer: fullerene blends with the modified polymer; fullerene blends with the 1,8-Diiodooctane (DIO) additive. Overall, the strain in the system was almost twice as low in the presence of additive. The inclusion of additive was found to help in stabilizing the system and, in particular, the π–π packing of the donor polymer chains.

Keywords: conjugated polymer and blends; in situ GIXD; additive; structure; strain

---

1. Introduction

One of the important research directions in the field of organic electronics is the development of stretchable/flexible electronics, which has attracted a lot of attention with a huge potential for numerous practical applications in different fields and daily life [1–3]. With the recent development of the flexible organic electronics, the performances of devices based on organic semiconducting molecules and polymers, such as organic light-emitting diodes (OLEDs), organic field effect transistors (OFETs) and organic solar cells (OSCs), have remarkably increased [4–8].

OSCs can be fabricated on large areas using low-cost roll-to-roll manufacturing techniques and, thus, represent an interesting alternative to conventional silicon solar cells [7,8], especially for nomadic applications and everyday connected objects, but also to construction, transport, urban furniture, etc. Recently, stretchable OSCs have demonstrated high potential as an energy source for the domain of flexible electronics [9–11]. The photoactive layer of an OSC is based on bulk heterojunctions, created by blending conjugated polymers and small molecule acceptors that provide low bandgap light absorption, high power conversion efficiency (PCE), appropriate energy level positions and quite high carrier mobilities [12,13]. In the last decade, tremendous efforts have been focused on the synthesis of new low bandgap donor polymers (e.g., PffBT4T-2OD polymer, commonly called PCE11) [14,15] and non-fullerene organic acceptors. The PCE of polymer-based solar cells has significantly increased and reached 16.5% in 2019 [16]. In most cases, additives, such as 1,8–diiodooctane (DIO), are used to improve nanoscale morphology and, thus, the performance of the organic solar cells [17,18]. DIO is a high boiling point additive used generally to dissolve $PC_{71}BM$ aggregates selectively [19], which improves the miscibility of fullerene molecules in PCE11. The additive slows down the formation of fullerene aggregates during the drying of the polymer blend, which makes the penetration of fullerene molecules between the chains of the donor polymer easier [20]. This effect was observed on the blend $TQ1:PC_{71}BM$ (TQ1:Poly[[2,3-bis(3-octyloxyphenyl)-5,8-quinoxalinediyl]-2,5-thiophenediyl]) [21] and on the $PCE11:PC_{71}BM$ mixture [19].

There are a lot of scientific efforts aimed at understanding and improving the microstructure of polymer-based OSCs, such as solution optimization [22,23], employing low volatile additives [24–26] and thermal annealing [27]. Moreover, the direct observation of film crystallization/formation during solvent evaporation gives insight into the phase separation in blends and allows correlating observed phenomenon to electrical properties [28], thus controlling polymer chain crystallization and fullerene molecule aggregation as a function of the blend concentration. In parallel, several methods have been developed with compact instrumentations, capable of tracking the crystallization process of semiconducting polymers and blends at synchrotron radiation facilities [8,29–31]

Taking into account this tremendous progress in understanding the structural evolution during the solution processing [31,32], the development of flexible and intrinsically stretchable organic conductors is very promising for the realization of emerging novel devices [33].

The commonly used approach is to probe thin films after stretching where the deformation of materials at a fixed value of strain can induce the anisotropy of the structural and morphological properties [34,35]. During the stretching process itself, the molecular structure and crystallinity can be strongly modified [36]. Moreover, if a phase transition takes place, it can dramatically influence the structural and electrical properties of the thin film [37]. Stretchable or wearable working devices rely on the knowledge and reproducibility of the evolution of electronic and optical properties under mechanical load [10,38,39]. It is worth noting that electrical transport measurements can also be a sensitive probe for strain-induced defects [40]. In our recent work, in situ grazing incidence X-ray diffraction (GIXD) studies were carried out to measure the structural parameters of pristine PCE11 polymer under uniaxial tensile load [41]. It was found that, after 15% of stretching a partial strain, relaxation takes place together with massive crack propagation. To go further towards relevant materials for OPV application, we expand on in situ GIXD metrology to the PCE11 blends with a focus on the impact of blending fullerenes inside the polymer on the mechanical properties. Furthermore, we put a particular focus on the role of additives under an applied strain in the pre-cracking regime, as additives not only improve device performances, but may also impact the nanoscale morphology and organization of the blended layers.

2. Materials and Methods

PCE11 (PffBT4T-2OD) (batch: YY11246CB) and $PC_{71}BM$ (99.5%) were purchased from 1-Material Inc. (Dorval, Quebec, Canada). PDMS stretchable substrate was prepared by spin coating of Sylgard 184 Silicone, obtained from Dow Corning (SAMARO, Lyon, France) using the ratio elastomer:hardener (10:1). After the degreasing of the polydimethylsiloxane (PDMS), using acetone, ethanol and water in an ultrasonic bath for 15 min for each bath and drying by argon flux gas, PDMS was fixed on a glass support, and the surface was activated by UV-ozone treatment at 80 °C for 10 min to improve the coatability.

The coating operation was carried out under an argon atmosphere inside a glove box. The blended layers with a thickness of 300 nm were spin coated at 1000 rpm from a chlorobenzene

(CB):dichlorobenzene (DCB) (1:1) mixed solution of PCE11:PC$_{71}$BM with a mixture ratio of 1:1.2 wt.% (33 mg/mL in total). For layers with 1,8-Diiodooctane (DIO), 3% DIO was added to the solution one hour before the coating.

The temperature of the polymer inks (with and without DIO) was kept at 110 °C and the substrate was heated at 100 °C during the spin coating operation. After coating, the layer was dried at 100 °C for 15 min and then the sample was immediately removed after drying.

Pure PCE11 layers were spin coated on glass and PDMS substrates at 1000 rpm from a CB solution of 15 mg/mL PCE11 concentration, the PCE11 ink was agitated at 110 °C for more than one hour in a glove box.

The structural properties of the thin flexible films were investigated by GIXD technique at two different beamlines:

(i)   DiffAbs beamline of SOLEIL synchrotron (Saint-Aubin, France) using a wide-area 2D XPAD detector (560 × 960 pixels of 130 μm) [42]. The measurements are recorded from the XPAD detector at different α_f ranges in the out-of-plane direction, as shown in Figure 1 (Positions 1 and 2). The distance sample-detector was 450 mm.

(ii)   BL9 beamline of DELTA synchrotron radiation facility at TU Dortmund, Dortmund, Germany, using a 2D image plate (MAR345) with a resolution of 100 μm/pixel [43]. The distance sample-detector was 394 mm.

For both experiments, X-ray photon energy of 15 keV was employed and a grazing incidence angle of 0.07° was chosen to probe the scattered signal from films. A photodiode point detector was used for aligning the sample at each stretching position.

A specially designed in situ stretching chamber allowed us to stretch thin films in a controlled way with a given amount of applied stretch and number of steps, as shown in Figure 1a. For each stretching step, the thin films were aligned prior to recording the GIXD patterns. The scattered signal was collected using a 2D detector. The sample surface was placed nearly horizontal, inclined by an incident angle $\alpha_i = 0.07°$. The exit angle was denoted α_f at the azimuth angle φ. Figure 1a shows the out-of-plane (q_z) and in-plane (q_{xy}) directions of the two-dimensional GIXD pattern. The molecular structure of PCE11 and PC$_{71}$BM molecules, and the schematic representation of the two main preferential orientations of the polymer chains (i.e., edge-on and face-on configurations with the backbone plane oriented perpendicular and parallel to the substrate, respectively), are depicted in Figure 1b.

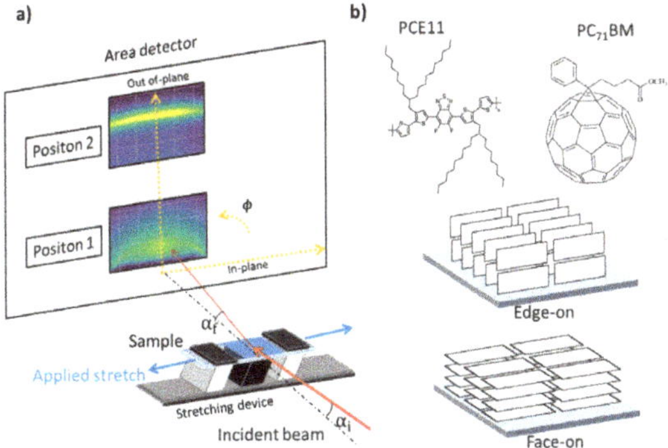

Figure 1. (a) Schematic view of the experimental setup for the GIXD experiments. (b) Molecular structure of PCE11 and PC$_{71}$BM—schematic representation of edge-on and face-on orientations.

3. Results and Discussions

3.1. Structural Properties before Stretching

Typically, spin coated semi-crystalline polymers favor edge-on orientation with the π–π stacking and conjugated backbone directions parallel to the substrate, whereas blending the donor polymer with fullerene acceptors leads to the co-existence of both edge-on and face-on orientations [19,20,44].

Our recent study on the structure of pristine PCE11 revealed that spin-coated layers on both rigid and stretchable substrates, are mostly edge-on oriented with more pronounced order and crystallinity for those coated on glass [41]. The present work is focused on PCE11 blends with fullerene molecules investigated via in situ GIXD under tensile stretching.

A 2D GIXD pattern for pristine PCE11 coated on glass is shown in Figure 2a. For pristine PCE11 film, the 2D pattern shows the edge-on orientation with the lamellar stacking perpendicular to the glass substrate. Such orientation results in a strong h00 series along the out-of-plane direction, and for in-plane direction, a pronounced 010 peak associated with π–π stacking, as shown in Figure 2a. The position of the 100 lamellar peak is at 3 nm^{-1} corresponding to a spacing of 2.09 nm, while the π–π stacking peak is centered at 17.5 nm^{-1} corresponding to a π–π spacing of 0.36 nm, which is in good agreement with the literature [15,45]. Initial GIXD patterns (before stretching) for the blends on PDMS are shown in Figure 2b,c. Interestingly, the lamellar stacking is flipped by 90°, resulting in dominating face-on orientation. In this case the lamellar 100 peak is mainly visible in-plane at 2.9 nm^{-1} (lamellar spacing of 2.17 nm) whereas 010 peaks appear only in the out-of-plane direction at 17.5 nm^{-1}, as shown in Figure 2b,c. Additionally, broad PDMS and PC$_{71}$BM halos, respectively, centered at 8.5 nm^{-1} and 13 nm^{-1} are visible.

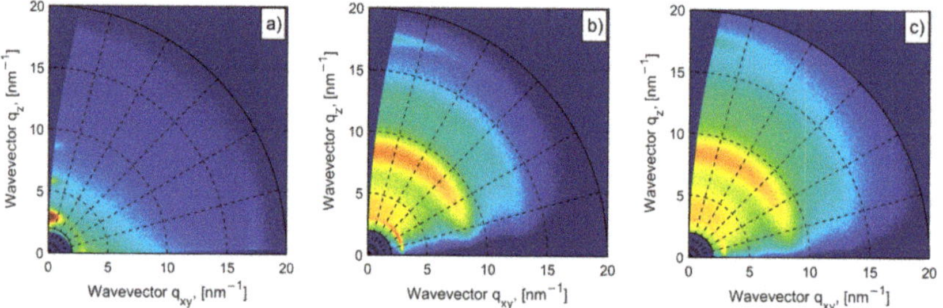

Figure 2. Two-dimensional GIXD patterns of (**a**) pristine PCE11 film coated on glass substrate and PCE11:PC$_{71}$BM blend (**b**) with DIO, (**c**) without DIO, coated on PDMS substrates.

Assuming the homogeneous distribution of lamellar spacings, the full width at half-maximum (FWHM) is mainly related to the finite domain size, as characterized by the coherence length, which is inversely proportional to the FWHM, according to the Scherrer equation [46]. Calculated domain sizes for each sample are given in Table 1. The FWHM corresponding to the in-plane 100 peak of pristine PCE11 is almost twice broader than for the PCE11:PC$_{71}$BM blends. Similar behavior was observed comparing pure PCE11 polymer with PCE11:PCBM blends on hard supports [45]. In this study, the increase in the size of ordered domains was related to the presence of PCBM molecules in the PCE11 ordered phases, forming less perfect but larger area ordered domains.

Furthermore, our study is focused on the polymer:fullerene blends on the flexible substrates. To understand the role of the additive on the crystalline organization of the PCE11 and PC$_{71}$BM blends, we have compared the line profiles extracted from 2D patterns, as shown in Figure 2b,c, along to the out-of-plane (q_z) and in-plane (q_{xy}) cuts. These profiles are shown in Figure 3a,b, respectively, and confirm dominating face-on orientation. The FWHM of the in-plane 100 peak for blend with DIO is ~1.2 larger than that for blend without DIO (same for the out-of-plane 010 peak), indicating smaller ordered domains, even though the degree of the π–π packing is improved after the addition of DIO, as shown in Table 1. This might also be due to the existence of bigger $PC_{71}BM$ molecules in the blend without DIO, forming less crystalline but larger area PCE11-ordered domains, as reported in [45]. With the addition of DIO, $PC_{71}BM$ aggregates dissolve better [19], leading to better crystallized but smaller PCE11 ordered domains.

Table 1. (100) and (010) domain sizes of pure PCE11 film, coated on glass, and PCE11:$PC_{71}BM$ films (with and without DIO additive) coated on PDMS.

Polymer	100 Domain Size (nm)		010 Domain Size (nm)	
	OOP	IP	OOP	IP
Pristine PCE11	12.9	9.6	4.46	4.22
Blend with DIO	11.3	11	6.75	-
Blend without DIO	13.56	13.2	8.1	-

A comparison of the azimuthal distribution for the blends processed with and without DIO is shown in Figure 3c,d, where the blend with DIO reveals more pronounced face-on orientation.

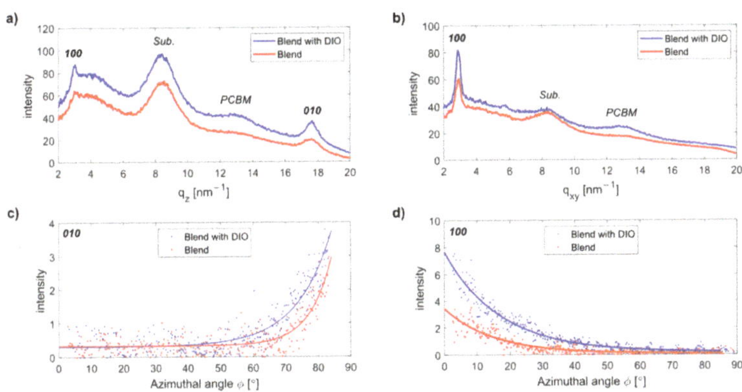

Figure 3. Out-of-plane (**a**) and in-plane (**b**) line profiles for PCE11:$PC_{71}BM$ blend (red circles) and PCE11:$PC_{71}BM$ blend with DIO (blue circles) on the PDMS; the azimuthal profiles for the 010 (**c**) and 100 (**d**) peaks, respectively.

The ordering of the donor polymer is a very important parameter affecting the charge transport and interconnectivity within the blends, thus improving the solar cell performance [47]. The face-on crystallization of PCE11 chains is favorable for the charge transport in the direction across the active layer [15]. We observe here that face-on orientation as shown, for example, by the intensity of $\pi-\pi$ stacking peak 010 is more pronounced for the blend film with DIO. Similar findings were reported in [48], supporting the idea that DIO is beneficial to the crystallinity of PCE11 chains in blends, resulting in better charge transport.

3.2. In Situ Structural Studies Under Stretching

In this section, we employ in situ GIXD metrology to monitor the microstructure of the polymer:fullerene blends upon stretching. Because of the increased scattering background on the flexible support, which makes it difficult to resolve the PCBM halo, we are focusing on the most intense PCE11 peaks. We compared the structural changes of the PCE11:$PC_{71}BM$ and PCE11:$PC_{71}BM$ with DIO blends upon stretching under grazing incidence geometry. The samples were uniaxially stretched during a tensile test starting from 1% with steps of 2% up to 7% and finally reaching 10%. For each step, the GIXD patterns have been recorded for the same angle of incidence of 0.07°. The maximum intensity of the in-plane 100 peak for the blended sample without additive was about two-times lower than for the one with DIO. To compensate the geometric effects and experimental constraints, we assume that the PDMS support provides an isotropic intensity distribution and the scattering ring from PDMS is independent of the azimuthal angle. Therefore, each pattern has been normalized on the corresponding PDMS ring. The normalized data show different trends for the blends processed with and without DIO upon stretching. Figure 4a shows the strain evolution (ε), calculated from Equation (1), of the π–π stacking distance d corresponding to the out-of-plane 010 peak position for the blends without and with DIO.

$$\varepsilon[\%] = 100(d - d_0)/d_0 \tag{1}$$

Strain evolution appears clearly different for both blends—the PCE11:PC$_{71}$BM processed without DIO shows an almost monotonous increase in tensile perpendicular strain up to 0.8% upon stretching. For the blend processed with DIO, the trend is opposite, resulting in a compressive perpendicular strain upon stretching.

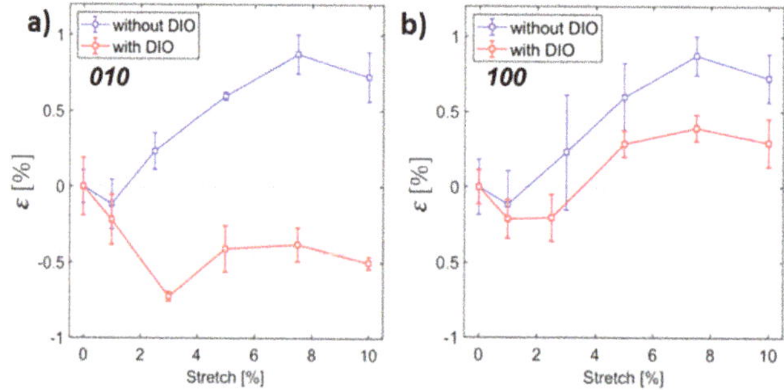

Figure 4. The strain evolution of the out-of-plane 010 π–π spacing (**a**) and in-plane 100 lamellar spacing (**b**) as a function of the applied stretch for the PCE11:PC$_{71}$BM blends without (blue) and with DIO (red).

Figure 4b shows the strain evolution of the polymer lamellar distance, corresponding to the in-plane 100 peak, for the blends with and without DIO. The strain evolution is almost similar for both blends. For the PCE11:PC$_{71}$BM blend, there is an almost monotonous increase in tensile strain up to 0.8% upon stretching (the same order of magnitude found in a previous study for pristine PCE11 [41]), whereas for the blend with DIO, this increase is smaller by a factor of two. Comparing the features for the 010 and 100 peaks, their behaviors are different—the PCE11:PC$_{71}$BM blend exhibits expansion in both directions. On the other hand, the PCE11:PC$_{71}$BM blend with DIO shows a compressive out-of-plane strain. A possible stabilization of the systems during first few steps of the stretching might be a reason for the higher compressive strain at 3% of stretching.

The FWHM of the 010 peak for the blend with DIO remains unchanged upon stretching. For the blend without DIO we found a monotonous increase in the FWHM, reaching a similar value as that of the PCE11:PC$_{71}$BM blend with DIO (not shown here). Similar to the 010 peak, the 100 FWHM for the PCE11:PC$_{71}$BM blend without DIO remains unchanged upon stretching. For the blend with DIO we found a monotonous decrease in the FWHM, reaching a similar value as that of the PCE11:PC$_{71}$BM blend.

From our recent findings on pristine PCE11 coated on PDMS [41], the initial orientation of PCE11 chains is mostly edge-on. In the present work, we found that blending PCE11 with PC$_{71}$BM molecules leads to the reversed orientation (face-on). This face-on orientation becomes more pronounced after the addition of DIO. The switch in orientation from edge-on to face-on is clearly observed to influence the mechanical behavior of PCE11 chains. When the chains are edge-on oriented, tensile tests revealed that in the out-of-plane direction, the chains undergo increasing compressive strain (negative values of strains) until 10% of stretching, in agreement with the Poisson effect. On the other hand, the face-on oriented chains are undergoing expansion (positive values of strains) along both directions.

4. Conclusions

In comparison to pristine PCE11 films with dominating edge-on orientation, polymer:fullerene blends have shown a 90° switch with preferential face-on orientation. For the PCE11:PC$_{71}$BM blends, the inclusion of the DIO additive further improves the microstructure and enhances face-on orientation.

In situ studies during tensile testing of PCE11:PC$_{71}$BM processed with and without additives showed different mechanical behavior. In the case of the PCE11:PC$_{71}$BM blend, only tensile strain has been observed, for the blend with DIO, we found a compressive out-of-plane strain associated with the π–π conjugation. It is also worth noting that the blends with DIO were almost by a factor of two less strained under the same stretching conditions. The implications of these findings, based on the inclusion of additives, could be beneficial for the flexibility and stability of organic photovoltaic devices. These in situ X-ray scattering studies during mechanical testing allow for the direct correlation of the structural and mechanical properties of organic materials for flexible electronics.

Author Contributions: Conceptualization, M.Y.A., S.E., J.A. and S.G.; methodology, M.Y.A., S.E., M.C.B., D.D. and S.G.; software, M.Y.A., D.K. and C.M.; validation, S.E. and S.G.; formal analysis, M.Y.A., D.K. and C.M.; investigation, M.Y.A., S.E., C.M., O.T. and S.G.; resources, J.A., D.T. and C.M.; data curation, M.Y.A., D.K., D.T. and C.M.; writing—original draft preparation, A.M.Y. and S.G.; writing—review and editing, S.E., J.A., O.T. and S.G.; visualization, M.Y.A., D.K. and S.G.; supervision, S.E., M.C.B. and D.D.; project administration, S.E., M.C.B. and S.G.; funding acquisition, S.E., M.C.B., O.T. and S.G. All authors have read and agreed to the published version of the manuscript.

Funding: We are grateful for financial support from the DAAD-PROCOPE (project No. 57211900, Campus France 35484SE) and PHC-Tassili (Project No. 17MDU994).

Acknowledgments: We acknowledge the BL9 beamline scientists at the DELTA synchrotron (Dortmund, Germany) and Synchrotron SOLEIL (Saint-Aubin, France) for supplying the beam-time for these experiments.

Conflicts of Interest: The authors declare no conflict of interest.

References

1. Liu, Y.; Wang, H.; Zhao, W.; Zhang, M.; Qin, H.; Xie, Y. Flexible, Stretchable Sensors for Wearable Health Monitoring: Sensing Mechanisms, Materials, Fabrication Strategies and Features. *Sensors* **2018**, *18*, 645. [CrossRef] [PubMed]

2. Sirringhaus, H.; Brown, P.J.; Friend, R.H.; Nielsen, M.M.; Bechgaard, K.; Langeveld-Voss, B.M.W.; Spiering, A.J.H.; Janssen, R.; Meijer, E.W.; Herwig, P.; et al. Two-dimensional charge transport in self-organized, high-mobility conjugated polymers. *Natural* **1999**, *401*, 685–688. [CrossRef]

3. McCulloch, I.; Heeney, M.; Bailey, C.; Genevičius, K.; Macdonald, I.; Shkunov, M.; Sparrowe, D.; Tierney, S.; Wagner, R.; Zhang, W.; et al. Liquid-crystalline semiconducting polymers with high charge-carrier mobility. *Nat. Mater.* **2006**, *5*, 328–333. [CrossRef] [PubMed]

4. Sirringhaus, H. 25th Anniversary Article: Organic Field-Effect Transistors: The Path Beyond Amorphous Silicon. *Adv. Mater.* **2014**, *26*, 1319–1335. [CrossRef] [PubMed]

5. Hösel, M.; Dam, H.F. Development of Lab-to-Fab Production Equipment Across Several Length Scales for Printed Energy Technologies, Including Solar Cells. *Energy Technol.* **2015**, *3*, 293–304. [CrossRef]

6. Oh, J.Y.; Kim, S.; Baik, H.-K.; Jeong, U. Conducting Polymer Dough for Deformable Electronics. *Adv. Mater.* **2015**, *28*, 4455–4461. [CrossRef]

7. Brabec, C.J.; Gowrisanker, S.; Halls, J.J.M.; Laird, D.; Jia, S.; Williams, S.P. Polymer-Fullerene Bulk-Heterojunction Solar Cells. *Adv. Mater.* **2010**, *22*, 3839–3856. [CrossRef]

8. Søndergaard, R.R.; Hösel, M.; Angmo, D.; Larsen-Olsen, T.T.; Krebs, F.C. Roll-to-roll fabrication of polymer solar cells. *Mater. Today* **2012**, *15*, 36–49. [CrossRef]

9. O'Connor, T.F.; Zaretski, A.; Savagatrup, S.; Printz, A.; Wilkes, C.D.; Diaz, M.I.; Sawyer, E.J.; Lipomi, D.J. Wearable organic solar cells with high cyclic bending stability: Materials selection criteria. *Sol. Energy Mater. Sol. Cells* **2016**, *144*, 438–444. [CrossRef]

10. Jinno, H.; Fukuda, K.; Xu, X.; Park, S.; Suzuki, Y.; Koizumi, M.; Yokota, T.; Osaka, I.; Takimiya, K.; Someya, T. Stretchable and waterproof elastomer-coated organic photovoltaics for washable electronic textile applications. *Nat. Energy* **2017**, *2*, 780–785. [CrossRef]

11. Kim, T.; Kim, J.-H.; Kang, T.E.; Lee, C.; Kang, H.; Shin, M.; Wang, C.; Ma, B.; Jeong, U.; Kim, T.-S.; et al. Flexible, highly efficient all-polymer solar cells. *Nat. Commun.* **2015**, *6*, 8547. [CrossRef]
12. Etxebarria, I.; Guerrero, A.; Albero, J.; Garcia-Belmonte, G.; Palomares, E.; Pacios, R. Inverted vs standard PTB7:PC70BM organic photovoltaic devices. The benefit of highly selective and extracting contacts in device performance. *Org. Electron.* **2014**, *15*, 2756–2762. [CrossRef]
13. Nam, S.; Hahm, S.G.; Han, H.; Seo, J.; Kim, C.; Kim, H.; Marder, S.R.; Ree, M.; Kim, Y. All-Polymer Solar Cells with Bulk Heterojunction Films Containing Electron-Accepting Triple Bond-Conjugated Perylene Diimide Polymer. *ACS Sustain. Chem. Eng.* **2015**, *4*, 767–774. [CrossRef]
14. Ma, W.; Yang, G.; Jiang, K.; Carpenter, J.H.; Wu, Y.; Meng, X.; McAfee, T.; Zhao, J.; Zhu, C.; Wang, C.; et al. Influence of Processing Parameters and Molecular Weight on the Morphology and Properties of High-Performance PffBT4T-2OD:PC71BM Organic Solar Cells. *Adv. Energy Mater.* **2015**, *5*, 1501400. [CrossRef]
15. Liu, Y.; Zhao, J.; Li, Z.; Mu, C.; Ma, W.; Hu, H.; Jiang, K.; Lin, H.; Ade, H.; Yan, H. Aggregation and morphology control enables multiple cases of high-efficiency polymer solar cells. *Nat. Commun.* **2014**, *5*, 5293. [CrossRef]
16. Cui, Y.; Yao, H.; Zhang, J.; Zhang, T.; Wang, Y.; Hong, L.; Xian, K.; Xu, B.; Zhang, S.; Peng, J.; et al. Over 16% efficiency organic photovoltaic cells enabled by a chlorinated acceptor with increased open-circuit voltages. *Nat. Commun.* **2019**, *10*, 2515. [CrossRef]
17. Liao, H.-C.; Ho, C.-C.; Chang, C.-Y.; Jao, M.-H.; Darling, S.; Su, W.-F. Additives for morphology control in high-efficiency organic solar cells. *Mater. Today* **2013**, *16*, 326–336. [CrossRef]
18. Pearson, A.J.; Hopkinson, P.E.; Couderc, E.; Domanski, K.; Abdi-Jalebi, M.; Greenham, N.C. Critical light instability in CB/DIO processed PBDTTT-EFT:PC 71 BM organic photovoltaic devices. *Org. Electron.* **2016**, *30*, 225–236. [CrossRef]
19. Zhao, J.; Zhao, S.; Xu, Z.; Qiao, B.; Huang, D.; Zhao, L.; Li, Y.; Zhu, Y.; Wang, P. Revealing the Effect of Additives with Different Solubility on the Morphology and the Donor Crystalline Structures of Organic Solar Cells. *ACS Appl. Mater. Interfaces* **2016**, *8*, 18231–18237. [CrossRef]
20. Güldal, N.S.; Berlinghof, M.; Kassar, T.; Du, X.; Jiao, X.; Meyer, M.; Ameri, T.; Osvet, A.; Li, N.; Destri, G.L.; et al. Controlling additive behavior to reveal an alternative morphology formation mechanism in polymer: Fullerene bulk-heterojunctions. *J. Mater. Chem. A* **2016**, *4*, 16136–16147. [CrossRef]
21. Kim, Y.; Yeom, H.R.; Kim, J.Y.; Yang, C. High-efficiency polymer solar cells with a cost-effective quinoxaline polymer through nanoscale morphology control induced by practical processing additives. *Energy Environ. Sci.* **2013**, *6*, 1909. [CrossRef]
22. Li, M.; Gao, K.; Wan, X.; Zhang, Q.; Kan, B.; Xia, R.; Liu, F.; Yang, X.; Feng, H.; Ni, W.; et al. Solution-processed organic tandem solar cells with power conversion efficiencies >12%. *Nat. Photon.* **2016**, *11*, 85–90. [CrossRef]
23. Lee, K.H.; Schwenn, P.; Smith, A.R.; Cavaye, H.; Shaw, P.E.; James, M.; Krueger, K.B.; Gentle, I.R.; Meredith, P.; Burn, P.L. Morphology of All-Solution-Processed "Bilayer" Organic Solar Cells. *Adv. Mater.* **2010**, *23*, 766–770. [CrossRef] [PubMed]
24. Zhou, Y.; Gu, K.L.; Gu, X.; Kurosawa, T.; Yan, H.; Guo, Y.; Koleilat, G.I.; Zhao, D.; Toney, M.F.; Bao, Z. All-Polymer Solar Cells Employing Non-Halogenated Solvent and Additive. *Chem. Mater.* **2016**, *28*, 5037–5042. [CrossRef]
25. Jo, J.; Kim, S.-S.; Na, S.-I.; Yu, B.-K.; Kim, D.-Y. Time-Dependent Morphology Evolution by Annealing Processes on Polymer:Fullerene Blend Solar Cells. *Adv. Funct. Mater.* **2009**, *19*, 866–874. [CrossRef]
26. Min, J.; Kwon, O.K.; Cui, C.; Park, J.-H.; Wu, Y.; Park, S.Y.; Li, Y.; Brabec, C.J. High performance all-small-molecule solar cells: Engineering the nanomorphology via processing additives. *J. Mater. Chem. A* **2016**, *4*, 14234–14240. [CrossRef]
27. Nguyen, L.H.; Hoppe, H.; Erb, T.; Güneş, S.; Gobsch, G.; Sariciftci, N.S. Effects of Annealing on the Nanomorphology and Performance of Poly(alkylthiophene):Fullerene Bulk-Heterojunction Solar Cells. *Adv. Funct. Mater.* **2007**, *17*, 1071–1078. [CrossRef]
28. Radchenko, E.S.; Anokhin, D.V.; Gerasimov, K.L.; Rodygin, A.I.; Rychkov, A.A.; Shabratova, E.D.; Grigorian, S.; Ivanov, D.A. Impact of the solubility of organic semiconductors for solution-processable electronics on the structure formation: A real-time study of morphology and electrical properties. *Soft Matter* **2018**, *14*, 2560–2566. [CrossRef]

29. Sanyal, M.; Schmidt-Hansberg, B.; Klein, M.; Colsmann, A.; Munuera, C.; Vorobiev, A.; Lemmer, U.; Schabel, W.; Dosch, H.; Barrena, E. In Situ X-Ray Study of Drying-Temperature Influence on the Structural Evolution of Bulk-Heterojunction Polymer-Fullerene Solar Cells Processed by Doctor-Blading. *Adv. Energy Mater.* **2011**, *1*, 363–367. [CrossRef]

30. Logothetidis, S. In situ characterization of organic electronic materials using X-ray techniques. In *Handbook of Flexible Organic Electronics*; Woodhead Publishing: Cambridge, UK, 2014; pp. 217–226. [CrossRef]

31. Güldal, N.S.; Kassar, T.; Berlinghof, M.; Ameri, T.; Osvet, A.; Pacios, R.; Destri, G.L.; Unruh, T.; Brabec, C.J. Real-time evaluation of thin film drying kinetics using an advanced, multi-probe optical setup. *J. Mater. Chem. C* **2016**, *4*, 2178–2186. [CrossRef]

32. Gu, X.; Reinspach, J.; Worfolk, B.; Diao, Y.; Zhou, Y.; Yan, H.; Gu, K.; Mannsfeld, S.; Toney, M.F.; Bao, Z. Compact Roll-to-Roll Coater for in Situ X-ray Diffraction Characterization of Organic Electronics Printing. *ACS Appl. Mater. Interfaces* **2016**, *8*, 1687–1694. [CrossRef] [PubMed]

33. Yao, S.; Zhu, Y. Nanomaterial-Enabled Stretchable Conductors: Strategies, Materials and Devices. *Adv. Mater.* **2015**, *27*, 1480–1511. [CrossRef]

34. O'Connor, B.; Kline, R.J.; Conrad, B.; Richter, L.J.; Gundlach, D.; Toney, M.F.; Delongchamp, D.M. Anisotropic Structure and Charge Transport in Highly Strain-Aligned Regioregular Poly(3-hexylthiophene). *Adv. Funct. Mater.* **2011**, *21*, 3697–3705. [CrossRef]

35. Mun, J.; Wang, G.-J.N.; Oh, J.Y.; Katsumata, T.; Lee, F.L.; Kang, J.; Wu, H.-C.; Lissel, F.; Rondeau-Gagné, S.; Tok, J.B.-H.; et al. Effect of Nonconjugated Spacers on Mechanical Properties of Semiconducting Polymers for Stretchable Transistors. *Adv. Funct. Mater.* **2018**, *28*, 1804222. [CrossRef]

36. Lu, C.; Lee, W.-Y.; Gu, X.; Xu, J.; Chou, H.-H.; Yan, H.; Chiu, Y.-C.; He, M.; Matthews, J.R.; Niu, W.; et al. Effects of Molecular Structure and Packing Order on the Stretchability of Semicrystalline Conjugated Poly(Tetrathienoacene-diketopyrrolopyrrole) Polymers. *Adv. Electron. Mater.* **2016**, *3*, 1600311. [CrossRef]

37. Grigorian, S.; Escoubas, S.; Ksenzov, D.; Duche, D.; Aliouat, M.; Simon, J.-J.; Bat-Erdene, B.; Allard, S.; Scherf, U.; Pietsch, U.; et al. A Complex Interrelationship between Temperature-Dependent Polyquaterthiophene (PQT) Structural and Electrical Properties. *J. Phys. Chem. C* **2017**, *121*, 23149–23157. [CrossRef]

38. Scenev, V.; Cosseddu, P.; Bonfiglio, A.; Salzmann, I.; Severin, N.; Oehzelt, M.; Koch, N.; Rabe, J.P. Origin of mechanical strain sensitivity of pentacene thin-film transistors. *Org. Electron.* **2013**, *14*, 1323–1329. [CrossRef]

39. Salari, M.; Joodaki, M.; Mehregan, S. Experimental investigation of tensile mechanical strain influence on the dark current of organic solar cells. *Org. Electron.* **2018**, *54*, 192–196. [CrossRef]

40. Cordill, M.J.; Glushko, O.; Kreith, J.; Marx, V.; Kirchlechner, C. Measuring electro-mechanical properties of thin films on polymer substrates. *Microelectron. Eng.* **2015**, *137*, 96–100. [CrossRef]

41. Aliouat, M.Y.; Escoubas, S.; Benoudia, M.C.; Ksenzov, D.; Duché, D.; Bènevent, E.; Videlot-Ackermann, C.; Ackermann, J.; Thomas, O.; Grigorian, S. In situ measurements of the structure and strain of a π-conjugated semiconducting polymer under mechanical load. *J. Appl. Phys.* **2020**, *127*, 045108. [CrossRef]

42. Gallard, M.; Amara, M.S.; Putero, M.; Burle, N.; Guichet, C.; Escoubas, S.; Richard, M.-I.; Mocuta, C.; Chahine, R.R.; Bernard, M.; et al. New insights into thermomechanical behavior of GeTe thin films during crystallization. *Acta Mater.* **2020**, *191*, 60–69. [CrossRef]

43. Krywka, C.; Sternemann, C.; Paulus, M.; Javid, N.; Winter, R.; Al-Sawalmih, A.; Yi, S.; Raabe, D.; Tolan, M. The small-angle and wide-angle X-ray scattering set-up at beamline BL9 of DELTA. *J. Synchrotron Radiat.* **2007**, *14*, 244–251. [CrossRef] [PubMed]

44. Li, N.; Perea, J.D.; Kassar, T.; Richter, M.; Heumueller, T.; Matt, G.J.; Hou, Y.; Güldal, N.S.; Chen, H.; Chen, S.; et al. Abnormal strong burn-in degradation of highly efficient polymer solar cells caused by spinodal donor-acceptor demixing. *Nat. Commun.* **2017**, *8*, 14541. [CrossRef]

45. Zhang, C.; Heumueller, T.; Gruber, W.; Almora, O.; Du, X.; Ying, L.; Chen, J.; Unruh, T.; Cao, Y.; Li, N.; et al. Comprehensive Investigation and Analysis of Bulk-Heterojunction Microstructure of High-Performance PCE11:PCBM Solar Cells. *ACS Appl. Mater. Interfaces* **2019**, *11*, 18555–18563. [CrossRef] [PubMed]

46. Scherrer, P. Bestimmung der Größe und der inneren Struktur von Kolloidteilchen mittels Röntgenstrahlen. *Nachr. Ges. Wiss. Göttingen* **1918**, *1918*, 98–100. Available online: http://eudml.org/doc/59018 (accessed on 20 June 2020).

47. Treat, N.D.; Chabinyc, M.L. Phase Separation in Bulk Heterojunctions of Semiconducting Polymers andFullerenes for Photovoltaics. *Annu. Rev. Phys. Chem.* **2014**, *65*, 59–81. [CrossRef] [PubMed]
48. Liu, C.-M.; Su, Y.-W.; Jiang, J.-M.; Chen, H.-C.; Lin, S.-W.; Su, C.-J.; Jeng, U.-S.; Wei, K.-H. Complementary solvent additives tune the orientation of polymer lamellae, reduce the sizes of aggregated fullerene domains, and enhance the performance of bulk heterojunction solar cells. *J. Mater. Chem. A* **2014**, *2*, 20760–20769. [CrossRef]

Article

Lattice Strain Evolutions in Ni-W Alloys during a Tensile Test Combined with Synchrotron X-ray Diffraction

Tarik Sadat [1,2,*], Damien Faurie [1], Dominique Thiaudière [3], Cristian Mocuta [3], David Tingaud [1] and Guy Dirras [1]

[1] LSPM-CNRS UPR3407, 99 Avenue Jean-Baptiste Clément, Université Sorbonne Paris Nord, 93430 Villetaneuse, France; faurie@univ-paris13.fr (D.F.); david.tingaud@univ-paris13.fr (D.T.); dirras@lspm.cnrs.fr (G.D.)

[2] Laboratoire d'Automatique, de Mécanique et d'Informatique Industrielles et Humaines (LAMIH), UMR CNRS 8201, Université Polytechnique Hauts-de-France, F-59313 Valenciennes, France

[3] Synchrotron SOLEIL, L'orme des Merisiers, Saint Aubin BP 48, 91192 Gif-Sur-Yvette, France; dominique.thiaudiere@synchrotron-soleil.fr (D.T.); cristian.mocuta@synchrotron-soleil.fr (C.M.)

* Correspondence: tarik.sadat@uphf.fr

Received: 31 July 2020; Accepted: 8 September 2020; Published: 11 September 2020

Abstract: Ni and Ni(W) solid solution of bulk Ni and Ni-W alloys (Ni-10W, Ni-30W, and Ni-50W) (wt%) were mechanically compared through the evolution of their {111} X-ray diffraction peaks during in situ tensile tests on the DiffAbs beamline at the Synchrotron SOLEIL. A significant difference in terms of strain heterogeneities and lattice strain evolution occurred as the plastic activity increased. Such differences are attributed to the number of brittle W clusters and the hardening due to the solid solution compared to the single-phase bulk Ni sample.

Keywords: metallic composites; synchrotron X-ray diffraction; Ni; Ni-W alloys

---

1. Introduction

Because of their good mechanical properties such as high hardness and wear resistance [1], nickel-tungsten (Ni-W) alloys are competitive materials that can be used to replace chrome deposits. In addition, they have better magnetic [2], tribological [3], corrosion [4–6], and electrical [7] properties. These alloys are generally produced by electrodeposition (ED) [8–17], magnetron co-sputtering [18], mechanical alloying [19], sintering processes [1,7], and thermal plasma-processes [20]. Due to diffusion of the body-centered cubic (bcc) tungsten phase (W) in the face-centered cubic (fcc) nickel one (Ni), a Ni(W) solid solution is commonly obtained by such processes [7–21]. The dissolution of tungsten atoms in the nickel lattice causes a shift of the fcc Bragg peaks towards lower scattering angles [7,13,19,22,23]. Based on Vegard's law [24], there is a linear relationship between the lattice parameter of the Ni(W) fcc phase and the W content in the solid solution [1,7,13,19,25]. Recently, the production of a bulk (Ni + W) composite-like microstructure by blending controlled amounts of high-purity Ni and W powder particles was successfully achieved by spark plasma sintering (SPS) [1,7,26,27]. Synchrotron X-ray diffraction using a high brilliance source is widely employed to analyze bulk metal samples. The mapping of the distribution of the precipitate microstructure [28], the investigation of the dynamic interactions during solidification of alloys [29], and the characterization of the evolution of phases during aging treatments [30] are some well-known applications. To determinate the influence of the Ni(W) solid solution versus the W phase on the macroscopic deformation of Ni-W alloys and in order to investigate the strain distributions in a Ni-30W (wt%) alloy sintered by SPS, in situ X-ray diffraction (XRD) experiments were performed in [27]. Five separate domains were identified to describe the whole mechanical behavior better. It has been clearly shown that during uniaxial tensile deformation, the cracks propagated inside W aggregates and were stopped at the Ni(W)/W interface at a macroscopic strain of 5% (corresponding to about 620 MPa) [27]. To better understand the intragranular heterogeneities between the two phases, we compare in this continued work, crystal lattice strains in Ni(W) and Ni phases, for different W amounts (Ni-10W, Ni-30W, and Ni-50W (wt%)). The Ni(W) {111} solid solution strain evolution during a uniaxial tensile test is studied and compared to that of the pure Ni {111} phase by using the same in situ tensile tests combined with XRD. In addition, we analyzed the plastic and fracture behavior using in situ monitoring of the full width at half maximum (FWHM), which has not been done previously for these systems to the best of our knowledge.

2. Materials and Methods

Bulk samples were sintered by the SPS technique. This is a fast-process sintering and high-temperature technique that provides a fast heating and cooling rate with a short consolidation time and controllable pressure [28]. A uniaxial load is applied to graphite dye that contains the powder. It is heated by the Joule effect via a pulsed current [29]. The details on the SPS parameters are published elsewhere [7].

The in situ experiments were carried out at the French radiation synchrotron facility (SOLEIL) on the Diffabs beamline with a six-circle diffractometer. A Deben^{TM} tensile stage was used to perform the tensile tests. Tensile tests were conducted using a step-by-step loading procedure at room temperature at a strain rate of 2×10^{-3} s^{-1}. The Digital Image Correlation (DIC) technique was used to obtain an accurate value of the sample macroscopic strain. Aramis software [30] was used to analyze the images. The monochromatic beam's energy was fixed at 18 keV, corresponding to a wavelength of about 0.68 Å. The size of the beam was set to 300×242 μm^2, and the incidence angle was fixed at 10°. A scintillator was used to adjust the sample height with respect to the beam at each deformation step (the vertical position is tailored by keeping half of the direct beam intensity). An XPAD-S140 two-dimensional detector [31] was mounted at about 64 cm from the sample surface to acquire the 560×240 px^2 2D diffractograms. A picture of the whole set-up can be found in a previous paper [27]. For each experimental data point, the corresponding lattice strain ε is calculated using the unloaded state as the reference one:

$$\varepsilon = \ln\left(\frac{d_{hkl}}{d_0}\right) = \ln\left(\frac{\sin(\theta_0)}{\sin(\theta_{hkl})}\right)$$

where d_{hkl} and d_0 are the interplanar distances of the respective loaded and unloaded states and θ_{hkl} and θ_0 the angular positions of the considered diffraction peak (hkl) in the respective loaded and unloaded states. The strain perpendicular to the tensile direction was analyzed. In this study, the peaks shifted towards the higher Bragg angle position during the tensile test, which means that the interplanar distance d_{hkl} decreased due to the macroscopic deformation.

3. Results and Discussion

As already discussed in [7], the Ni-30W and Ni-50W alloys are made of fine-grained multi-crystalline clusters of W (average grain size of about 0.5~0.8 μm) surrounded by randomly oriented grains of Ni(W) (average grain size between 3.9 μm and 8.4 μm depending on the initial amount of W). The Ni-10W was found to be composed of the sole Ni(W) phase (no W cluster). It was established that the average grain size of the Ni(W) phase within the alloy decreased significantly with an increasing amount of W. The bulk Ni exhibited coarse grains with an average grain size of 19.2 μm [7]. To illustrate the microstructure of the alloys, Ni-50W Electron backscatter diffraction (EBSD) phase map is presented in Figure 1 where the Ni(W) solid solution and W clusters appear in red and in green, respectively (other phase maps can be found in [7]).

Figure 1. EBSD phase map of the Ni-50W alloy; the Ni(W) solid solution and W clusters appear in red and in green, respectively. (For interpretation of the references to color in this figure legend, the reader is referred to the web version of the article [7]).

Applied stresses as a function of applied strains (measured by DIC) are presented in Figure 2. As seen, the Ni-30W alloy displays the best combination of both ultimate tensile strength (UTS) (800 MPa) and uniform strain. As expected, the bulk Ni is the sample that deformed the most but was also the one that highlighted the lowest value of UTS (400 MPa). On the contrary, the Ni-50W exhibited the highest UTS (919 MPa) at the expense of ductility. Indeed, it is well known that the addition of W leads to an increase in UTS at the expense of the uniform elongation.

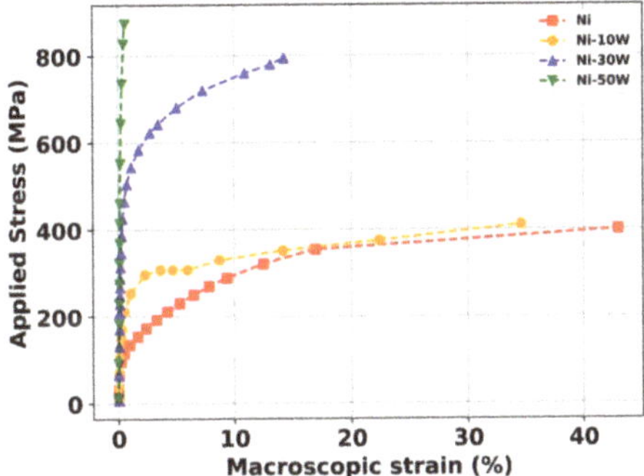

Figure 2. Applied stress as a function of the macroscopic strain of the bulk Ni and Ni-W alloys.

Figure 3a illustrates the lattice strain evolution of the Ni {111} and Ni(W) {111} of the bulk Ni and Ni-W alloys as a function of the applied strain computed by the DIC technique. Figure 3b displays the applied stresses as a function of the lattice strains of Ni {111} or Ni(W) {111} of the bulk Ni and different Ni-W alloys. It is interesting to notice that the elastic domain is well established for all the tracked lattice strains. Indeed, the linearity of the curves occurs at the beginning of the deformation and lasts more or less as the applied stresses increase. It is worth noting that the lattice strains are negative due to the fact that only the crystallographic orientations perpendicular to the tensile tests are probed in this work. Regarding the Ni-10W alloy and the bulk Ni, the curves are characterized by three or four distinct domains:

(i)   From 0 MPa to 75 MPa (for the Ni) and 180 MPa (for the Ni-10W), the lattice strains increase linearly with the applied stresses. This is in accordance with the elastic domain at the macroscopic scale highlighted in Figure 2.

(ii)   At some points, from 75 MPa to 170 MPa (for the bulk Ni) and from 180 MPa to 290 MPa (for the Ni-10W alloy), a sudden change in lattice strain occurs.

(iii)   From 170 MPa to 350 MPa (for the bulk Ni) and from 290 to 350 MPa (for the Ni-10W alloy), a very slight variation of the crystal lattice occurs.

(iv)   The last points are associated with a sharper increase, in absolute value, of the crystal lattices, most probably due to the sample's striction.

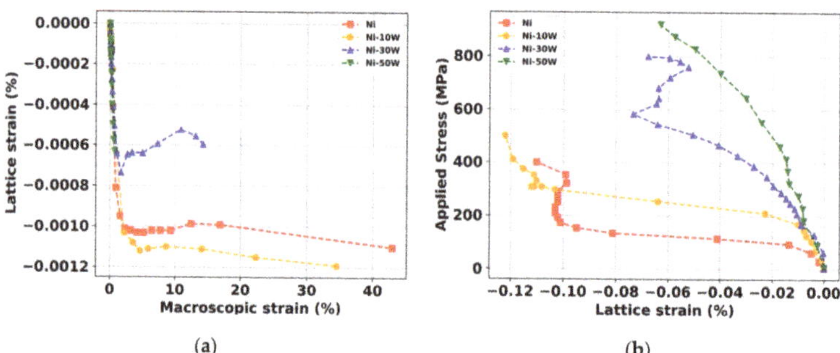

Figure 3. (**a**) Ni {111} or Ni(W) {111} lattice strain as a function of the macroscopic strain of bulk Ni and Ni-W alloys. (**b**) Applied stress as a function of the Ni {111} or Ni(W) {111} lattice strain of the bulk Ni and Ni-W alloys.

Domain (ii) is quite surprising. We did not find in the literature such variations of about 0.1% for polycrystalline materials with micrometric grain size distribution [31,32]. The lattice strain is generally smoother beyond the elastic domain (of the order of 0.1%). This effect might be attributed to the internal microstructure of our materials designed by SPS. Indeed, the disorientation of the grain boundaries of the Ni and Ni(W) phases has been characterized by electron back scatter diffraction (EBSD) analysis [7]. A relatively significant fraction number of $\Sigma = 3$ boundaries ($\Sigma 3$) was observed, showing an ideal misorientation angle of 60° around the <111> axis and including coherent {111} and incoherent {112} twin boundaries. For example, the fraction number of $\Sigma 3$ boundaries was equal to 40.5% and 52.4% for the bulk Ni and the Ni-10W alloy, respectively. It has been reported, in the recent literature, that after the mechanical deformation, at the post mortem state, a significant decrease of the initial fraction of $\Sigma 3$ boundaries occurs. Such a phenomenon is due to an intense dislocation activity and interaction with those $\Sigma 3$ boundaries [33]. This phenomenon seems to be more severe in the case of pure Ni as compared to the Ni(W) phase. It is remarkable that such evolutions are imperceptible at the macroscopic scale presented in Figure 2.

Regarding the Ni-30W alloy, which has been studied in previous work [27], five separate regions are identified in the evolution of the lattice strain. Furthermore, as it has been demonstrated in [27], the cracks propagating inside W aggregates are stopped at the Ni(W)/W interface for a macroscopic strain equal to 5% (corresponding to about 620 MPa). We now turn to the Ni-50W alloy, which presents a different lattice strain evolution of the Ni(W) {111} orientation. Here, we can only distinguish two domains:

(i) An elastic domain up to 400 MPa of the applied macroscopic stress;

(ii) An elastoplastic domain up to 900 MPa of the applied macroscopic stress and, finally, the sample's failure.

The applied stress as a function of the lattice strain of the Ni(W) {111} and the W {110} orientations of the Ni-50W alloy is presented in Figure 4. The brittle fracture is well established in both cases. Around 870 MPa, an inversion of the curve occurs; a load-transfer seems then to take place close to the failure of the sample from Ni(W) to the W phase.

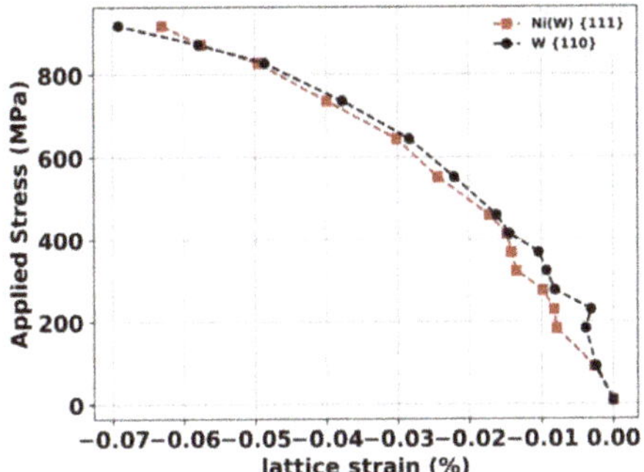

Figure 4. Applied stress as a function of the lattice strain of the Ni(W) {111} and W {110} orientations of the Ni-50W alloy.

To gain more insight into the influence of the W amount, we present in Figure 5 the evolution of the FWHM of the Ni or Ni(W) {111} orientation in the unloaded state as a function of this amount (from 0 to 50 wt.%). A linear increase of the FWHM is clearly observed. Such an evolution is in accordance with the grain size decreasing of the Ni and Ni(W) phases with the amount of W [7].

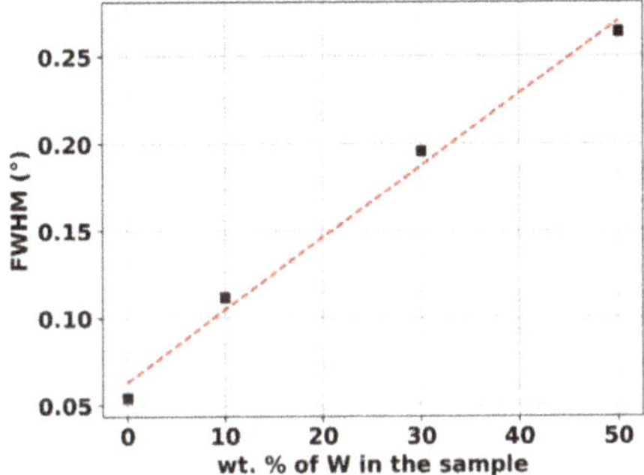

Figure 5. Evolution of the full width at half maximum (FWHM) of the Ni {111} or Ni(W) {111} as function of the weight percentage of W in the sample of the bulk Ni and Ni-W alloys in the unloaded state.

Finally, the evolutions of the FWHM with the macroscopic strain were analyzed. Normalization with the unloaded state value was first performed, and the as-obtained curves are presented in Figure 6.

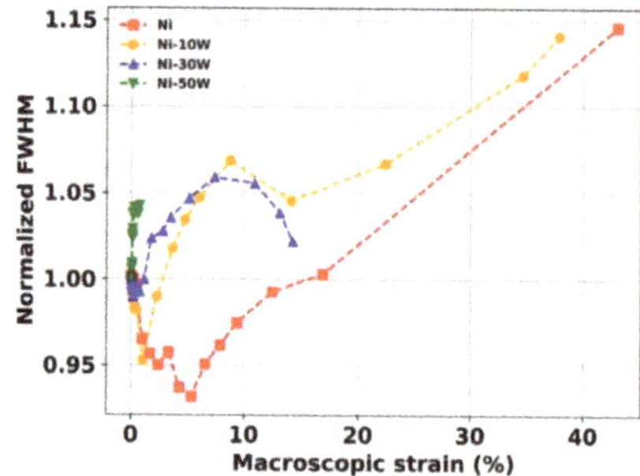

Figure 6. Normalized FWHM of the Ni {111} and Ni(W) {111} as a function of the macroscopic strain of the bulk Ni and Ni-W alloys.

As can be seen, the FWHM evolves in the same way for Ni and the Ni-10W alloy. Indeed, in the elastic domain and at the beginning of the plastic domain activity, a decrease is observed (up to 5% for the Ni and 2% for the Ni-10W) which can be attributed to decreased strain heterogeneities inside those materials. Such a phenomenon might be attributed to a relaxation of internal stresses and induced by the tensile test. Another explanation might be the decreasing of the $\Sigma 3$ boundary fraction leading to an increase of the coherent diffraction domains and finally, an increase of the FHWMs. After that, the FWHMs reach a minimum and then seem to follow an increasing tendency with the applied strain up to failure. This increase of the FWHM is synonymous with an increase in deformation heterogeneities due to the dislocation activity, which increases with the plastic deformation, which is classical in the case of coarse-grained materials [34].

The Ni-30W alloy has a more singular behavior. At the beginning of the test, the FWHM decreases up to 1% of applied strain, which might also be related to the decreasing of the $\Sigma 3$ boundary fraction in the Ni(W) phase during the tensile test. Such a decrease was quantified from post mortem EBSD analysis in [7]. After that, the FHWM increases because of dislocation activity during plastic deformation, and then, from 7% applied strain, it decreases until failure. Indeed, the crack propagation from the brittle W aggregates to the Ni(W) matrix induces a mean stress relaxation in the Ni(W) phase that explains this decrease [27].

The FWHM evolution of the Ni-50W alloy is much different. An increase at the beginning of the macroscopic deformation (contrarily to the other ones) is first observed. It is then followed by a plateau, difficult to distinguish at this scale. In this sample, with the highest amount of the bcc phase, elastic interactions might be stronger between the phases than in the Ni-30W alloy. Moreover, the cc phase's stresses are more pronounced due to a higher theoretical Young's modulus (400 GPa), which can explain such dissimilarity. Regardless, this sample is globally fragile because the W-phase is continuously distributed in the material, which allows the cracks to propagate quickly throughout the material [7].

4. Conclusions

Ni and Ni-xW alloys (x varying between 10, 30, and 50 wt.%) were mechanically deformed in situ under synchrotron XRD. Differences in terms of lattice strains of the Ni {111} or Ni(W) {111} crystallographic orientations were identified. A significant increase in lattice strains between 75–170 MPa (for the bulk Ni) and 180–290 MPa (for the Ni-10W) occurred. Such a phenomenon might be correlated to a decrease of the $\Sigma 3$ boundary fraction during the tensile tests. A load transfer between the Ni(W) phase and the W one was observed in both Ni-30W and Ni-50W alloys. The full width at half maximum (FWHM) of the samples decreased (up to 5% mechanical deformation for the Ni and 2% for the Ni-10W one), which can be attributed to a decrease of strain heterogeneities inside those materials. These in situ observations based on X-ray diffraction are relevant to target in situ transmission electron microscopy (TEM) analysis to observe the dislocations and interaction with the Σ3 boundaries and the crack propagation.

Author Contributions: Conceptualization, T.S., D.F., D.T. (David Tingaud) and G.D.; methodology, T.S., D.F., D.T. (David Tingaud) and G.D.; formal analysis, T.S., D.F. and C.M.; investigation, T.S., D.F., D.T. (David Tingaud) and G.D.; resources, C.M. and D.T. (Dominique Thiaudière); data curation, T.S., C.M., D.T. (Dominique Thiaudière) and D.F.; writing—original draft preparation, T.S. and D.F.; writing—review and editing, T.S., D.F. and G.D.; supervision, D.F., D.T. (David Tingaud) and G.D. All authors have read and agreed to the published version of the manuscript.

Funding: This research received no external funding.

Acknowledgments: The authors would like to thank SOLEIL synchrotron for allocating beamtime for the experiments. P. Joly is acknowledged for excellent technical support during the experimental campaign.

Conflicts of Interest: The authors declare no conflict of interest. The funders had no role in the design of the study; in the collection, analyses, or interpretation of data; in the writing of the manuscript, or in the decision to publish the results.

References

1. Genç, A.; Kaya, P.; Ayas, E.; Ovecoglu, M.L.; Turan, S. Microstructural evolution of mechanically alloyed and spark plasma sintered Ni-W alloy matrix composites. *J. Alloys Compd.* **2013**, *571*, 159–167. [CrossRef]

2. Elias, L.; Hegde, A.C. Effect of magnetic field on corrosion protection efficacy of Ni-W alloy coatings. *J. Alloys Compd.* **2017**, *712*, 618–626. [CrossRef]

3. Yu, J.; Wang, Y.; Zhao, X.; Li, Q.; Qiao, Q.; Zhao, J.; Zhai, S. Wear resistance of ni-based alloy coatings. *Adv. Mater. Sci. Eng.* **2019**, *2019*, 1–7. [CrossRef]

4. Chianpairot, A.; Lothongkum, G.; Schuh, C.A.; Boonyongmaneerat, Y. Corrosion of nanocrystalline Ni-W alloys in alkaline and acidic 3.5wt.% NaCl solutions. *Corros. Sci.* **2011**, *53*, 1066–1071. [CrossRef]

5. Druga, J.; Kašiarová, M.; Dobročka, E.; Zemanová, M. Corrosion and tribological properties of nanocrystalline pulse electrodeposited Ni–W alloy coatings. *Trans. Inst. Met. Finish.* **2017**, *95*, 39–45. [CrossRef]

6. Sriraman, K.R.; Ganesh Sundara Raman, S.; Seshadri, S.K. Corrosion behaviour of electrodeposited nanocrystalline Ni-W and Ni-Fe-W alloys. *Mater. Sci. Eng. A* **2007**, *460–461*, 39–45. [CrossRef]

7. Sadat, T.; Dirras, G.; Tingaud, D.; Ota, M.; Chauveau, T.; Faurie, D.; Vajpai, S.; Ameyama, K. Bulk Ni-W alloys with a composite-like microstructure processed by spark plasma sintering: Microstructure and mechanical properties. *Mater. Des.* **2016**, *89*, 1181–1190. [CrossRef]

8. Iwasaki, H.; Higashi, K.; Nieh, T.G. Tensile deformation and microstructure of a nanocrystalline Ni-W alloy produced by electrodeposition. *Scr. Mater.* **2004**, *50*, 395–399. [CrossRef]

9. Jinlong, L.; Zhuqing, W.; Tongxiang, L.; Suzuki, K.; Hideo, M. Effect of tungsten on microstructures of annealed electrodeposited Ni-W alloy and its corrosion resistance. *Surf. Coat. Technol.* **2018**, *337*, 516–524. [CrossRef]

10. Indyka, P.; Beltowska-Lehman, E.; Tarkowski, L.; Bigos, A.; García-Lecina, E. Structure characterization of nanocrystalline Ni-W alloys obtained by electrodeposition. *J. Alloys Compd.* **2014**, *590*, 75–79. [CrossRef]

11. Lee, S.; Choi, M.; Park, S.; Jung, H.; Yoo, B. Mechanical properties of electrodeposited Ni-W thin films with alternate W-Rich and W-Poor multilayers. *Electrochim. Acta* **2015**, *153*, 225–231. [CrossRef]

12. Nasu, T.; Sakurai, M.; Kamiyama, T.; Usuki, T.; Uemura, O.; Tokumitsu, K.; Yamasaki, T. Structural comparison of M-W (M = Fe, Ni) alloys produced by electrodeposition and mechanical alloying. *Mater. Sci. Eng. A* **2004**, *375–377*, 163–170. [CrossRef]

13. Argañaraz, M.P.Q.; Ribotta, S.B.; Folquer, M.E.; Zelaya, E.; Llorente, C.; Ramallo-López, J.M.; Benítez, G.; Rubert, A.; Gassa, L.M.; Vela, M.E.; et al. The chemistry and structure of nickel-tungsten coatings obtained by pulse galvanostatic electrodeposition. *Electrochim. Acta* **2012**, *72*, 87–93. [CrossRef]

14. Schloßmacher, P.; Yamasaki, T. Structural analysis of electroplated amorphous-nanocrystalline Ni-W. *Microchim. Acta* **2000**, *313*, 309–313. [CrossRef]

15. Schuh, C.A.; Nieh, T.G.; Iwasaki, H. The effect of solid solution W additions on the mechanical properties of nanocrystalline Ni. *Acta Mater.* **2003**, *51*, 431–443. [CrossRef]

16. Ong, C.Y.A.; Blackwood, D.J.; Li, Y. The effects of W content on solid-solution strengthening and the critical Hall-Petch grain size in Ni-W alloy. *Surf. Coat. Technol.* **2019**, *357*, 23–27. [CrossRef]

17. Zhu, L.; Younes, O.; Ashkenasy, N.; Shacham-Diamand, Y.; Gileadi, E. STM/AFM studies of the evolution of morphology of electroplated Ni/W alloys. *Appl. Surf. Sci.* **2002**, *200*, 1–14. [CrossRef]

18. Kurz, S.J.B.; Ensslen, C.; Welzel, U.; Leineweber, A.; Mittemeijer, E.J. The thermal stability of Ni-Mo and Ni-W thin films: Solute segregation and planar faults. *Scr. Mater.* **2013**, *69*, 65–68. [CrossRef]

19. Genç, A.; Oveçoglu, M.L.; Baydogan, M.; Turan, S. Fabrication and characterization of Ni-W solid solution alloys via mechanical alloying and pressureless sintering. *Mater. Des.* **2012**, *42*, 495–504. [CrossRef]

20. Na, H.; Park, J.W.; Choi, H.; Cho, Y.S. Radio frequency thermal plasma-processed Ni-W nanostructures for printable microcircuit electrodes. *Mater. Des.* **2020**, *191*, 108590. [CrossRef]

21. Cao, W.; Marvel, C.; Yin, D.; Zhang, Y.; Cantwell, P.; Harmer, M.P.; Luo, J.; Vinci, R.P. Correlations between microstructure, fracture morphology, and fracture toughness of nanocrystalline Ni-W alloys. *Scr. Mater.* **2016**, *113*, 84–88. [CrossRef]

22. Mi, S. Processing, structure and properties of Ni-W alloys fabricated by mechanical alloying and hot-isostatic pressing. *Scr. Mater.* **1998**, *38*, 637–644. [CrossRef]

23. Genç, A.; Lütfi Oveçoglu, M. Characterization investigations during mechanical alloying and sintering of Ni-W solid solution alloys dispersed with WC and Y2O3 particles. *J. Alloys Compd.* **2010**, *508*, 162–171. [CrossRef]

24. Vegard, V.L. Die konstitution der mischkristalle und die raumffillung der atome. *J. Mater. Sci.* **1921**, *1*, 79–90. [CrossRef]

25. Genç, A.; Ayas, E.; Oveçoglu, M.L.; Turan, S. Fabrication of in situ Ni(W)-WC nano composites via mechanical alloying and spark plasma sintering. *J. Alloys Compd.* **2012**, *542*, 97–104. [CrossRef]

26. Sadat, T.; Hocini, A.; Lilensten, L.; Faurie, D.; Tingaud, D.; Dirras, G. Data on the impact of increasing the W amount on the mass density and compressive properties of Ni-W alloys processed by spark plasma sintering. *Data Br.* **2016**, *7*, 1405–1408. [CrossRef]

27. Sadat, T.; Faurie, D.; Tingaud, D.; Mocuta, C.; Thiaudière, D.; Dirras, G. Fracture behavior of Ni-W alloy probed by in situ synchrotron X-ray diffraction. *Mater. Lett.* **2019**, *239*, 116–119. [CrossRef]

28. Malard, B.; De Geuser, F.; Deschamps, A. Microstructure distribution in an AA2050 T34 friction stir weld and its evolution during post-welding heat treatment. *Acta Mater.* **2015**, *101*, 90–100. [CrossRef]

29. Wang, B.; Tan, D.; Lee, T.L.; Khong, J.C.; Wang, F.; Eskin, D.; Connolley, T.; Fezzaa, K.; Mi, J. Ultrafast synchrotron X-ray imaging studies of microstructure fragmentation in solidification under ultrasound. *Acta Mater.* **2018**, *144*, 505–515. [CrossRef]

30. Callegari, B.; Oliveira, J.P.; Aristizabal, K.; Coelho, R.S.; Brito, P.P.; Wu, L.; Schell, N.; Soldera, F.A.; Mücklich, F.; Pinto, H.C. In-situ synchrotron radiation study of the aging response of Ti-6Al-4V alloy with different starting microstructures. *Mater. Charact.* **2020**, *165*, 110400. [CrossRef]

31. Neil, C.J.; Wollmershauser, J.A.; Clausen, B.; Tomé, C.N.; Agnew, S.R. Modeling lattice strain evolution at finite strains and experimental verification for copper and stainless steel using in situ neutron diffraction. *Int. J. Plast.* **2010**, *26*, 1772–1791. [CrossRef]

32. Clausen, B.; Lorentzen, T.; Leffers, T. Self-consistent modeling of the plastic deformation of f.c.c polycristals and its implications for diffraction measurements of internal stresses. *Acta Mater.* **1998**, *9*, 3087–3098. [CrossRef]

33. Dutel, G.D.; Tingaud, D.; Langlois, P.; Dirras, G. Nickel with multimodal grain size distribution achieved by SPS: Microstructure and mechanical properties. *J. Mater. Sci.* **2012**, *47*, 7926–7931. [CrossRef]

34. Fan, G.J.; Li, L.; Yang, B.; Choo, H.; Liaw, P.K.; Saleh, T.A.; Clausen, B.; Brown, D.W. In situ neutron-diffraction study of tensile deformation of a bulk nanocrystalline alloy. *Mater. Sci. Eng. A* **2009**, *506*, 187–190. [CrossRef]

 *materials* 

Article

Transformation of Ammonium Azide at High Pressure and Temperature

Guozhao Zhang [1,2], Haiwa Zhang [2], Sandra Ninet [2], Hongyang Zhu [3], Keevin Beneut [2], Cailong Liu [4], Mohamed Mezouar [5], Chunxiao Gao [1,*] and Frédéric Datchi [2,*]

[1] State Key Laboratory of Superhard Materials, Jilin University, Changchun 130012, China; z2012gz@163.com
[2] Institut de Minéralogie, de Physique des Matériaux et de Cosmochimie (IMPMC), Sorbonne Université, CNRS UMR 7590, MNHN, 4 Place Jussieu, F-75005 Paris, France; haiwa.zhang@outlook.com (H.Z.); sandra.ninet@sorbonne-universite.fr (S.N.); keevin.beneut@sorbonne-universite.fr (K.B.)
[3] School of Physics and Electronic Engineering, Linyi University, Linyi 276005, China; hongyangzhu@jlu.edu.cn
[4] Shandong Key Laboratory of Optical Communication Science and Technology, School of Physical Science and Information Technology of Liaocheng University, Liaocheng 252059, China; cailong_liu@jlu.edu.cn
[5] European Synchrotron Radiation Facility, BP 220, F-38043 Grenoble CEDEX, France; mezouar@esrf.fr
* Correspondence: gaocx@jlu.edu.cn (C.G.); frederic.datchi@sorbonne-universite.fr (F.D.)

Received: 6 August 2020; Accepted: 11 September 2020; Published: 15 September 2020

Abstract: The compression of ammonium azide (AA) has been considered to be a promising route for producing high energy-density polynitrogen compounds. So far though, there is no experimental evidence that pure AA can be transformed into polynitrogen materials under high pressure at room temperature. We report here on high pressure (P) and temperature (T) experiments on AA embedded in N_2 and on pure AA in the range 0–30 GPa, 300–700 K. The decomposition of AA into N_2 and NH_3 was observed in liquid N_2 around 15 GPa–700 K. For pressures above 20 GPa, our results show that AA in N_2 transforms into a new crystalline compound and solid ammonia when heated above 620 K. This compound is stable at room temperature and on decompression down to at least 7.0 GPa. Pure AA also transforms into a new compound at similar P–T conditions, but the product is different. The newly observed phases are studied by Raman spectroscopy and X-ray diffraction and compared to nitrogen and hydronitrogen compounds that have been predicted in the literature. While there is no exact match with any of them, similar vibrational features are found between the product that was obtained in AA + N_2 with a polymeric compound of N_9H formula.

Keywords: high energy-density materials; high pressure and temperature; Raman spectroscopy; X-ray diffraction; ammonium azide; polynitrogen compounds

1. Introduction

At ambient conditions, nitrogen forms one of the strongest triple bonds found in nature in the diatomic molecule N_2. When submitted to high pressures ($P > 110\,GPa$) and temperatures ($T > 2000\,K$) the N_2 molecular crystal transforms into the extended, single-bonded cubic gauche phase of nitrogen (cg-N) [1,2]. cg-N is a high energy density material (HEDM), with a much larger energy density than conventional explosives, due to the large energy difference between single and triple bonds. Such a HEDM would be an ideal source of clean energy to be used, for example, as propellant. Unfortunately, cg-N is not recoverable at ambient conditions, since it reverts to molecular N_2 below 42 GPa at 300 K.

Various groups have considered adding elements that could enhance the stability of nitrogen oligomers to increase the stability of polynitrogen materials at ambient conditions. Hydronitrogen (N_xH_y) compounds have been given particular attention, when considering that their high mass ratio of nitrogen makes them attractive for HEDM and that hydrogen could effectively passivate polymeric structures at ambient conditions [3–5]. Various poly-hydronitrogen compounds have been indeed predicted under high pressure, not yet observed at ambient conditions [3,4,6,7]. Depending on the composition, these structures contain infinite one-dimensional (1D)-polymeric chains (N_2H, N_3H, N_4H), infinite 2D-polymeric chains (N_9H_4), or pentazole (N_5H) molecules. Experiments [3,6,8–10] on N_2-H_2 mixtures observed that when compressed over 50 GPa, the two molecules react and form N-H bonds suggesting the presence of N_xH_y oligomers. However, their identification has been made difficult by the disordered nature of the product(s), which results in broad Raman and IR peaks and diffuse X-ray diffraction.

Ammonium azide (AA) has been considered as a promising precursor to poly-hydronitrogen. Indeed, the double bond of the azide ion N_3^- is weaker than the triple bond of N_2 and, thus, easier to break. Computer simulations have predicted the polymerization of nitrogen in various forms in AA compressed to pressures in excess of 60 GPa [11–13]. So far, experimental works have investigated AA compressed to 85 GPa at room temperature and did not report any evidence of polymerization [12,14]. Synthesizing polynitrogen compounds from the sole compression of AA would thus require extreme pressures in the 100 GPa range and is, therefore, not a practical route. Recently, a theoretical work has predicted that crystals of NH_4N_5 and N_5H, both featuring all-nitrogen cyclic pentazole ions (N_5^-) could be synthesized at high pressure [15]. In particular, compressing NH_4N_3 with N_2 above 12.5 GPa was proposed as a route for obtaining NH_4N_5. In this respect, it is worth mentioning the recent synthesis of the pentazolate salts CsN_5 [16] and LiN_5 [17] obtained by the reaction of the respective metal or metal azide with nitrogen at high pressures and temperatures. When compared with metal pentazolate salts, NH_4N_5 and N_5H would present a higher mass ratio of nitrogen, an environmentally friendly clean HEDM, and lower *P–T* conditions of synthesis, making it more practical for large production and more likely to recover the product at ambient conditions.

Below, we report the results of our investigations on the AA + N_2 and pure AA systems at pressures up to 30 GPa and temperatures up to 700 K, while using Raman spectroscopy and X-ray diffraction. A transformation of the AA sample is found at 620 K for pressures above 20 GPa in both cases, but the recovered product is different. The Raman spectrum and X-ray diffraction pattern of the new phases have been measured up to 25 GPa at room temperature. We find no evidence for the predicted NH_4N_5 nor HN_5 compounds. Comparison is made with other N_xH_y compounds predicted stable at high *P*; while there is no exact match with any of them, similar vibrational features are found between the product obtained in AA + N_2 with a polymeric compound of N_9H formula.

2. Materials and Methods

2.1. Experimental Methods

AA powder samples were synthesized by a metathetical reaction of NaN_3 and NH_4NO_3, as reported in previous studies [18]. A membrane diamond anvil cell (MDAC) was used to generate high pressures [19], equipped with diamond anvils of culet size of 300 µm. Re gaskets of initial thickness of 200 µm were preindented to a thickness of 35 µm, and a hole of diameter of 95 µm was laser-drilled at the center of the indent in order to serve as the sample chamber. A piece of AA sample of lateral size 40 to 50 µm was loaded in the gasket hole and embedded in N_2 by cryogenic loading of N_2, in the same fashion as our previous study using liquid argon [14]. The Raman spectrum of the AA sample after loading was compared to those that were measured in our previous study [14] and confirmed the absence of impurities, as shown in Figure S1 of the Supporting Information file (SI).

In Raman experiments, pressure was determined using the ruby and SrB_4O_7:Sm^{2+} fluorescent sensors, according to calibrations from Refs. [20,21], respectively. In XRD experiments, pressure was determined either from the measured volume of a small piece of gold powder or of solid nitrogen, while using the equations of state of Refs. [22,23], respectively.

We adopted an external heating method to heat the sample, using a cylindrical resistive heater (Watlow France, Asnières-sur-seine, France) whose internal diameter and length fit the outer dimensions of the cell, ensuring homogeneous heating. The heater is connected to a power unit and it is temperature controlled using a feedback loop. A K-type thermocouple is placed in contact with the cell close to the sample chamber in order to determine the sample temperature. The accuracy of temperature measurement is about 5 K.

Raman scattering was excited by the 514.5 nm line of an argon laser, collected by a confocal in-house optical set-up and dispersed by a HR460 (Horiba Jobin Yvon, Longjumeau, France) spectrometer of focal length 460 mm onto a water-cooled CCD camera (Andor Technology Ltd, Belfast, U. K.). The diameter of the laser spot on the sample, focused by 20X Mitutoyo objective, was about 2 μm. The Raman spectra were recorded in the range of 70 to 4800 cm^{-1} using a 1200 lines/mm grating.

XRD experiments were done in the angular dispersive mode using monochromatic X-rays at beamline ID27 of the European Radiation Synchrotron Facility (ESRF, Grenoble, France). The X-ray beam of wavelength 0.3738 Å was focussed to a spot size of 2.5 × 3 μm FWHM. Diffracted X-rays were collected by a bidimensional SX-165 CCD detector (marXperts, Norderstedt, Germany). The sample distance and position of the X-ray beam were calibrated while using a CeO_2 powder standard. The analysis and refinement of XRD patterns were performed using the Fullprof software suite [24].

2.2. Theoretical Methods

Structural optimization and vibrational calculations were performed using, respectively, the first-principle plane-wave pseudo-potential density functional theory (DFT) and density functional perturbation theory (DFPT) [25,26], as implemented in the CASTEP code [27]. The ultrasoft pseudopotentials were used with exchange and correlation effects described by the generalized-gradient-approximation (GGA) [28] of Perdew–Burke–Ernzerhof. van der Waals corrections were implemented using the Tkatchenko–Scheffler method [29], and the value of the two scheme parameters sR and d were set at 0.94 and 20.0, respectively. The kinetic energy cutoff and Monkhorst-Pack k-point meshes (5 × 5 × 2 meshes) for structural optimization were set at 898 eV and 0.07 A^{-1}, respectively. The self-consistent energy convergence criterium was set to less than 5.0×10^{-6} eV/atom, and the maximal force, stress and displacement set to be 0.01 eV/Å, 0.02 GPa and 5.0×10^{-4} Å, respectively. The Reflex powder diffraction module in the material studio software was used in order to calculate the XRD patterns.

2.3. High P-T Transformation of AA in N_2

2.3.1. Raman Experiments

We first investigated AA embedded in N_2 compressed up to 30 GPa at room temperature using Raman spectroscopy. We observed no sign of phase transition in the Raman spectrum, the latter being consistent with previous measurements on pure AA and AA embedded in argon [12,14]. We then prepared three samples at pressures of about 15 GPa, 20 GPa, and 30 GPa, and studied them as a function of temperature in the range 300–700 K. The followed *P–T* paths are illustrated in Figure S2 of the SI.

At $P{\sim}15$ GPa, the Raman spectrum was collected at 100 K steps and showed no phase transition to 600 K (see Figure S3 of the SI). At 700 K, nitrogen became liquid and the AA sample decomposed quickly and dissolved into liquid nitrogen. Only the Raman band of the N_2 molecule stretching could be clearly observed at this temperature. Upon cooling back to 300 K, N-H stretch modes in the range 3150–3450 cm^{-1} were also detected at several positions, which are compatible with the Raman spectrum of ammonia (see Figure S4 of SI). The reasons why NH_3 was not detected at high temperature likely comes from its dilution in the N_2 liquid, the small AA sample volume, which was loaded as compared to that of N_2 and weaker Raman intensity at high *T*. At 13 GPa–300 K, N_2 and NH_3 were solid and phase separated (hence, the inhomogeneous aspect of the sample shown in panel (a4) of Figure S3), and a stronger Raman signal from the NH_3 solid was recorded in sample regions, where NH_3 crystallized. Although not tested, we consider it unlikely that H_2 was formed in the decomposition of AA, since (1) none were formed at higher pressures (see below) and (2) NH_3 is more stable than $N_2 + H_2$ at this pressure [4,30,31]. Thus, we conclude that AA is not stable and decompose into N_2 and NH_3 in the presence of liquid N_2 at 15 GPa. For this reason, we took care not to cross the N_2 melting line in the two subsequent heating runs at 20 GPa and 30 GPa. The heating rate was in these cases 50 K/step. Since the experimental results for these two pressures are identical (see Figures S5–S7 of the SI), we only report below the results obtained on heating at 30 GPa.

Upon heating the sample initally at 30 GPa at 300 K, the pressure slowly reduced to 26.5 GPa at 600 K. At 600 K–26.5 GPa, the Raman spectrum of the $AA + N_2$ exhibits no significant change from the one at 380 K–29.4 GPa, apart from the expected *P-T* induced peak shifts and slight peak broadening, as can be seen from Figure 1. When the temperature was raised to 650 K, the Raman spectrum of the sample suddenly changed, especially in the frequency ranges colored in gray in Figure 1a. The sample light absorption also increased at this temperature [see the photographs in Figure 1b]. After holding for 10 min., some Raman peaks disappeared or weakened. After half an hour, the Raman spectrum of the sample no longer varied; it also qualitatively stayed the same when the sample was cooled down to room temperature at the same pressure.

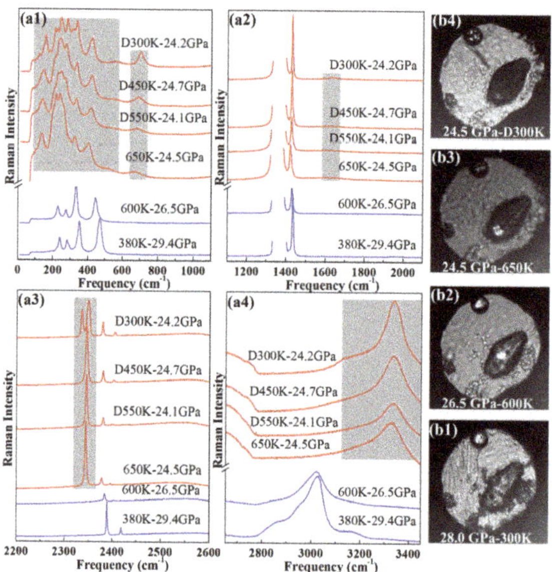

Figure 1. (**a**) Raman spectrum of $AA + N_2$ up to 25 GPa at different temperatures. The gray colored bands emphasize the frequency windows where major changes are observed in the sample Raman spectrum. "D" indicates that the measurement was performed on decreasing temperature (**b**) Photographs of the sample chamber at different pressures and temperatures.

A comparison of the sample Raman spectra before and after the high pressure heat treatment reveals major changes both in the lattice and internal modes. In particular, the N-H stretch bands discontinuously shifted by $\sim$300 cm^{-1} and new bands appeared in the frequency regions of N≡N stretch (2340 cm^{-1}) and N=N bend ($\sim$700 cm^{-1}). The Raman spectra taken at different positions in the sample chamber (see Figure S8 of the SI) show that the reaction product is homogeneous, and the sharp molecular and lattice Raman bands strongly indicate that it is crystalline. We also note that a large amount of nitrogen remained around the sample after the transition. No evidence for the formation of H_2 was found in the Raman spectra (see Figure S9 of the SI). As seen below, the reaction product is actually most likely composed of two phases, ammonia solid and an unknown phase referred to as phase A in the following.

The Raman spectrum of the reacted sample was collected upon decompression at 300 K (see Figure S9 of the SI). All of the modes observed after the high *P-T* treatment are conserved down to 7.1 GPa. Below this pressure, no measurement could be made, as lowering the force in the DAC resulted in a rapid jump to ambient pressure. When opened to air, the products synthesized in the sample chamber and nitrogen quickly volatilized. Before its complete decomposition, we could only detect the Raman spectrum characteristic of pure AA. This shows either that the high *P-T* reaction is reversible at a pressure between ambient and 7.1 GPa at room temperature, or that a small remnant of AA was still present and remained after all other products volatilized. This amount must be small as AA was not detected either by Raman or XRD after the reaction. AA itself eventually decomposed into volatile components and disappeared from the sample chamber after a few minutes.

Figure 2 shows the evolution of the Raman mode frequencies with pressure at 300 K, and compares them to those of the pure AA [14], N_2 [32], NH_3 [33,34] and hydrazine (N_2H_4) [35] solids obtained from literature. We recall that NH_3 (and possibly heavier, unidentified azanes) and N_2H_4 (below 10 GPa on decompression) were identified as reaction products of $N_2 + H_2$ mixtures at high pressure [3,6,8,9], which justifies the present comparison. The higher frequency modes correspond to N-H stretching, and the deconvolution of the observed band reveals at least five peaks in the 3100–3400 cm^{-1} range. It can be seen that four of these N-H peaks nearly coincide with those of the pure ammonia solid reported in Refs. [33,34]; the small frequecy differences are likely due to the different sample temperature, which was 50 K in Ref. [34]. There is also a good match between the observed modes in the 1620–1700 cm^{-1} region with the reported torsion modes of NH_3, and the same is true for the lattice mode around 500 cm^{-1} at 20 GPa, which corresponds to the A^e mode of NH_3 [34]. The frequency region below 400 cm^{-1} in phase A is composed of many peaks and is more difficult to analyze in details. These observations strongly suggest that ammonia is one of the reaction products and separated in solid form from the other products. The disappearance of the N-H stretch and bending modes of NH_4^+ also show that these ions disappeared and probably converted to NH_3.

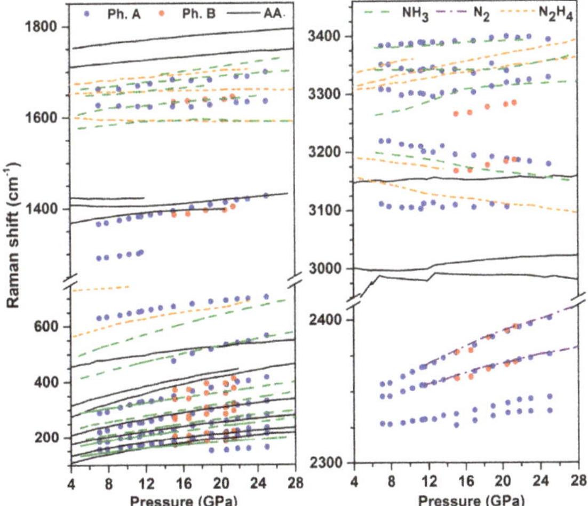

Figure 2. Frequency evolution with pressure of the Raman peaks of the products of the AA transformation at high *P-T*. The measurements were made on decompression at 300 K. The experimental data for phases A and B are represented by blue and red solid dots, respectively. The different colored lines represent literature data for the Raman modes of pure AA (black solid lines) [14], NH_3 [33,34] (green dashed lines), N_2 [32] (purple dashed lines), and hydrazine [35] (N_2H_4, orange dashed lines) solids.

The evolution with pressure of the mode at 1420 cm^{-1} at 20 GPa is very similar to that of the N=N stretch in AA, indicating that phase A contains species with N=N double bonds. We note that in AA, the N=N stretch mode is in Fermi resonance with the bending mode of NH$_4^+$, which is no longer the case in phase A, supporting again that NH$_4^+$ have disappeared in this phase. Four Raman bands are observed in the N≡N stretch frequency range; two of them perfectly match those of pure solid N$_2$, and the remaining two, which are located at lower frequencies (by about 20 to 40 cm^{-1}), more likely belong to phase A. Finally, the band at ca. 700 cm^{-1} at 26 GPa cannot be assigned to N$_2$ or NH$_3$ solids and, thus, also belongs to phase A.

2.3.2. XRD Experiments

We performed XRD measurements on the same sample as above from 25 GPa to 12 GPa at 300 K on decompression in order to explore the structure of phase A. Thanks to the small X-ray beam, XRD patterns could be collected at different positions in the sample chamber, either on the reacted sample or at a location which only contained pure solid nitrogen. Because the AA sample is surrounded by N$_2$, diffraction peaks from solid N$_2$ are also expected in the XRD pattern of the reacted sample. The latter at 20.1 GPa–300 K is shown in Figure 3a. Diffraction peaks of solid N$_2$ (ϵ-N$_2$ above 17 GPa and δ^*-N$_2$ below) and ammonia (phases V above 12 GPa and IV below) could be identified in the pattern of the reacted sample (see Figure S11 of the SI). The cell parameters and volume of N$_2$ and NH$_3$ obtained from the Le Bail refinement of these patterns are compared to literature [32,36] in Figure S12 of the SI, showing very good agreement in both cases. This confirms that pure solid ammonia is produced in the reaction.

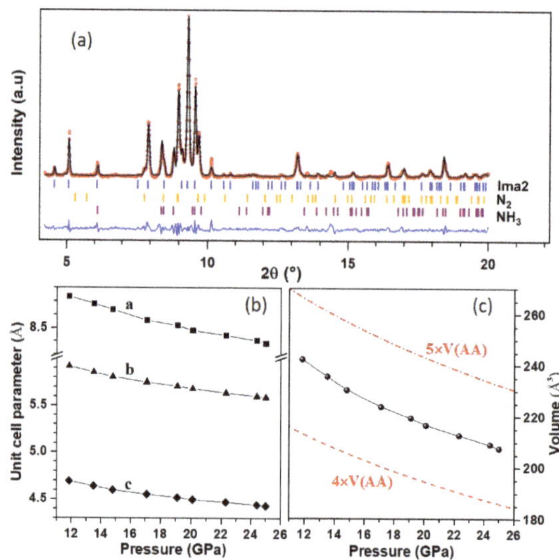

Figure 3. XRD pattern of the AA + N$_2$ sample at 20.1 GPa–300 K after the high *P-T* transformation. Panel (**a**) shows the three-phase Le Bail fit of the pattern using the *Ima2* unit cell for phase A, ϵ-N$_2$ (*R-3c*) and NH$_3$-V (*P2$_1$2$_1$2$_1$*). The background has been subtracted for easier viualization. The red circles are experimental data, the black line is the Le Bail fit and the blue line is the fit residual. Ticks indicate the position of the Bragg peaks for each phase. The bottom panels show the unit cell parameters (**b**) and volume (**c**) of the *Ima2* cell. The latter may be compared to the dashed and dash-dotted lines, representing, respectively, four and five times the volume of AA-II (*P2/c*).

Searches for the unit cell of phase A were performed with the list of non-indexed peaks in the XRD pattern at 20.1 GPa while using the DICVOL06 software [37]. This returned two possible solutions:

first, an orthorhombic cell with most probable space group *Ima2*, cell parameters $a = 8.475(9)$ Å, $b = 5.679(4)$ Å, $c = 4.490(5)$ Å and volume $V = 216.1(4)$ Å^3; second, a monoclinic cell of most probable space group *P2/m*, cell parameters $a = 4.738(3)$ Å, $b = 3.524(1)$ Å, $c = 4.249(2)$ Å, $\beta = 93.75(3)°$, and volume $V = 70.8(1)$ Å^3. The variation with pressure of the orthorhombic cell parameters and volume are shown in Figure 3b,c, and those of the monoclinic cell in Figure S13 of the SI. These parameters vary continuously with pressure, confirming that phase A is stable in this pressure range. Figure 3c also compares the volume of phase A to that of the AA-II (*P2/c*) phase reported in Ref. [14]. It may be seen that the volume of the *Ima2* cell is bracketed by four and five times the volume of AA-II. If we assume that phase A is composed of the same formula units as AA-II and four times as much units, the volume reduction at the transition would be about 12%, which seems rather large. This supports the fact that phase A is composed of different moieties than AA. It can also be observed that the compressibility of phase A is similar to that of AA-II, which suggests that phase A is a molecular solid.

2.4. Transformation of Pure AA

2.4.1. Raman Experiments

We carried out high-temperature Raman experiments on pure AA in order to determine whether the formation of phase A requires the participation of nitrogen. Starting from a sample at 26.2 GPa at 300 K, the temperature was increased in several steps. The measured Raman spectra are shown in Figure 4. Upon heating to 600 K, the sample pressure slowly reduced to 20 GPa. As for the AA + N$_2$ system, we found no evidence for a phase transition in AA below 600 K and around 20 GPa. When the temperature was further increased to 620 K, the pressure in the sample chamber reduced to 17.4 GPa and a clear change of the Raman spectrum took place: first, the Raman band peaked at 277 cm^{-1} split and new Raman bands appeared at 332 cm^{-1} and 2359 cm^{-1}; second, the N-H stretch bands discontinuously shifted to higher frequencies. After increasing temperature to 650 K, no more change was detected in the Raman spectrum or on cooling down the sample to 300 K. The visual images presented in Figure 4b also show a loss in white light transmission of the sample occurring concomitantly as the changes in Raman spectrum, starting from one side of the sample chamber at 620 K, and completing by 650 K.

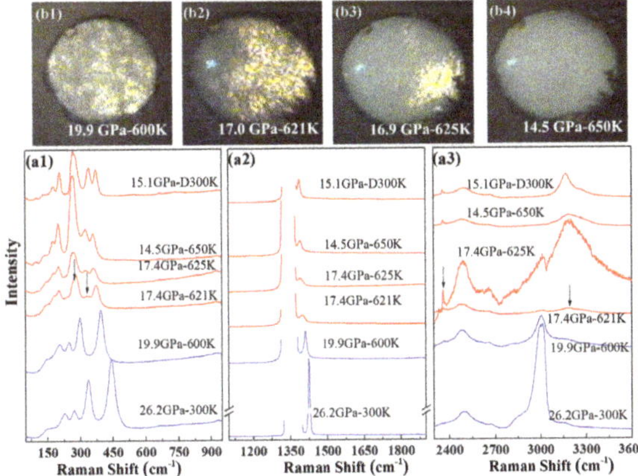

Figure 4. High *P-T* transformation of pure AA. (**a**) Raman spectra collected before (blue lines) and during (red lines) the transformation (621–650 K). The uppermost spectrum was collected after cooling back to 300 K. The arrows in the figure indicate the positions of new peaks. "D" indicates that the measurement was performed on decreasing temperature; (**b**) Photographs of the sample chamber before (600 K) and during the transformation (621–650 K).

Thus, pure AA transforms to a new phase above 620 K which is referred hereafter as phase B. The sharp lattice and internal modes of the Raman spectrum of phase B are in favor of a crystalline compound. Further Raman measurements were performed on decompression at 300 K from 21.4 GPa to 15.1 GPa, and the peak frequencies as a function of pressure are plotted in Figure 2.

The N-H stretch band can be deconvoluted into two peaks located at 3183 cm^{-1} and 3281 cm^{-1} at 20.5 GPa–300 K, and whose frequency increases with pressure. A single N-H bending mode is detected at 1639 cm^{-1}. These N-H vibration frequencies largely differ from those of NH_4^+ in AA, indicating that NH_4^+ ions disappeared in the high *P-T* transformation. They are in the same frequency range as those of NH_3, but, unlike for phase A, the number of modes and their frequencies do not match those of pure solid NH_3, which suggests that the latter is not formed. There is also no good match with the Raman modes of N_2H_4. As for AA + N_2, we found no evidence for the formation of H_2 in the Raman spectra (see Figure S10 of the SI).

The two peaks at 2368 cm^{-1} and 2392 cm^{-1} at 20.5 GPa coincide, both in frequency and pressure shift, with the N≡N stretch of the pure N_2 solid, strongly indicating that N_2 molecules were formed at the transition and separated from phase B. The Raman peak at 1395 cm^{-1} at 20.5 GPa is close to, and shifts similarly with pressure as the N=N stretching mode of AA and phase A, indicating that phase B also contains species with doubly bonded nitrogen atoms. As in phase A, no Fermi resonance is observed between this mode and the N-H bending of NH_4^+, thus confirming that NH_4^+ are no longer present in phase B. Further comparison of phases A and B will be presented in the next section.

2.4.2. XRD Experiments

The XRD pattern of phase B was measured at 24.8 GPa and 15.9 GPa. In both patterns, peaks from solid N_2 either in the ϵ or the δ^* phase could be identified, thus confirming the formation of solid N_2 in the transformation. The remaining peaks in the pattern at 15.9 GPa were used for unit cell search using DICVOL06 [37]. Here again, two solutions were found compatible with the data; the first one is an orthorhombic unit cell of most probable space group *Pnna* with $a = 12.799(3)$ Å, $b = 8.957(3)$ Å, $c = 4.724(1)$ Å and volume $541.6(3)$ Å^3 at 15.9 GPa; the second one is a monoclinic cell of most probable space group $P2/m$, $a = 8.254$ Å, $b = 6.722(1)$ Å, $c = 5.356(1)$ Å, $\beta = 91.99(1)°$, and volume $296.95(1)$ Å^3. The Le Bail refinement using the *Pnna* cell is shown in Figure 5, and the one for the $P2/m$ structure is given in Figure S14 of the SI. The orthorhombic cell better fits the data, and is thus the preferred solution. The volume ratio with respect to AA-II is 10.7, which suggests at least ten times more formula units. However, this large volume could also signal that phase B is not a single phase.

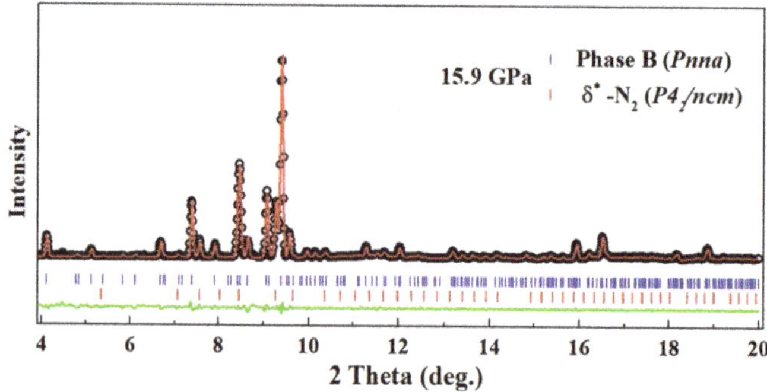

Figure 5. XRD pattern of the pure AA sample at 15.9 GPa–300 K after the high *P-T* transformation. The background has been subtracted for easier visualization. Experimental data are shown by red circles. The black line is a two-phase Le Bail fit of the pattern using the *Pnna* unit cell for phase A and δ^*-N_2 ($P4_2/ncm$). The blue line is the fit residual. Ticks indicate the position of the Bragg peaks of each phase.

3. Discussion

The experimental results that are presented above show that AA ceases to be stable and undergoes a transformation for $P > 20$ GPa and $T > 620$ K, whether it is or not embedded in N_2. However, the transformation gives a different product in the presence of N_2, judging both from the Raman spectra and XRD patterns. It may be conjectured that phase A is obtained in two steps, first from the transformation of AA into phase B, and second from the reaction of phase B with N_2. Below, we first compare the vibrational and structural properties of phases A and B, and then compare the two with predicted stable N_xH_y compounds in the literature.

3.1. Comparison of Phases A and B

The comparison of the Raman spectra of phases A and B at 20.5 GPa at 300 K, reveal significant differences, as seen in Figures 2 and 6. First, as noted above, there are clear differences in the number and frequency vs. pressure evolution of the N-H stretch bands. In phase A, this band is composed at least of five peaks, but only 1 of them, located at $\sim$3110 cm^{-1} and nearly invariant with pressure was assigned to the new phase, as the 4 others match those of solid NH_3. As a matter of fact it is possible that other peaks overlapping with those of solid NH_3, are present but not resolved. In phase B, two N-H stretch modes are observed which differ in frequency and pressure shift from those of phase A. In both phases though, we found that these modes are not compatible with the N-H stretch of the NH_4^+ ion, which signals their disappearance in the transformation and likely conversion into ammonia molecules or other species with N-H radicals.

Another marked difference in the Raman spectra resides in the Raman bands at $\sim$700 cm^{-1} and 2320–2340 cm^{-1} in phase A, which are not present in phase B. The first one is difficult to assign as it may arise from several possible vibrational motions. One the one hand, the frequency of this mode is similar to that of the N-N stretching vibration in the extended covalent nitrogen solid cg-N [1]; however, the Raman spectrum and compressibility of phase A suggests that it is a hydronitrogen molecular solid, so the N-N stretch would rather be expected around 1100–1150 cm^{-1} as in hydrazine [35]. The latter also presents a Raman mode close to 700 cm^{-1} at 25 GPa (see Figure 2), which corresponds to a torsion of the N-N bond through rocking of the NH_2 units. Examining the vibrational modes that were obtained by our calculations for several N_xH_y compounds (see below), we also found that this mode is compatible with other types of motion such as the torsion (in N_3H and N_5H) or the bending (in N_6 and N_8) of N-N-N units. The second band at 2320–2340 cm^{-1}, composed of two peaks, is located in the range of frequency of the $N\equiv N$ stretch of the N_2 molecule and thus likely comes from the presence of this molecule inside the crystal structure of phase A. N_2 molecules are also formed in the transformation of pure AA, but, in this case, separate from phase B. This suggests that the N_2 molecules are not interstitial in phase A, and that they are not formed in sufficient amount in the transformation of pure AA to reach the required stoichiometry of phase A. This would confirm that excess N_2 is required to obtain phase A from AA. It is also interesting to note that a shifted N_2 Raman vibron was observed in the reaction products of compressed N_2-H_2 mixtures [8–10], and assigned to trapped N_2 molecules.

3.2. Comparison with Predicted N_xH_y Compounds

In order to establish the composition of the two newly formed phases A and B resulting from the transformation of AA at high P–T, we compared their Raman spectra and XRD patterns to those of poly-hydronitrogen N_xH_y compounds predicted as stable in the literature. We limited the comparison to compounds which have been predicted as stable below 60 GPa. Furthermore, we only present below results for compounds with a nitrogen/hydrogen content ratio equal or superior to AA. As a matter of fact, to the best of our knowledge, only two H-rich compounds besides ammonia have been reported so far to be stable in calculations below 60 GPa, of respective formula NH_4 and NH_5 [4]. Both contain H_2 molecules and, as shown in Figure S10, have strong associated Raman peaks that are not observed in our experiments. The XRD the patterns were obtained from the optimized structure at 20 GPa, and the Raman spectra were computed using density-functional perturbation theory, as described in the Methods section. Figure 6 shows the comparison for the Raman spectra, and Figure S15 of the SI compares the XRD patterns.

N_4H_4. A compound of same stoichiometry as AA, but composed of tetrazene (NH_2-N-N-NH_2) molecules was predicted by computational studies to be stable above 36 GPa [4,11]. Our previous study showed that this compound is less stable than AA-II in calculations below 102 GPa, and it is not observed in experiments up to 85 GPa at 300 K [14]; however, the transition may be facilitated at high temperature. The present calculations were made using the $P2_1/c$ structure that is reported in Ref. [4]. The comparison of the computed Raman spectra with phases A and B show strong differences, especially in the 1000–1600 cm^{-1} region, and the same is true for the XRD pattern; we may thus safely rule out the formation of this compound in our experiments.

NH_4N_5. As stated in the introduction, a motivation of this work was the theoretical prediction [15] that ammonium azide will react with N_2 and transform into ammonium pentazolate (NH_4N_5) above 12.5 GPa. The recent synthesis of the metal pentazolate salts CsN_5 [16] and LiN_5 [17] from the direct reaction of, respectively, CsN_3 and Li with N_2 at high P-T gave credit to this prediction. According to Ref. [15], the stable phase of NH_4N_5 at this pressure is of *Pbcm* symmetry and it is composed of hydrogen-bonded NH_4^+ and cyclo-N_5^- ions. In the DFPT calculated Raman spectrum of this structure at 20 GPa, the distinctive features of th N_5^- ion are the breathing (1240 cm^{-1}), bending (758 cm^{-1}), and deformation (1090–1125 cm^{-1}) vibrations. The frequencies of these modes in NH_4N_5 obtained from our calculations compare very well with those measured in LiN_5 [17] and CsN_5 [16]. There is no corresponding mode in the Raman spectra of phases A and B, apart maybe for the peak of phase A at 692 cm^{-1}, but the predicted strong breathing mode at 1240 cm^{-1} is definitely not observed, as seen in Figure 6. Thus, we may conclude that NH_4N_5 has not been formed in our experiments and that the synthesis of this compound at high P-T is not as straightforward as for the metal pentazolate salt. The reaction might require a higher pressure or temperature to overcome the kinetic barrier, however the decomposition of AA in liquid N_2 observed in this work limits the temperature range of investigation to below the melting line of N_2.

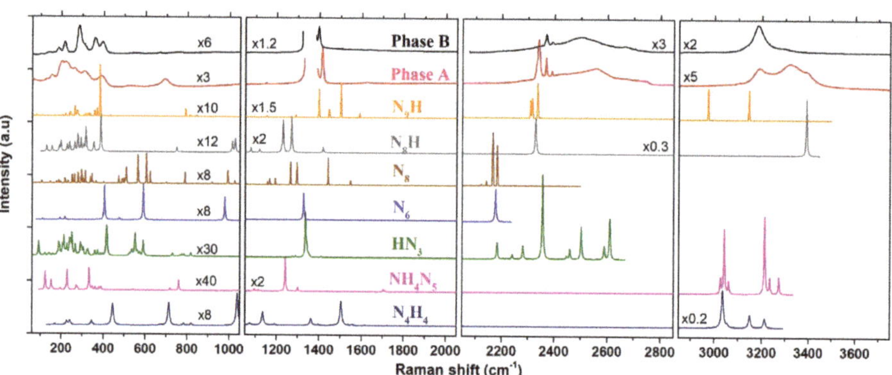

Figure 6. Comparison between the experimental Raman spectra of phases A and B at 20.5 GPa–300 K, and the theoretical ones of N_4H_4 ($P2_1/c$), NH_4N_5 (*Pbcm*), HN_3 (*Cc*), N_6 (*C2/m*), N_8 (*P1*), N_8H (*P-1*) and N_9H (*P1*) at 20 GPa–0 K, computed by DFPT. The structures used for the Raman calculations are from Ref. [15] for NH_4N_5, Ref. [38] for N_6, Ref. [39] for N_8, Ref. [40] for HN_3, Ref. [4] for N_4H_4 N_8H and Ref. [5] for N_9H. The frequency range was divided into four panels for easier visualization. The vertical scale of each panel is the same but individual scale factors were applied as indicated to some spectra in order to increase the visibility of peaks.

HN$_3$. The disappearance of NH$_4^+$ ions in the high *P–T* transformation of AA + N$_2$ and the observation of NH$_3$ as a reaction product could indicate that proton transfers from NH$_4^+$ to N$_3^-$ occur during the transformation to yield HN$_3$. To the best of our knowledge, the crystal structure of the HN$_3$ solid has not been investigated at high pressure, but theoretical studies [4,5] have predicted a *P*2$_1$/*c* structure composed of long chains with N-N bond lengths in the range of single and double bonds. The XRD pattern and Raman spectrum of this structure reported in Ref. [5] are very different from that observed, so we did not consider it. Instead, we assumed that the crystal structure remains the same as that determined at ambient pressure [40]. The latter, of *Cc* symmetry, consists of 16 HN$_3$ molecules per unit cell connected to each other by N-H$\cdots$N hydrogen bonds. We note that this structure is dynamically stable at 20 GPa in our calculations, since all of the vibrational modes have positive frequencies. As seen in Figure 6, the calculated Raman spectrum of the *Cc* HN$_3$ solid also largely differ from those of phases A and B; in particular, the N-H stretching modes are located at much lower frequencies (2350–2610 cm^{-1}) and the strongest N=N stretch peak (1336 cm^{-1}) is also 70 cm^{-1} below. Although there remains uncertainty regarding the correct structure of HN$_3$ at 20 GPa, we thus consider that it is not a good candidate for the reaction products of AA + N$_2$.

N$_6$ and N$_8$. Recent theoretical studies have predicted the existence of two new allotropes of nitrogen in the form of N$_6$ [38] and N$_8$ [39] molecules. Both molecules exhibit several resonnance structures mixing single, double, and triple nitrogen bonds. The predicted crystal structures for N$_6$ (*C2/m*) and N$_8$ (*P1*) were found more stable than the cg-N solid below 20 GPa. The experimental synthesis of the N$_8$ solid was recently claimed by compressing hydrazinium azide [(N$_2$H$_5$)(N$_3$)] above 40 GPa, decomposing to N$_2$ below 20 GPa [41]. The theoretical Raman spectrum of the N$_8$ crystal, computed here at 20 GPa while using the structure reported in Hirshberg et al. [39], contains many (96) active modes due to the low symmetry of this structure. The Raman spectra of phases A and B are much simpler than that of N$_8$ and many predicted peaks for this solid are not observed. Moreover, the predicted frequency range for the N≡N stretch modes (2120–2180 cm^{-1}) in N$_8$ is lower than that observed in both phases A and B. The N$_6$ crystal has a higher symmetry (*C2/m*) and thus a simpler Raman spectrum, yet the Raman peaks occur at similar frequencies as in N$_8$ and, thus, differ from those of phases A and B. Thus, we may safely conclude that these species are not formed in the transformation of AA. We also note that the calculated Raman spectrum of N$_8$ largely differ from that measured in Duwal et al. [41] (in particular, for the N≡N stretch observed around 2380 cm^{-1}), casting doubts that these authors actually produced N$_8$ by compression of (N$_2$H$_5$)(N$_3$).

N$_8$H and N$_9$H. N$_8$H (*P-1*) and N$_9$H (*P1*) are two hydronitrogen compounds that emerged from theoretical structural searches, in Refs. [4,5], respectively. These two compounds have in common the presence of N$_2$ molecules in their crystal structure, which we suspect to be the case in phase A. N$_8$H is also composed of N$_5$H moieties, whereas N$_9$H contains infinite zig-zag chains of N and H atoms. The computed Raman spectra for these two solids at 20 GPa display intense peaks around 2330 cm^{-1} coming from N$_2$ molecules. Interstingly, there is a good match with the spectrum of phase A, supporting our intuition that N$_2$ molecules are part of the crystal structure of phase A. N$_9$H also presents Raman peaks which matches well with the N=N and N-H stretch of phases A and B; however, some strong predicted peaks, such as those at 1500 cm^{-1} and 2973 cm^{-1}, are not observed, and the XRD pattern of the *P1* N$_9$H solid does not match those measured for phases A and B (Figure S15).

4. Conclusions

In conclusion, the present experimental investigation of ammonium azide at high pressure and temperature showed, first, that this compound decomposes in liquid N$_2$ at 13 GPa–700 K, into N$_2$ and NH$_3$; second, that it transforms to new phases for *P* > 20 GPa and *T* > 620–650 K. This transformation was observed both in pure AA and in AA embedded in N$_2$, but the final products are different in the two cases. In pure AA, a new crystalline phase, called phase B, is formed mixed with solid N$_2$. For AA + N$_2$, we determined that the products are more likely composed of a new crystalline phase, called phase A, which is mixed with solid ammonia and the excess N_2 solid. Phases A and B could both be recovered at ambient temperature and were found stable to at least 7 GPa (A) and 15 GPa (B). The XRD patterns of these solids have been collected on varying pressure at 300 K, showing well defined crystalline peaks in the two cases. Candidate structures have been disclosed with larger unit cell volumes as compared to the AA solid. To establish the content of these solids, the Raman spectrum was measured over the 7–25 GPa range. A common feature of the two phases appears to be the loss of ammonium ions and the presence of doubly-bonded nitrogen species. In addition, phase A presents Raman bands assigned to $N{\equiv}N$ stretch and $N{=}N$ bend which are not observed in phase B. The first one are more likely the signature of N_2 molecules in the crystal structure. The Raman spectra and XRD patterns were finally compared to nitrogen and hydronitrogen compounds predicted in the literature as stable at high pressure. No correspondence was found for any of the considered compounds; however, similar vibrational features were found with the N_9H solid, composed of N_2 molecules and infinite zig-zag chains. This study thus shows that new hydronitrogen compounds may be discovered in a moderate *P-T* range. Future work will aim to establish the nature and structure of the newly produced crystalline solids.

Supplementary Materials: The following are available online at http://www.mdpi.com/1996-1944/13/18/4102/s1, Figure S1: Raman spectrum of the AA sample embedded in N_2 at 6 GPa–300 K after loading in the DAC; Figure S2: Experimental paths mapped in the nitrogen phase diagram; Figure S3: Photographs of the sample chamber and Raman spectrum of AA + N_2 along path 1; Figure S4: Raman spectrum at different positions in the sample chamber at 12.6 GPa–300 K after the decomposition of AA in liquid N_2 at 700 K; Figure S5: Raman spectrum of AA + N_2 collected along path 2; Figure S6: Raman spectrum of phase A upon heating to 650 K around 20 GPa; Figure S7: Comparison of Raman spectra in paths 2 and 3; Figure S8: Raman spectrum of phase A at different positions in the sample chamber; Figure S9: Raman spectrum of phase A upon decompression at 300 K; Figure S10: Extended Raman spectra of phase B and comparison with NH_4 and NH_5; Figure S11: XRD pattern (a) and image (b) of phase A (a); Figure S12: Measured cell parameters and volume of N_2 and NH_3 solids mixed with phase A as a function of pressure at 300 K; Figure S13: XRD pattern of the AA + N_2 sample at 20.1 GPa–300 K after the high *P-T* transformation; Figure S14: XRD pattern of the pure AA sample at 15.9 GPa–300 K after the high *P-T* transformation; Figure S15: Comparison between the experimental XRD patterns of phases A at 20.1 GPa–300 K and Phase B at 15.9 GPa–300 K, with the theoretical ones of N_4H_4, N_6, N_8, N_8H, N_9H, HN_3 and NH_4N_5 at 20 GPa–0 K.

Author Contributions: Conceptualization, F.D.; Formal analysis, G.Z., S.N. and F.D.; Funding acquisition, S.N., C.L., C.G. and F.D.; Investigation, G.Z., H.Z. (Haiwa Zhang), S.N., K.B., M.M. and F.D.; Methodology, S.N. and F.D.; Project administration, F.D.; Resources, H.Z.(Hongyang Zhu); Supervision, C.G. and F.D.; Writing—original draft, G.Z. and F.D.; Writing—review & editing, G.Z., H.Z. (Haiwa Zhang), S.N., H.Z. (Hongyang Zhu), C.L., M.M., C.G. and F.D. All authors have read and agreed to the published version of the manuscript.

Funding: This work was supported by the Chinese Scholarship Council through the allocation of a scholarship to G.Z., the National Natural Science Foundation of China (Grant Nos. 11674404, 11874174, and 11774128), the National Science Foundation of Shandong Province (ZR2018JL003), the Introduction and Cultivation Plan of Youth Innovation Talents for Universities of Shandong Province, and the French Agence Nationale de la Recherche under grant ANR-15-CE30-0008-01 (SUPERICES).

Acknowledgments: We gratefully acknowledge the European Synchrotron Radiation Facility for the allocation of beamtime, and the spectroscopy platform of the IMPMC for access to the Raman spectroscopy equipment.

Conflicts of Interest: The authors declare no conflict of interest.

Abbreviations

The following abbreviations are used in this manuscript:

P	Pressure
T	Temperature
HEDM	High energy-density material
AA	Ammonium azide
FWHM	Full width at half maximum
DFT	Density functional theory
DFPT	Density functional perturbation theory

References

1. Eremets, M.; Gavriliuk, A.; Trojan, I.; Dzivenko, D.; Boehler, R. Single-Bonded Cubic Form of Nitrogen. *Nat. Mater.* **2004**, *3*, 558. [CrossRef] [PubMed]
2. Lipp, M.; Klepeis, J.; Baer, B.; Cynn, H.; Evans, W.; Iota, V.; Yoo, C. Transformation of Molecular Nitrogen to Nonmolecular Phases at Megabar Pressures by Direct Laser Heating. *Phys. Rev. B* **2007**, *76*, 14113. [CrossRef]
3. Goncharov, A.; Holtgrewe, N.; Qian, G.; Hu, C.; Oganov, A.; Somayazulu, M.; Stavrou, E.; Pickard, C.; Berlie, A.; Yen, F. Backbone NxH Compounds at High Pressures. *J. Chem. Phys.* **2015**, *142*, 214308. [CrossRef] [PubMed]
4. Qian, G.R.; Niu, H.; Hu, C.H.; Oganov, A.; Zeng, Q.; Zhou, H.Y. Diverse Chemistry of Stable Hydronitrogens, and Implications for Planetary and Materials Sciences. *Sci. Rep.* **2016**, *6*, 25947. [CrossRef]
5. Batyrev, I.G. Modeling of Extended N-H Solids at High Pressures. *J. Phys. Chem. A* **2017**, *121*, 638–647. [CrossRef]
6. Wang, H.; Eremets, M.I.; Troyan, I.; Liu, H.; Ma, Y.; Vereecken, L. Nitrogen Backbone Oligomers. *Sci. Rep.* **2015**, *5*, 13239. [CrossRef]
7. Yu, H.; Duan, D.; Tian, F.; Liu, H.; Li, D.; Huang, X.; Liu, Y.; Liu, B.; Cui, T. Polymerization of Nitrogen in Ammonium Azide at High Pressures. *J. Phys. Chem. C* **2015**, *119*, 25268–25272. [CrossRef]
8. Spaulding, D.K.; Weck, G.; Loubeyre, P.; Datchi, F.; Dumas, P.; Hanfland, M. Pressure-Induced Chemistry in a Nitrogen-Hydrogen Host-Guest Structure. *Nat. Commun.* **2014**, *5*, 5739. [CrossRef]
9. Laniel, D.; Svitlyk, V.; Weck, G.; Loubeyre, P. Pressure-induced chemical reactions in the $N_2(H_2)_2$ compound: From the N_2 and H_2 species to ammonia and back down into hydrazine. *Phys. Chem. Chem. Phys.* **2018**, *20*, 4050–4057. [CrossRef]
10. Turnbull, R.; Donnelly, M.E.; Wang, M.; Peña-Alvarez, M.; Ji, C.; Dalladay-Simpson, P.; kwang Mao, H.; Gregoryanz, E.; Howie, R.T. Reactivity of Hydrogen-Helium and Hydrogen-Nitrogen Mixtures at High Pressures. *Phys. Rev. Lett.* **2018**, *121*. [CrossRef] [PubMed]
11. Hu, A.; Zhang, F. A Hydronitrogen Solid: High Pressure Ab Initio Evolutionary Structure Searches. *J. Phys. Condens. Matter* **2010**, *23*, 22203. [CrossRef] [PubMed]
12. Crowhurst, J.C.; Zaug, J.M.; Radousky, H.B.; Steele, B.A.; Landerville, A.C.; Oleynik, I.I. Ammonium Azide under High Pressure: A Combined Theoretical and Experimental Study. *J. Phys. Chem. A* **2014**, *118*, 8695–8700. [CrossRef] [PubMed]
13. Yedukondalu, N.; Vaitheeswaran, G.; Anees, P.; Valsakumar, M.C. Phase stability and lattice dynamics of ammonium azide under hydrostatic compression. *Phys. Chem. Chem. Phys.* **2015**, *17*, 29210–29225. [CrossRef]
14. Zhang, G.; Zhang, H.; Ninet, S.; Zhu, H.; Liu, C.; Itié, J.P.; Gao, C.; Datchi, F. Crystal Structure and Stability of Ammonium Azide Under High Pressure. *J. Phys. Chem. C* **2020**, *124*, 135–142. [CrossRef]
15. Steele, B.A.; Oleynik, I.I. Pentazole and Ammonium Pentazolate: Crystalline Hydro-Nitrogens at High Pressure. *J. Phys. Chem. A* **2017**, *121*, 1808–1813. [CrossRef]
16. Steele, B.A.; Stavrou, E.; Crowhurst, J.C.; Zaug, J.M.; Prakapenka, V.B.; Oleynik, I.I. High-Pressure Synthesis of a Pentazolate Salt. *Chem. Mater.* **2017**, *29*, 735–741. [CrossRef]
17. Laniel, D.; Weck, G.; Gaiffe, G.; Garbarino, G.; Loubeyre, P. High-Pressure Synthesized Lithium Pentazolate Compound Metastable under Ambient Conditions. *J. Phys. Chem. Lett.* **2018**, *9*, 1600–1604. [CrossRef]
18. Frierson, W.J.; Browne, A.W. Preparation of Ammonium Trinitride from Dry Mixtures of Sodium Trinitride and an Ammonium Salt. *J. Am. Chem. Soc.* **1934**, *56*, 2384. [CrossRef]
19. Letoullec, R.; Pinceaux, J.P.; Loubeyre, P. The membrane diamond anvil cell: A new device for generating continuous pressure and temperature variations. *Int. J. High Press. Res.* **1988**, *1*, 77–90. [CrossRef]
20. Dewaele, A.; Torrent, M.; Loubeyre, P.; Mezouar, M. Compression curves of transition metals in the Mbar range: Experiments and projector augmented-wave calculations. *Phys. Rev. B* **2008**, *78*, 104102. [CrossRef]
21. Datchi, F.; Dewaele, A.; Loubeyre, P.; Letoullec, R.; Godec, Y.L.; Canny, B. Optical pressure sensors for high-pressure–high-temperature studies in a diamond anvil cell. *High Press. Res.* **2007**, *27*, 447–463. [CrossRef]
22. Takemura, K.; Dewaele, A. Isothermal equation of state for gold with a He-pressure medium. *Phys. Rev. B* **2008**, *78*, 104119. [CrossRef]
23. Olijnyk, H. High pressure X-ray diffraction studies on solid N_2 up to 43.9 GPa. *J. Chem. Phys.* **1990**, *93*, 8968–8972. [CrossRef]

Materials **2020**, *13*, 4102

24. Rodriguez-Carvajal, J. Recent advances in magnetic structure determination by neutron powder diffraction. *Phys. B* **1993**, *192*, 55–69. [CrossRef]

25. Baroni, S.; De Gironcoli, S.; Dal Corso, A.; Giannozzi, P. Phonons and Related Crystal Properties from Density-Functional Perturbation Theory. *Rev. Mod. Phys* **2001**, *73*, 515. [CrossRef]

26. Refson, K.; Tulip, P.; Clark, S. Variational Density-Functional Perturbation Theory for Dielectrics and Lattice Dynamics. *Phys. Rev. B* **2006**, *73*, 155114. [CrossRef]

27. Segall, M.; Lindan, P.; Probert, M.; Pickard, C.; Hasnip, P.; Clark, S.; Payne, M. First-Principles Simulation: Ideas, Illustrations and the CASTEP Code. *J. Phys. Condens. Matter* **2002**, *14*, 2717. [CrossRef]

28. Perdew, J.; Burke, K.; Ernzerhof, M. Generalized Gradient Approximation Made Simple. *Phys. Rev. Lett.* **1996**, *77*, 3865. [CrossRef]

29. Tkatchenko, A.; Scheffler, M. Accurate Molecular van der Waals Interactions from Ground-State Electron Density and Free-Atom Reference Data. *Phys. Rev. Lett.* **2009**, *102*, 73005. [CrossRef]

30. Ninet, S.; Datchi, F. High pressure–high temperature phase diagram of ammonia. *J. Chem. Phys.* **2008**, *128*, 154508. [CrossRef]

31. Queyroux, J.A.; Ninet, S.; Weck, G.; Garbarino, G.; Plisson, T.; Mezouar, M.; Datchi, F. Melting curve and chemical stability of ammonia at high pressure: Combined X-ray diffraction and Raman study. *Phys. Rev. B* **2019**, *99*. [CrossRef]

32. Olijnyk, H.; Jephcoat, A.P. Vibrational Dynamics of Isotopically Dilute Nitrogen to 104 GPa. *Phys. Rev. Lett.* **1999**, *83*, 332. [CrossRef]

33. Ninet, S. Propriétés Structurales et Vibrationnelles de la Glace D'ammoniac Sous Pression. Ph.D. Thesis, University of Toulouse, Paris, France, 2006.

34. Ninet, S.; Datchi, F.; Saitta, A.M.; Lazzeri, M.; Canny, B. Raman spectrum of ammonia IV. *Phys. Rev. B* **2006**, *74*, 104101. [CrossRef]

35. Jiang, S.; Huang, X.; Duan, D.; Zheng, S.; Li, F.; Yang, X.; Zhou, Q.; Liu, B.; Cui, T. Hydrogen Bond in Compressed Solid Hydrazine. *J. Phys. Chem. C* **2014**, *118*, 3236–3243. [CrossRef]

36. Datchi, F.; Ninet, S.; Gauthier, M.; Saitta, A.M.; Canny, B.; Decremps, F. Solid ammonia at high pressure: A single-crystal X-ray diffraction study to123GPa. *Phys. Rev. B* **2006**, *73*, 174111. [CrossRef]

37. Boultif, A.; Louër, D. Powder pattern indexing with the dichotomy method. *J. Appl. Crystallogr.* **2004**, *37*, 724–731. [CrossRef]

38. Greschner, M.J.; Zhang, M.; Majumdar, A.; Liu, H.; Peng, F.; Tse, J.S.; Yao, Y. A New Allotrope of Nitrogen as High-Energy Density Material. *J. Phys. Chem. A* **2016**, *120*, 2920–2925. [CrossRef]

39. Hirshberg, B.; Gerber, R.B.; Krylov, A.I. Calculations predict a stable molecular crystal of N8. *Nat. Chem.* **2013**, *6*, 52–56. [CrossRef]

40. Evers, J.; Göbel, M.; Krumm, B.; Martin, F.; Medvedyev, S.; Oehlinger, G.; Steemann, F.X.; Troyan, I.; Klapötke, T.M.; Eremets, M.I. Molecular Structure of Hydrazoic Acid with Hydrogen-Bonded Tetramers in Nearly Planar Layers. *J. Am. Chem. Soc.* **2011**, *133*, 12100–12105. [CrossRef]

41. Duwal, S.; Ryu, Y.J.; Kim, M.; Yoo, C.S.; Bang, S.; Kim, K.; Hur, N.H. Transformation of hydrazinium azide to molecular N_8 at 40 GPa. *J. Chem. Phys.* **2018**, *148*, 134310. [CrossRef]

Article

Comparative Compressibility of Smectite Group under Anhydrous and Hydrous Environments

Yongmoon Lee [1], Pyosang Kim [2], Hyeonsu Kim [2] and Donghoon Seoung [2,*]

1   Department of Geological Sciences, Pusan National University, Busan 46241, Korea; lym1229@pusan.ac.kr
2   Department of Earth Systems and Environmental Sciences, Chonnam National University,
     Gwangju 61186, Korea; 197944@jnu.ac.kr (P.K.); 197942@jnu.ac.kr (H.K.)
*   Correspondence: dseoung@jnu.ac.kr; Tel.: +82-62-530-3452

Received: 19 July 2020; Accepted: 24 August 2020; Published: 27 August 2020

Abstract: High-pressure synchrotron X-ray powder diffraction studies of smectite group minerals (beidellite, montmorillonite, and nontronite) reveal comparative volumetric changes in the presence of different fluids, as pressure transmitting media (PTM) of silicone oil and distilled water for anhydrous and hydrous environments at room temperature. Using silicone oil PTM, all minerals show gradual contraction of unit-cell volumes and atomistic interplane distances. They, however, show abrupt collapse near 1.0 GPa under distilled water conditions due to hydrostatic to quasi-hydrostatic environmental changes of water PTM around samples concomitant with the transition from liquid to ICE-VI and ICE-VII. The degrees of volume contractions of beidellite, montmorillonite, and nontronite up to ca. 3 GPa are ca. 6.6%, 8.9%, and 7.5% with bulk moduli of ca. 38(1) GPa, 31(2) GPa, and 26(1) GPa under silicone oil pressure, whereas 13(1) GPa, 13(2) GPa, and 17(2) GPa, and 17(1) GPa, 20(1) GPa, and 21(1) GPa under hydrostatic and quasi-hydrostatic environments before and after 1.50 GPa, respectively.

Keywords: high-pressure; smectite; bulk moduli; anhydrous and hydrous environments; synchrotron X-ray powder diffraction; pressure-transmitting media

---

1. Introduction

Smectite is the most abundant two-dimensional phyllosilicate mineral group and comprises the most common natural materials, which occur by weathering, diagenesis, and hydrothermal alteration [1,2]. It is an important material for applications such as engineered barrier systems for nuclear waste storage and industrial catalysts [3,4]. Fundamental studies on the elastic properties of the smectite group have provided thermodynamic parameters that relate to the stability of materials [1,2].

Representative minerals in the smectite group are beidellite, montmorillonite, and nontronite stratified by 2:1 (T–O–T) layered framework and counterbalance cation surrounded by water molecules at expandable interlayers [5–7]. Specifically, the montmorillonite shows predominant changes of cation charge in the octahedral layer (mostly Si^{4+} at tetrahedral site; Al^{3+} and Mg^{2+} at octahedral sites), whereas beidellite (Al^{3+} in octahedral site) and nontronite (Fe^{3+} for octahedral site) show changes in the tetrahedral layer (Al^{3+} and Si^{4+} at tetrahedral sites) (Figure 1).

One of the most interesting and distinctive properties of smectite is swelling concomitant with interlayer hydration due to the influx of surrounding water molecules [8,9]. Recent studies of phyllosilicate clays under different pressure conditions reveal that the kaolinite family ($Al_2Si_2O_5(OH)_4$, kaolinite, and nacrite), stratified by 1:1 (T–O) layered framework, shows swelling and superhydration of the mono-water layer under water saturated environments at 5.75(1) GPa and 460(5) °C and, subsequently, transitions to a high-pressure phase (coesite, diaspore, and topaz–OH) under consecutive pressure and temperature conditions [10,11]. A synthetic Na–hectorite ($Na_{0.3}(Mg)_2(Si_4O_{10})(F)_2$-$xH_2O$),

2:1 (T–O–T) layered framework, shows anomalous swelling due to pressure-induced hydration at 2.2(1) GPa concomitant with bi- to tri-water layers increment [12].

In topologically anisotropic phases such as 2-dimensional phyllosilicates, including smectite clays, pressure-dependent unit-cell axes and volume changes are related to key structural directions defined by silicate framework sheets and interlayers. The comparative high-pressure study of smectites under anhydrous and hydrous environments is therefore important to provide compressibility that can be used with thermodynamic parameters to calculate key reactions for understanding their stabilities.

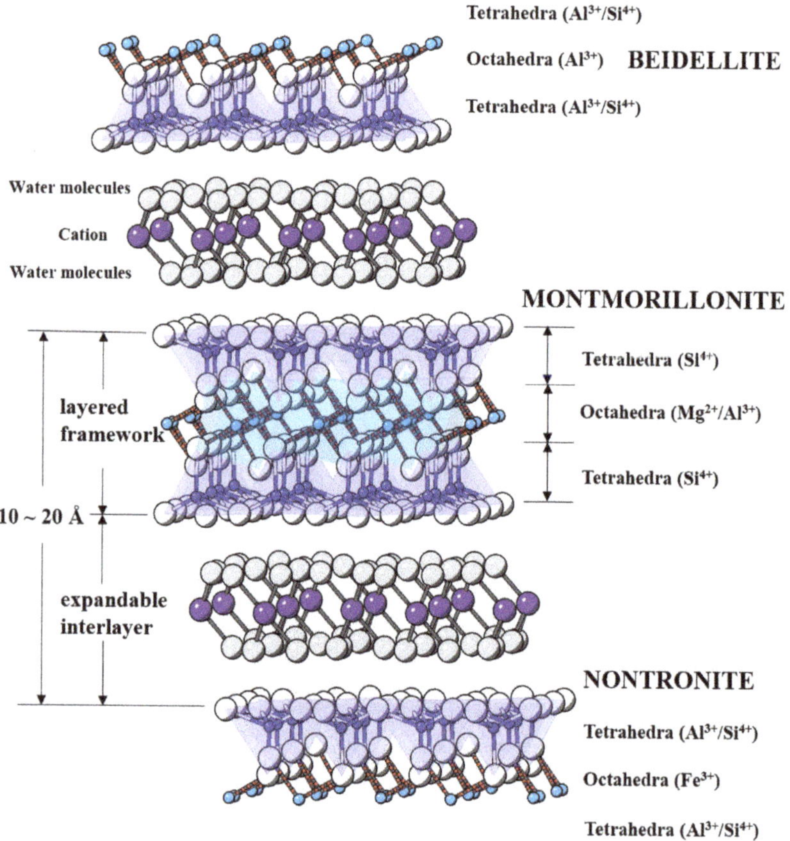

Figure 1. Schematic illustration of smectite structures represented by beidellite, montmorillonite, and nontronite. Blue and cyan colored circles represent the central cation in tetrahedra and octahedra comprising layered framework and white circles represent framework oxygen atoms, respectively. Purple and gray colored circles illustrate counterbalance interlayer cations and water molecules, respectively.

In this work, we studied the comparative compressibility of beidellite, montmorillonite, and nontronite through high-pressure synchrotron X-ray powder diffraction experiments from ambient condition up to nearly 4 GPa using diamond anvil cell (DAC) with two different pressure transmitting media (PTM) of silicone oil and distilled water for anhydrous and hydrous environments, respectively.

2. Materials and Methods

Natural smectite samples (beidellite from ID, USA., SBId-1, montmorillonite (Ca-rich) from Apache County, AZ, USA., SAz-2 from the Source Clays of the Clay Mineral Society, The Clay Mineral Society, Chantilly, VA, USA and nontronite from Uley Graphite Mine, Australian Graphite Pty Ltd, Port Lincoln, Australia, NAu-1 [13]) were ground and loaded inside a sample chamber with 200 μm diameter and 100 μm thickness at the center of a stainless steel gasket between two opposed diamond culets with 500 μm diameters.

In-situ high-pressure synchrotron X-ray powder diffraction on smectite at room temperature (25 °C) was performed at beamline 3D and 5A at Pohang Light Source II (PLS-II) at Pohang Accelerator Laboratory (PAL) in Korea. At both beamlines, the primary white beam from the bending magnet was monochromatized using Si(111) crystal DCM (double crystal monochromator), and a pin-hole was used to create an approximately 100 μm beam of monochromatic X-rays with a wavelength of 0.6888 Å. MAR345 IP detectors were used at both beamlines and the wavelength of the incident beam was calibrated using a LaB6 standard (SRM660c).

A modified 4-pin type diamond anvil cell (DAC) was used for the high-pressure experiments, equipped with two type-I diamond anvils (500 μm culet diameter) and tungsten carbide supports [14]. A stainless steel foil of 200 μm thickness was pre-indented to a thickness of about 100 μm and a 200 μm hole for the sample chamber was drilled by electro-discharge machine erosion. The powder sample of smectite (beidellite, montmorillonite, and nontronite) was placed in the sample chamber hole on a gasket with several ruby chips for in-situ pressure measurements.

Ambient pressure data of the smectite group (beidellite, montmorillonite, and nontronite) were collected on the dry powder sample for nonswelling conditions and the wet powder samples, which are saturated by water molecules for swelling environment, respectively. Subsequently, the silicone oil and distilled water were added to the sample chamber as a pressure-transmitting media, and then, the DAC was sealed to the first pressure point. The sample pressure is measured by detecting the shift in the R1 emission line of included ruby chips [15]. The samples were typically equilibrated for about 10 min. in the DAC at each measured pressure points. The pressure was increased in steps of 0.3–0.6 GPa up to 3–4 GPa. The FIT2D program (16-041) suite was used to integrate the diffraction images into diffraction patterns.

The pressure-dependent changes in the unit-cell lengths and volumes were derived from a series of whole-profile fitting procedures using the LeBail Method implemented EXPGUI program (1251) suite [16,17]. The background was fitted with a Chebyshev polynomial with 20 coefficients, and the pseudo-Voigt profile function proposed by Thompson et al. was used to model the observed Bragg reflections [18]. The derived bulk moduli from normalized volume (V/V_0) were calculated using the third order Birch–Murnaghan equation of state, Equation (1) [19]. The value of B' for bulk moduli calculation is fixed as "4" for natural materials approximation.

$$P = \frac{3}{2}B_0\left[\left(\frac{V_0}{V}\right)^{-7/3} - \left(\frac{V_0}{V}\right)^{-5/3}\left\{1 + \frac{3}{4}(B'_0 - 4)\left[\left(\frac{V_0}{V}\right)^{-2/3} - 1\right]\right\}\right] \tag{1}$$

3. Results and Discussion

Pressure-dependent changes in the measured in-situ synchrotron X-ray powder diffraction patterns of smectites (beidellite, montmorillonite, and nontronite) under anhydrous (silicone oil) and hydrous (distilled water) conditions are shown in Figures 2 and 3 and Tables S1–S3. At ambient pressure, smectites in a water-saturated environment showed noticeable extension of $d(001)$ up to ca. 19 Å, which was interpreted as interlayer expansion due to formation of tri-water layers (additional water layer intercalation), whereas $d(001)$ of all samples in a silicone oil environment was constant at ca. 15 Å, which is similar length of bi-water layers (Figure 4c,d) [20].

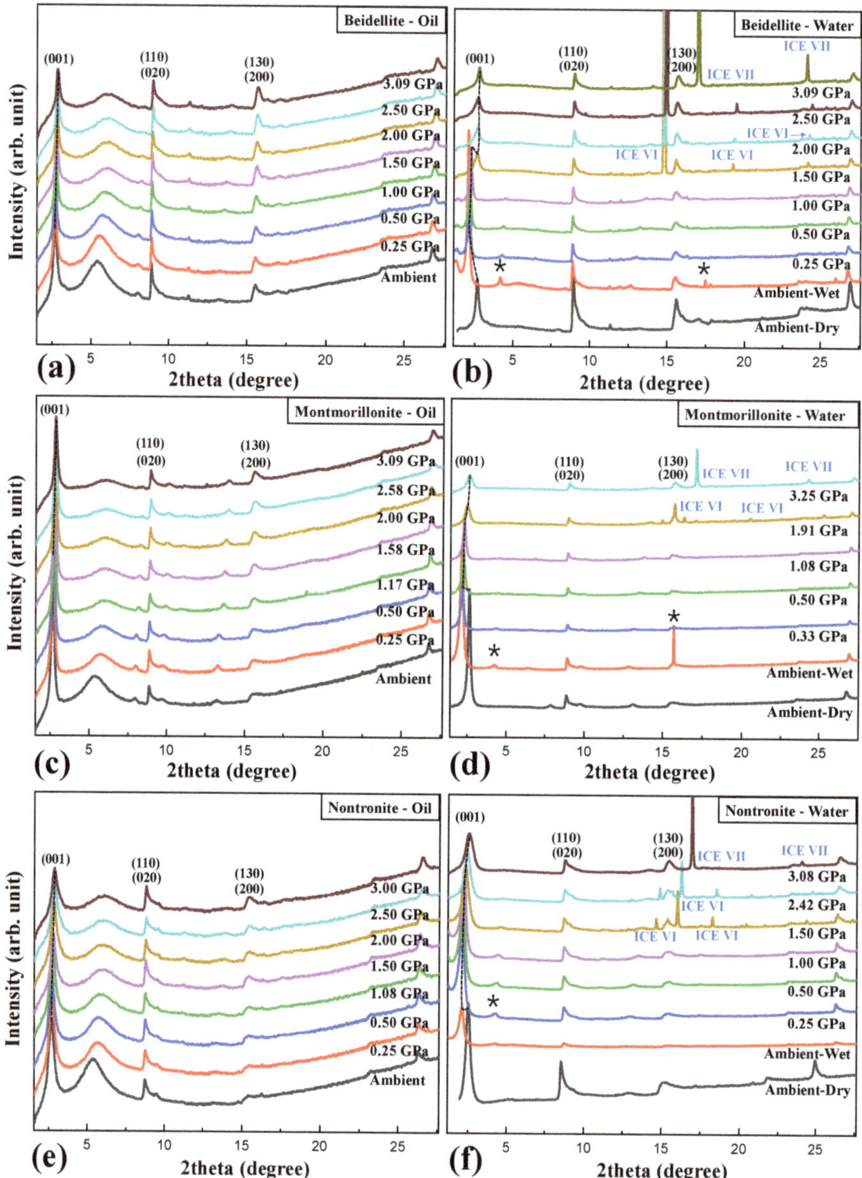

Figure 2. In-situ high-pressure synchrotron X-ray powder diffraction patterns of beidellite, montmorillonite, and nontronite in silicone oil and distilled water pressure transmitting media (PTM) for anhydrous and hydrous environments at room temperature. Each stacked pattern represents (**a**) beidellite oil, (**b**) beidellite water, (**c**) montmorillonite oil, (**d**) montmorillonite water, (**e**) nontronite oil, and (**f**) nontronite water, respectively. Selected Miller indices are shown. The reflections of impurities are marked as asterisks. The humps near 5.5 to 6 degrees in 2-theta in silicone oil data indicate short-range ordering of silicon oil PTM.

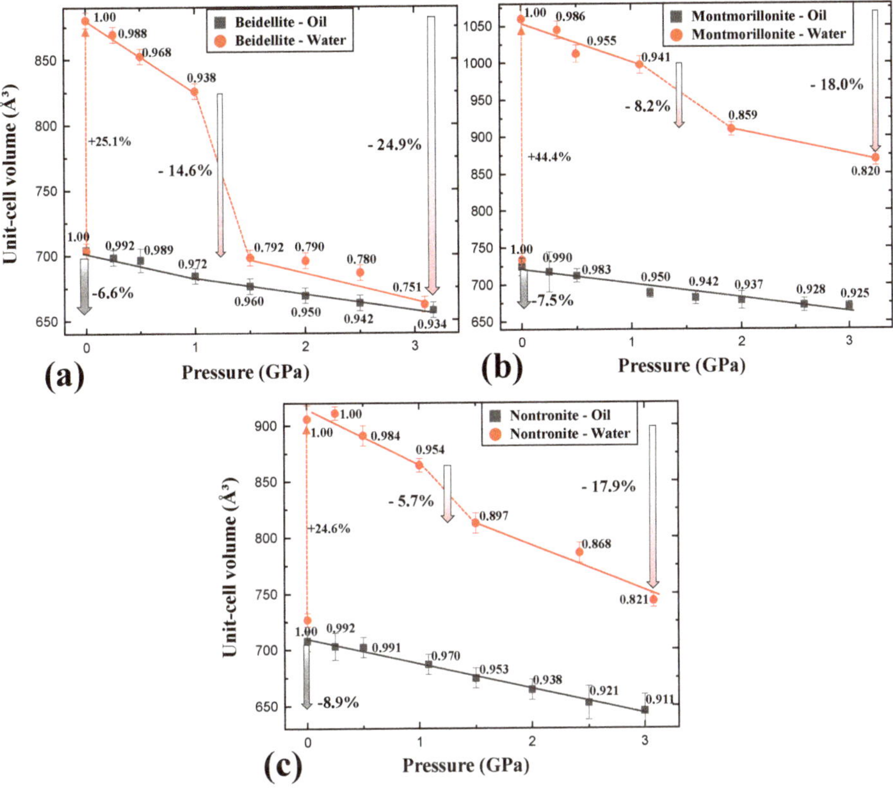

Figure 3. Pressure-dependent unit-cell volume changes of (**a**) beidellite, (**b**) montmorillonite, and (**c**) nontronite in silicone oil and distilled water PTMs for anhydrous and hydrous environments, respectively, at room temperature. Normalized volumes are shown on data points. The dashed and bold arrows show percent changes.

In the case of beidellite ($Na_{0.3}Al_2(Si, Al)_4O_{10}(OH)_2 \cdot nH_2O$) in the presence of silicone oil, all diffraction peaks gradually shifted to higher 2-theta showing normal compression behavior up to 3.09 GPa concomitant with contraction of unit-cell volume by 6.6% Figures 2a and 3a), whereas (001) reflection shows an anomalous shift at 1.50 GPa in distilled water PTM (Figure 2b). At this pressure, the unit-cell volume of beidellite abruptly contracted ca. 14.6% and showed discontinuity of volume contraction behavior due to the changing environment around the sample from a hydrostatic to a quasi-hydrostatic condition by the transition of water PTM to ICE-VI (Figure 3a). Subsequently, the (001) reflection contracted gradually up to the final pressure of ca. 3 GPa with a volume contraction of ca. 24.9% Figures 2b and 3a). Upon the change to quasi-hydrostatic condition, the FWHM (full with at half maximum) of (001) reflections also abruptly increased up to ca. 255% before and after 1.50 GPa accompanying consecutive peak shifting to a higher 2-theta angle, whereas the FWHM of (001) reflections, in case of silicone oil runs, showed gradual increments within 20% (Figure 5a,b) [21,22]. Bulk modulus, B_0 (GPa), of beidellite was calculated to be 38(1) GPa under a silicone oil environment, 13(1) GPa under hydrostatic water pressure, and 17(1) GPa under quasi-hydrostatic water pressure, respectively (Figure 4a,b and Table 1).

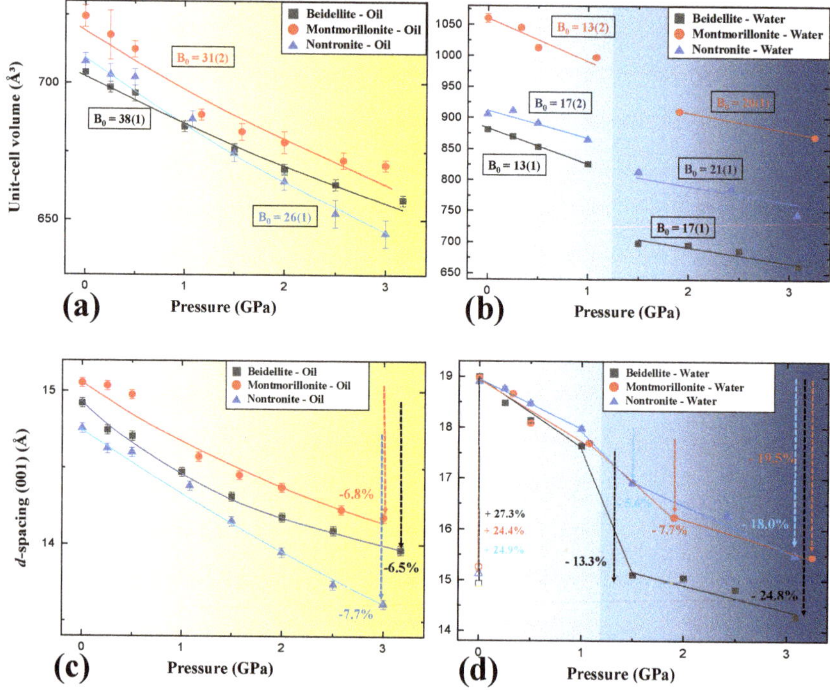

Figure 4. Comparative bulk-moduli of smectite under (**a**) silicone oil and (**b**) distilled water PTMs conditions at room temperature. Pressure-dependent interplane (001) distance changes of smectite under (**c**) silicone oil and (**d**) distilled water PTM conditions. The dashed arrows show percent changes.

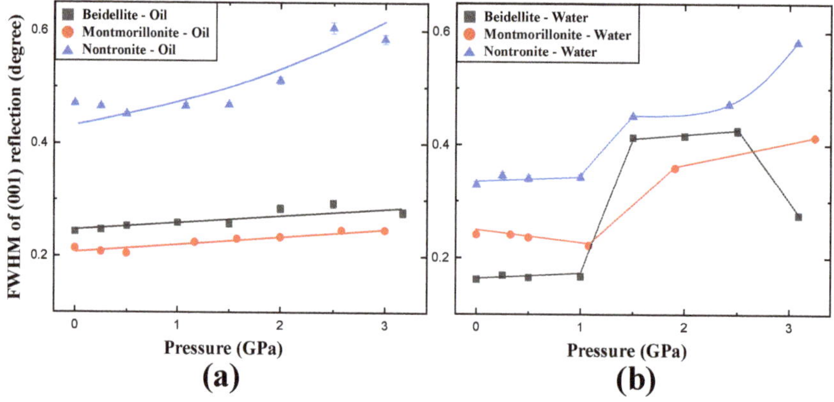

Figure 5. Pressure-dependent FWHM changes of the interplane (001) Bragg reflections of beidellite, montmorillonite, and nontronite under (**a**) anhydrous and (**b**) hydrous environments, respectively.

Unit-cell volumes of montmorillonite ($Ca_{0.3}(Al, Mg)_2(Si_4O_{10})(OH)_2 \cdot nH_2O$) and nontronite ($Ca_{0.5}Fe_4(Si, Al)_4O_{10}(OH)_2 \cdot nH_2O$) in silicone oil PTM gradually decreased with contraction of 8.9% and 7.5% up to 3.0 GPa, respectively (Figure 3b,c). Similarly to beidellite, which showed an anomalous peak shift with drastic volume contraction of 14.6% at 1.50 GPa under water saturated PTM condition, montmorillonite and nontronite showed discontinuous behaviors with 5.7% and 8.2% of drastic volume contraction near 1.5 GPa, where the hydrostatic environment started to break down and concomitant with existence of ICE-VI and VII in diffraction patterns (Figures 2 and 3). Bulk moduli, B_0 (GPa) of montmorillonite and nontronite was calculated to 31(2) GPa, and 26(1) GPa under silicone oil environment, and 13(2) GPa and 17(2) GPa under hydrostatic water pressure, and 20(1) GPa and 21(1) GPa under quasi-hydrostatic pressure, respectively (Figure 4a,b and Table 1).

Overall, comparative compressional behaviors under two different PTM conditions suggest that beidellite, specifically, showed relative higher incompressibility than others under silicone oil PTM conditions, whereas lower under water PTM conditions (Table 1). Under the water PTM conditions, smectite group minerals showed compressional discontinuities accompanying the "hydrostatic to quasi-hydrostatic" environmental change around the sample by the transition of water PTM to ICE-VI near 1.5 GPa.

Table 1. Bulk moduli of beidellite, montmorillonite, and nontronite under silicone oil and distilled water PTMs (ESD's (estimated standard deviations) are in parentheses).

Bulk Modulus, B_0 (GPa)		Beidellite	Montmorillonite	Nontronite
Silicone-Oil		38(1)	31(2)	26(1)
distilled water	(hydrostatic)	13(1)	13(2)	17(2)
	(quasi-hydrostatic)	17(1)	20(1)	21(1)

We, therefore, undertook the calculation of the atomistic interplane distance of (001) plane to understand the relationship between interlayer changes of smectite group minerals and different fluid conditions as a function of pressure (Figure 4c,d). In silicone oil PTMs, the d(001) of beidellite, montmorillonite, and nontronite up to ca. 3 GPa gradually reduced ca. 6.5%, 6.8%, and 7.7% accompanying volume contraction of ca. 6.6%, 8.9%, and 7.5%, respectively. Montmorillonite, specifically, showed ca. 2.1% greater volume contraction behavior than d(001) contraction. Under the water PTM conditions, smectite minerals showed similar d(001) distances near 19 Å in wet conditions, showing interlayer expansion due to additional intercalation of water layer inside interlayer from water-saturated environments, whereas they showed about 22% differences in maximal unit-cell volumes, respectively (Figure 4a,d). Before 1.0 GPa, the interlayer of three smectites showed similar compressional behaviors by pressure. In consecutive increasing pressures over 1.50 GPa, however, beidellite, montmorillonite, and nontronite showed ca. 13.3%, 7.7%, and 5.6% of the abrupt collapse of interlayer distances, whereas they showed 14.6%, 5.7%, and 8.2% of unit-cell volume contraction.

4. Conclusions

This study establishes that beidellite, montmorillonite, and nontronite show different behaviors in anhydrous and hydrous PTM environments. All samples show modulated volume contraction by pressure and with bulk moduli of ca. 38(1) GPa, 31(2) GPa, and 26(1) GPa for beidellite, montmorillonite, and nontronite, respectively. When samples are exposed to water-saturated conditions, volume expansions are observed accompanying d(001) swelling behaviors up to 19 Å. In water PTM, they show bulk moduli of 13(1) GPa, 13(2) GPa, and 17(2) GPa for beidellite, montmorillonite, and nontronite, respectively, before 1.0 GPa. After 1.50 GPa, hydrous environments change to quasi-hydrostatic conditions due to the transition of liquid water PTM to ICE-VI and ICE-VII. We then observed further enhancement of bulk moduli for beidellite, montmorillonite, and nontronite to be 17(1) GPa, 20(1) GPa, and 21(1) GPa, respectively. From the results in change of bulk moduli before and after 1.50 GPa, we expect that water layers can interact with water molecules of PTM under the hydrostatic pressure conditions. After the change to quasi-hydrostatic conditions, however, the interaction might be prohibited due to the liquid-to-solid transition of water PTM and, therefore, they show normal volume contractions after 1.50 GPa. High-pressure spectroscopic investigations are underway to understand the detailed relationship between anomalous changes in bulk moduli and water interactions under hydrous and anhydrous environments.

Supplementary Materials: The following are available online at http://www.mdpi.com/1996-1944/13/17/3784/s1. Figure S1: Pressure-dependent changes of the interplane (001) distances of beidellite, montmorillonite, and nontronite in present of silicone-oil and distilled water PTMs. Table S1: Final refined unit-cell parameters, volumes, *d-spacing* of (001) plane, and FWHM of (001) reflections of beidellite under pressure conditions, at room temperature. Table S2: Final refined unit-cell parameters, volumes, *d-spacing* of (001) plane, and FWHM of (001) reflections of montmorillonite under pressure conditions, at room temperature. Table S3: Final refined unit-cell parameters, volumes, *d-spacing* of (001) plane, and FWHM of (001) reflections of nontronite under pressure conditions, at room temperature.

Author Contributions: Conceptualization, D.S.; methodology, P.K. and H.K.; formal analysis, P.K. and H.K.; data curation, P.K. and H.K.; writing—original draft preparation, D.S. and Y.L.; writing—review and editing, D.S. and Y.L.; visualization, D.S.; supervision, D.S.; project administration, D.S.; funding acquisition, D.S. and Y.L. All authors have read and agreed to the published version of the manuscript.

Funding: This research was funded by National Research Foundation of the Ministry of Science and ICT of Korean Government, grant number NRF-2019R1F1A106258, NRF-2017R1D1A1B03035418, NRF-2020R1C1C1013642, NRF-2019K1A3A7A09101574 and Chonnam National University Research Grant, 2019-0215 and Pusan National University Research Grant, 201902840001.

Acknowledgments: Experiments using synchrotron radiation were supported by Pohang Light Source II (PLS-II) at Pohang Accelerator Laboratory (PAL). We thank Taeyeol Jeon and Hyun-Hwi Lee for the support at beamline 3D and 5A.

Conflicts of Interest: The authors declare no conflict of interest.

References

1. Meunier, A. *Clays*; Springer: New York, NY, USA, 2005.
2. Bergaya, F.; Theng, B.K.G.; Lagaly, G. *Handbook of Clay Science*; Elsevier: Oxford, UK, 2005.
3. Ewing, R.C. Long-term storage of spent nuclear fuel. *Nat. Mater.* **2015**, *14*, 252–257. [CrossRef] [PubMed]
4. Hansen, E.L.; Hemmen, H.; Fonseca, D.M.; Coutant, C.; Knudsen, K.D.; Plivelic, T.S.; Bonn, D.; Fossum, J.O. Swelling transition of a clay induced by heating. *Sci. Rep.-UK* **2012**, *2*, 618. [CrossRef] [PubMed]
5. Bailey, S.W. Classification and structures of the micas. *Rev. Mineral. Geochem.* **1984**, *13*, 1–12.
6. Severmann, S.; Mills, R.A.; Palmer, M.R.; Fallick, A.E. The origin of clay minerals in active and relict hydrothermal deposit. *Geochim. Cosmochim. Acta* **2004**, *68*, 73–88. [CrossRef]
7. Güven, N. Bentonites—Clays for Molecular Enginerring. *Elements* **2009**, *5*, 89–92. [CrossRef]
8. Huggett, J.M. Clay Minerals. In *Reference Module in Earth System and Environmental Sciences*; Elsevier: Oxford, UK, 2015. [CrossRef]
9. Schulze, D.G. Clay Minerals. In *Encyclopedia of Soils in the Environment*; Elsevier: Oxford, UK, 2005; pp. 246–254.
10. Hwang, H.; Seoung, D.; Lee, Y.; Liu, Z.; Liemann, H.P.; Cynn, H.; Vogt, T.; Kao, C.C.; Mao, H.K. A role for subducted super-hydrated kaolinite in Earth's deep water cycle. *Nat. Geosci.* **2017**, *10*, 947–953. [CrossRef]
11. Hwang, H.; Choi, J.; Liu, Z.; Kim, D.-Y.; He, Y.; Celestian, A.J.; Vogt, T.; Lee, Y. Pressure induced hydration and formation of bilayer Ice in nacrite, a kaolin-group clay. *ACS Earth Space Chem.* **2019**, *4*, 183–188. [CrossRef]
12. You, S.; Kunz, D.; Stöter, M.; Kalo, H.; Putz, B.; Breu, J.; Talyzin, A.V. Pressure-induced water insertion in synthetic clays. *Angew. Chem. Int. Ed.* **2013**, *52*, 3891–3895. [CrossRef] [PubMed]
13. Keeling, J.L.; Raven, M.D.; Gates, W.P. Geology and characterization of two hydrothermal nontronites from weathered metamorphic rocks at the Uley Graphite Mine, Souoth Australia. *Clay. Clay Miner.* **2000**, *48*, 537–548. [CrossRef]
14. Mao, H.K.; Hemley, R.J. Experimental studies of the Earth's deep interior: Accuracy and versatility of diamond-anvil cells. *Philos. Trans. R. Soc. A* **1996**, *354*, 1315–1332.
15. Mao, H.-K.; Bell, P.M. *Carnegie Institute Year Book*; Carnegie Institute: Troy, MI, USA, 1979; Volume 78, p. 663.
16. Larson, A.C.; von Dreele, R.B. *General Structure Analysis System (GSAS)*; Report LAUR: Los Alamos, NM, USA, 1994; pp. 86–748.
17. Toby, B.H. EXPGUI, a graphical user interface for GSAS. *J. Appl. Crystallogr.* **2001**, *34*, 210–213. [CrossRef]

18. Thompson, P.; Cox, D.E.; Hastings, J.B. Rietveld refinement of Debye-Scherrer synchrotron X-ray data from Al_2O_3. *J. Appl. Crystallogr.* **1987**, *20*, 79–83. [CrossRef]
19. Birch, F. Finite elastic strain of cubic crystals. *Phys. Rev.* **1947**, *71*, 809–824. [CrossRef]
20. Ferrage, E. Investigation of the interlayer organization of water and ions in smectite from the combined use of diffraction experiments and molecular simulations. A review of methodology, applications, and perspectives. *Clay. Clay Miner.* **2016**, *64*, 348–373. [CrossRef]
21. Klotz, S.; Chervin, J.C.; Munsch, P.; Marchand, G.L. Hydrostatic limits of 11 pressure transmitting media. *J. Phys. D Appl. Phys.* **2009**, *42*, 075413. [CrossRef]
22. Angel, R.J.; Bujak, M.; Zhao, J.; Gatta, G.D.; Jacobsen, S.D. Effective hydrostatic limits of pressure media for high-pressure crystallographic studies. *J. Appl. Crystallogr.* **2007**, *40*, 26–32. [CrossRef]

Article

Pressure- and Temperature-Induced Insertion of N_2, O_2 and CH_4 to Ag-Natrolite

Donghoon Seoung [1], Hyeonsu Kim [1], Pyosang Kim [1] and Yongmoon Lee [2,*]

[1] Department of Earth Systems and Environmental Sciences, Chonnam National University, Gwangju 61186, Korea; dseoung@jnu.ac.kr (D.S.); 197942@jnu.ac.kr (H.K.); 197944@jnu.ac.kr (P.K.)
[2] Department of Geological Sciences, Pusan National University, Busan 46241, Korea
* Correspondence: lym1229@pusan.ac.kr; Tel.: +82-51-510-2254

Received: 13 July 2020; Accepted: 14 September 2020; Published: 15 September 2020

Abstract: This paper aimed to investigate the structural and chemical changes of Ag-natrolite ($Ag_{16}Al_{16}Si_{24}O_{80} \cdot 16H_2O$, Ag-NAT) in the presence of different pressure transmitting mediums (PTMs), such as N_2, O_2 and CH_4, up to ~8 GPa and 250 °C using in situ synchrotron X-ray powder diffraction and Rietveld refinement. Pressure-induced insertion occurs in two stages in the case of N_2 and O_2 runs, as opposed to the CH_4 run. First changes of the unit cell volume in N_2, O_2 and CH_4 runs are observed at 0.88(5) GPa, 1.05(5) GPa and 1.84(5) GPa with increase of 5.7(1)%, 5.5(1)% and 5.7(1)%, respectively. Subsequent volume changes of Ag-natrolite in the presence of N_2 and O_2 appear at 2.15(5) GPa and 5.24(5) GPa with a volume increase of 0.8(1)% and a decrease of 3.0(1)%, respectively. The bulk moduli of the Ag-NAT change from 42(1) to 49(7), from 38(1) to 227(1) and from 49(3) to 79(2) in the case of N_2, O_2 and CH_4 runs, respectively, revealing that the Ag-NAT becomes more incompressible after each insertion of PTM molecules. The shape of the channel window of the Ag-NAT changes from elliptical to more circular after the uptake of N_2, O_2 and CH_4. Overall, the experimental results of Ag-NAT from our previous data and this work establish that the onset pressure exponentially increases with the molecular size. The unit cell volumes of the expanded (or contracted) phases of the Ag-NAT have a linear relationship and limit to maximally expand and contract upon pressure-induced insertion.

Keywords: silver-exchanged natrolite; pressure-induced insertion; synchrotron X-ray diffraction; Rietveld refinement

---

1. Introduction

Zeolites—one of the most abundant microporous materials—have been widely studied as 3D functional materials and employed as sorbents, catalysts or gas separators due to their various pore sizes and ion-exchange and polar compound adsorption properties [1–4]. Over the last 60 years, numerous experiments under ambient and applied temperatures have addressed the fundamental behaviors of zeolites and their potential applications in human life and industry. Experiments under pressure conditions, performed in the last few decades, show zeolites exhibit elastic behaviors and pressure-induced anomalous expansion in response to adopted pressure [5]. However, numerous high-pressure studies of natrolite ($Na_{16}Al_{16}Si_{24}O_{80} \cdot 16 H_2O$) have recently been carried out using pressure-induced hydration (PIH) and pressure-induced insertion (PII) accompanying abnormal volume expansion under applied pressures. A notable quantity of studies reported potential for various applications, such as in sequestration of cations and molecules (e.g., Cs^+, Sr^{2+}, Pb^{2+} and CO_2) and trapping of noble gases (e.g., Ar, Kr and Xe), resulting in the insertion of chemical species by widening the window of natrolite pores using pressure; the noble gases then remain trapped by narrowing the channel opening after pressure release [6–13]. With respect to the crystal structure, the PIH and PII are a consequence of the auxetic behavior of the natrolite framework, which is visualized using a rotating-squares model of framework topology [14]. The natrolite framework is composed of a secondary building unit of T_5O_{10} (T = Al and Si), which is 3D corner sharing [15]. This unit consists of alternatively bridged Si- and Al-tetrahedra and forms a helical and elliptical channel along the c-axis [16]. The PIH phenomenon was first discovered by Lee et al. when pressurized in water containing PTM [17,18]. The first PIH occurs around 1.0 GPa to form the paranatrolite phase, ($Na_{16}Al_{16}Si_{24}O_{80} \cdot 24\,H_2O$) accompanying ~6.7% of unit cell volume expansion and subsequent water insertion with ~3.9% volume contraction occurs at approximately 1.2 GPa to form the super-hydrated natrolite phase ($Na_{16}Al_{16}Si_{24}O_{80} \cdot 32H_2O$). Natrolite shows reversible sequential phase transitions under pressure conditions and is irreversible under simultaneous pressure and temperature conditions. For example, Cs- or Pb-containing natrolite ($Cs_{16}Al_{16}Si_{24}O_{80} \cdot 16H_2O$ and $Pb_8Al_{16}Si_{24}O_{80} \cdot 16H_2O$, respectively) become pollucite ($CsAlSi_2O_6 \cdot H_2O$) after heating to 160 °C at 2 GPa and lawsonite ($Pb_4Al_8Si_8O_{28} \cdot 4H_2O$) after heating at 200 °C and 4.5 GPa, respectively. The pollucite and lawsonite maintain ~40 wt% of the remaining Cs^+ and Pb^{2+} cations and show low leaching rates due to tight coordinate bonding with the framework after irreversible phase transition. The pressure- and temperature-driven processes make natrolite a more suitable form for the sequestration of nuclear waste and as long-term storage material under ambient conditions [11,12].

Among the various cation-exchanged natrolites, the silver-exchanged form (Ag-NAT) absorbs water and CO_2 molecules at comparatively low pressures (0.4(1) GPa and 0.8(1) GPa, respectively), whereas natural natrolite absorbs both at 1.0(1) GPa. The onset pressure of pressure-induced insertion (PII) arises from the circular geometry of the channel window, and we have suggested that one of the possible materials for CO_2 storage under crustal conditions [9]. We investigated the pressure-induced insertion (PII) of N_2, O_2 and CH_4 gases inside microchannels of the Ag-NAT in order to explore potential material for (ir)reversible gas storage by controlling the pressure and temperature. Herein, we report the structural investigation of Ag-NAT in the presence of N_2, O_2 and CH_4 as PTMs under applied pressure using a Diamond Anvil Cell (DAC).

2. Materials and Methods

2.1. Sample Preparation

The Ag-NAT was prepared as described by Lee et al. [19]. The starting material, K-NAT, was prepared using a 4 M KNO_3 (ACS reagent grade from Sigma-Aldrich, St Louis, MO, USA) solution and a ground mineral natrolite ($Na_{16}Al_{16}Si_{24}O_{80} \cdot 16H_2O$, San Juan, Argentina, OBG International) in a 100:1 weight ratio. The mixture was stirred at 80 °C in a reflux system (SciLab Korea Co., LTD, Seoul, Korea) to minimize the loss of water. After 24 h, the solid was separated from the solution by vacuum filtration (SciLab Korea Co., LTD, South). The dried powder was used for the second and third exchange cycles under the same conditions. The final product was washed with deionized water and subsequently air-dried under ambient conditions. Over 99% K-exchange was confirmed by energy-dispersive X-ray spectroscopy (SUPRA25, Zeiss, Germany) (EDS). The Ag-form was prepared with a fully saturated $AgNO_3$ solution in the same sequence as above. Stoichiometric analyses of the Ag-NAT on the products were performed using EDS and confirmed that the silver cation was fully exchanged. To determine the amount of H_2O molecules, Thermogravimetry Analysis (TGA) was performed in the heating range of 25–800 °C at a heating rate of 10 °C/min under a N_2 atmosphere. The EDS and TGA results are summarized in Table S1 and Figure S1, respectively.

2.2. Synchrotron X-ray Powder Diffraction

High-pressure synchrotron X-ray powder diffraction experiments were performed at the 3D and 5A beamlines of the Pohang Accelerator Laboratory (PAL). The primary white beam from the bending magnet at 3D or the superconducting insertion device at 5A, was directed on a Si (111) crystal and sets of parallel slits were used to create monochromatic X-rays with a wavelength of 0.6888(1) Å and 0.6927(1) Å for 3D and 5A, respectively. The diffracted beam was collected using a MAR345 image plate detector (marXperts GmbH, Norderstedt, Germany) as a diffractometer. A LaB$_6$ standard (SRM 660c, National Institute of Standards and Technology, USA) was used to calibrate several factors. The calibrated factors were 339.2250(1) mm of sample to detector distance, 1726.827(1) of the X-pixel coordinate of the direct beam, 247.0732(1) of the Y-pixel coordinate of the direct beam, −38.3913(1)° of the rotation angle of the detector and 0.2233(1)° of the tilt angle of the detector.

2.3. Diamond Anvil-Cell Preparation

A modified piston–cylinder type DAC (Beijing Scistar Technology CO. LTD., Beijing, China) was used for the high-pressure experiments, equipped with type-I diamond anvils (Almax·easyLab, Ashford, UK) (culet diameter of 700 μm) and tungsten carbide supports. A stainless-steel foil of 250 μm thickness was pre-indented to a thickness of ~100 μm and holes with 300 μm diameter were obtained by electro-spark erosion. The powdered sample of the Ag-NAT was placed in the gasket hole along with some ruby chips for in situ pressure measurements. Ambient pressure data were collected first on the dry powder sample inside the DAC. Subsequently, O$_2$ (N$_2$ or CH$_4$) was added to the sample chamber as a hydrostatic PTM in a cryogenic environment of liquid N$_2$ temperature, and then the DAC was sealed at the first pressure point. The pressure of a sample in the DAC was measured by detecting the shift in the R1 emission line of included ruby chips (precision: ±0.05 GPa). The sample was typically equilibrated for approximately 10 min in the DAC at each measured pressure. DAC was heated at 110 °C, 150 °C, 200 °C or 250 °C for 1 h in a dry oven to maintain the hydrostatic pressure around the samples. DAC was then cooled to ambient conditions for 1 h and the pressure was measured again.

2.4. Structural Analysis

Pressure- and temperature-dependent changes in the unit cell lengths and volume were derived from a series of whole profile fitting procedures using the GSAS suite of programs in [20]. The background was fixed at selected points and the pseudo-Voigt profile function proposed by Thompson et al. was used to model the observed Bragg peaks, while a March–Dollase function [21] was used to account for the preferred orientation [22]. The structural model at the selected pressure was obtained using Rietveld refinement [20,23]. To reduce the number of parameters, isotropic displacement factors were refined by grouping the framework tetrahedral atoms, framework oxygen atoms and extra-framework species, respectively. Geometric soft-restraints on the T–O (T = Si, Al) and O–O bond distances of the tetrahedra were applied: the distances between Si–O and Al–O were restrained to target values of 1.620 ± 0.001 Å and 1.750 ± 0.001 Å, respectively and the O–O distances to 2.646 ± 0.005 Å for the Si-tetrahedra and 2.858 ± 0.005 Å for the Al-tetrahedra. In the final stage of the refinements, the weights of the restraints on the framework were maintained. Convergence was achieved by simultaneously refining all background and profile parameters, scale factor, lattice constants, two theta zero, preferred orientation function and the atomic positional and thermal displacement parameters. The final refined parameters are summarized in Table S2 and the selected bond distances and angles are listed in Table S3.

3. Results and Discussions

Pressure- and temperature-induced changes in the observed synchrotron X-ray diffraction patterns of the Ag-NAT in the presence of different PTM, N$_2$, O$_2$ or CH$_4$ are shown in Figure 1. The Bragg peaks of the initial phase are indexed to the orthorhombic space group *Fdd2* under ambient conditions [19]. In all cases except the CH$_4$ run, the Ag-NAT expands in two stages by applying pressure and temperature. The Bragg peak of (220) is obviously observed to shift to lower two theta angles in all the diffraction data when PII occurs, indicating that structural changes in the *ab*-plane are dominant. The Ag-NAT in the presence of N$_2$, the first expanded phase is observed at 0.88(5) GPa owing to the starting pressure-induced insertion (PII) of the N$_2$ molecule. The intensity of peaks that belong to the first expanded phase (space group: *Fdd2*) increases up to 1.14(5) GPa, and the ambient phase disappears after heating at 110 °C for an h. The second expanded phase is subsequently observed with a phase transition to monoclinic, *Cc*, at 2.15(5) GPa after heating at 200 °C for 1 h (Figure 1a). The second expanded phase gradually contracts without any further phase change due to pressure and temperature. In the case of O_2 run, the first expansion of the Ag-NAT is accompanied by a transition to monoclinic, *Cc*, at 1.05(5) GPa. The second expanded phase (orthorhombic: *Fdd2*) was observed at 3.84(5) GPa after heating at 250 °C for 1 h. This phase gradually contracted up to a final pressure of 8.12(5) GPa (Figure 1b). In the case of the CH_4 run, an expanded phase (space group: *Cc*) is observed at 1.43(5) GPa after heating at 150 °C for 1 h and exists up to 3.81(5) GPa (Figure 1c). In all cases, the Ag-NAT reversibly changes to the initial phase after pressure is released.

A series of whole-profile refinements reveals the details of the compressional changes of the unit cell lengths and volume of the Ag-NAT in the presence of different PTMs (Figure 2). When we convert to a non-conventional *Fd* setting for comparison with the *Fdd2* structure, we find that there are three distinct regions of unit cell parameter changes in the case of N_2 and O_2 as PTMs, while there are two regions of unit cell parameter changes in case of the CH_4 as PTM. In all cases, the *a*- and *b*-axes increase when the first PII occurs (in order of *a*- and *b*-axis, 4.1(1)% and 2.6(1)% at 0.88(5) GPa in the N_2 run; 3.6(1)% and 2.4(1)% at 1.05(5) GPa in the O_2 run; 4.1(1)% and 2.4(1)% at 1.84(5) GPa in the CH_4 run). Compared with the *a*- and *b*-axes, the *c*-axis slightly decreases at the pressure of the first PII (−1.0(1)% in the N_2 run; −0.5(1)% in the O_2 run; −0.8(1)% in the CH_4 run). This phenomenon of axes changing when PII occurs is related to expansion and becoming more circular in the channel window [9]. In the N_2 run, all axes slightly increase within ~0.5% at 2.15(5) GPa, the onset pressure of the second PII. In regard to O_2 run, all axes decrease at 5.24(5) GPa. Regarding our Rietveld refinement results and pressure of second PII, the increase of all axes at 2.15(5) GPa in the case of the N_2 run is dominantly reflected by the PII effect rather than axial contraction by pressure. A decrease in the axes at 5.24(5) GPa and second PII, in the case of O_2 run compressional effect by pressure is more dominant.

The unit cell volume of the Ag-NAT with the N_2 expands 5.7(1)% at 0.88(5) GPa and 0.8(1)% at 2.15(1) GPa, respectively. Except for the abrupt volume expansion at the pressures of the first and second PII (0.88(5) GPa and 2.15(5) GPa, respectively), the unit cell volume linearly decreases with increasing pressure. The volume changes are mainly caused by changes in the *a*- and *b*-axes. Related to our Rietveld refinement results, the number of 12.3 N_2 molecules per 80 framework oxygen (O_f) of the Ag-NAT is inserted into the void of the channel at 1.44(5) GPa, after the first PII. At 2.74(5) GPa, 16 N_2 molecules per 80 of are inserted by the second PII. To understand the relationship between molecules by PII and the compressibility of the Ag-NAT in the presence of N_2, O_2 and CH_4, we used the Birch-Murnaghan equation of state with second order and fixed the derivative of the bulk modulus (B_0) to 4. In the case of N_2, the bulk modulus (B_0) of the Ag-NAT is 42(1) GPa before the first PII occurs at 0.88(5) GPa. Between 0.88(5) GPa and 2.15(5) GPa, the bulk modulus of the Ag-NAT changes to 57(5) GPa. After the second PII occurs at 2.15(5) GPa, the bulk modulus is 49(7) GPa. The bulk modulus changes of the Ag-NAT reveal that the Ag-NAT becomes more incompressible due to insertion of N_2 molecules into the NAT framework (Figures 2d and 3). In the O_2 run, the unit cell volume of the Ag-NAT also linearly decreases with pressure except an abrupt expansion of 5.5(1)% at 1.05(5) GPa and contraction of 3.0(1)% at 5.24(5) GPa caused by the first and second PII, respectively. The numbers of 14.2 O_2 and 16 O_2 are inserted per 80 of at 2.51(5) GPa and 8.12(5) GPa, after the first and second PII, respectively. The bulk modulus of Ag-NAT in the presence of O_2 is 38(2) GPa before the first PII at 1.05(5) GPa. From 1.05(5) GPa to 5.24(5) GPa, the bulk modulus of the Ag-NAT increases up to 85(5) GPa. After the second PII at 5.24(1) GPa, the bulk modulus is 227(1) GPa. The highest bulk modulus of 227(1) among our results makes it possible to make fourteen coordinate bonds of O_2 molecule and framework oxygen (Figure 3f) compared to three to eight bonds are formed after PII in other models (Figure 3a–e). This means that the bonds of O_2 molecules after the second PII sustain the collapsible framework by pressure and therefore the structure becomes more incompressible. Similar to the case of N_2 run, the bulk modulus of the Ag-NAT in O_2 increases after each PII (Figures 2e and 3). In the case of CH_4, the volume increases 5.7(1)% in response to insertion of 8 CH_4 per 80 O_f at 1.84(5) GPa. Before and after 1.84(5) GPa, the bulk moduli are 55(3) GPa and 79(2) GPa, respectively.

After pressure is reduced to ambient pressure, the unit cell volumes of Ag-NAT in all cases recover to a similar volume of volume at ambient pressure (open symbols in Figure 2). Overall, the volume changes are accompanied by the PII and the bulk moduli increase after molecule uptake.

For a detailed understanding of the structural changes before and after pressure-induced insertion of each molecule, Rietveld models were established at selected pressure points in each run (Tables S2 and S3). All models are projected along the [001] direction in Figure 3, and the ambient model is from our previous study [19]. The extra-framework cation (EFC), Ag^+ and water molecules in the ambient model are located at the center and side of the NAT channel, respectively, and the extra-framework species show an ordered distribution. Silver cations have six-coordinated bonding with four framework oxygens and two water molecules. The geometry of the channel window is determined by measuring the chain rotation angle of the T_5O_{10} (T = Si and Al) secondary building unit, ψ and the elongation ratio between the lengths of the longest and shortest diagonal (L/S) of the eight-membered rings. The lower degree of the chain rotation angle and elongation ratio indicates a more circular shape of the channel window. The ψ value of the Ag-NAT at ambient pressure is $22.2(1)°$ and the angle decreases to $19.3(1)°$, $18.3(1)°$ and $20.2(1)°$ after first PII of N_2, O_2 and CH_4 molecules, respectively (Figure 3). In the N_2 run, the channel window became more circular with increasing amount of N_2 molecules (12.3 per 80 out of $\rightarrow$ 16 per 80 O_f) inside the channel after second PII ($19.3(1)°$ at 1.44(5) GPa $\rightarrow$ $18.9(1)°$ at 2.74(5) GPa in Figure 3b,c. The rotation angle of Ag-NAT with O_2 increases after the second PII (14.2 per 80 O_f $\rightarrow$ 16 per 80; $18.3(1)°$ at 2.51(5) GPa $\rightarrow$ $23.5(1)°$ at 8.12(5) GPa in Figure 3e,f. The changes in the elongation ratio are quite similar to the changes in the rotation angles under pressure. The ratio decreases from 2.37(1) at ambient to 2.10(1), 2.01(1) and 2.16(1) after the first PII of N_2, O_2 and CH_4 molecules, respectively. After the second PII, the ratio decreases to 2.02(1) in the N_2 run and increases to 2.52(1) in the O_2 run. Considering a comparatively high pressure of 8.12(5) GPa, the compressional force affects the channel window to become more elliptical rather than becoming circular by insertion and increasing the amount of O_2 molecules. The difference Fourier map in the channel shows that two unknown sites are adjacent with interatomic distances of ~1.1 Å and ~1.2 Å in all models of Ag-NAT-N_2 and Ag-NAT-O_2, respectively. We therefore assign unknown sites to N_2 and O_2 and understand that the chemical properties of guest molecules may be retained after inserting into the framework at high pressure.

In all cases, guest molecules were located near the Ag^+ cation and coordinated with Ag^+, water molecules and framework oxygen after insertion. The N_2 molecules are bonded with eight and seven framework oxygens after the first and second PII, respectively, in the case of N_2 (Figure 3b,c). The six bindings of O_2 and framework oxygen are formed after the first PII and the number of bonds is changed to fourteen after PII in the case of O_2 run (Figure 3e,f) due to sustaining the NAT framework at a comparatively high pressure of 8.12(5) GPa in our pressure range. Three framework oxygen atoms are connected by CH_4 molecules after PII occurs. When CH_4 is inserted at 2.62(5) GPa, the water molecule migrates to the opposite site of CH_4 and forms a coordinate bond with the Ag^+, whereas the atomic positions of the water molecules of the N_2 and O_2 models after the first and second PII are similar to those of the ambient model. In the CH_4 model, CH_4 molecules may push neighboring host water molecules by repulsion during the CH_4 occupies one site in the channel and then the water molecule subsequently migrates to the opposite site of CH_4 to balance the charge distribution inside the channel. Unlike the CH_4 model, the position of the host water molecules in the pressure models of the Ag-NAT with N_2 and O_2 are similar to the position of water molecules in the ambient model because the charge distribution balance is already satisfied due to N_2 and O_2 molecules at both sides of the channel. For a detailed understanding of the insertion of different guest molecules inside the Ag-NAT channel, high-pressure spectroscopy experiments were performed.

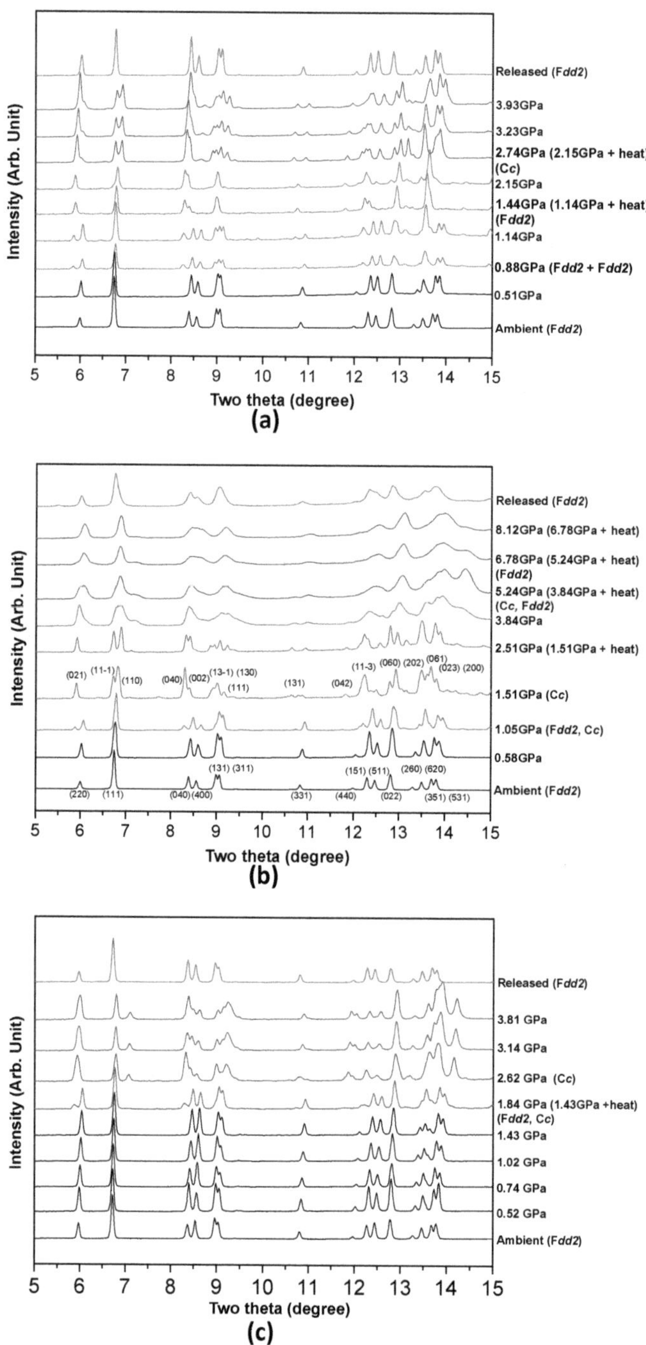

Figure 1. Pressure- and temperature-induced changes of the synchrotron X-ray powder diffraction patterns of Ag-NAT in presence of (**a**) N_2, (**b**) O_2 and (**c**) CH_4. Wavelength of X-ray w 0.6888(1) Å and 0.6927(1) Å in a case of N_2 (and O_2) run and CH_4 run, respectively.

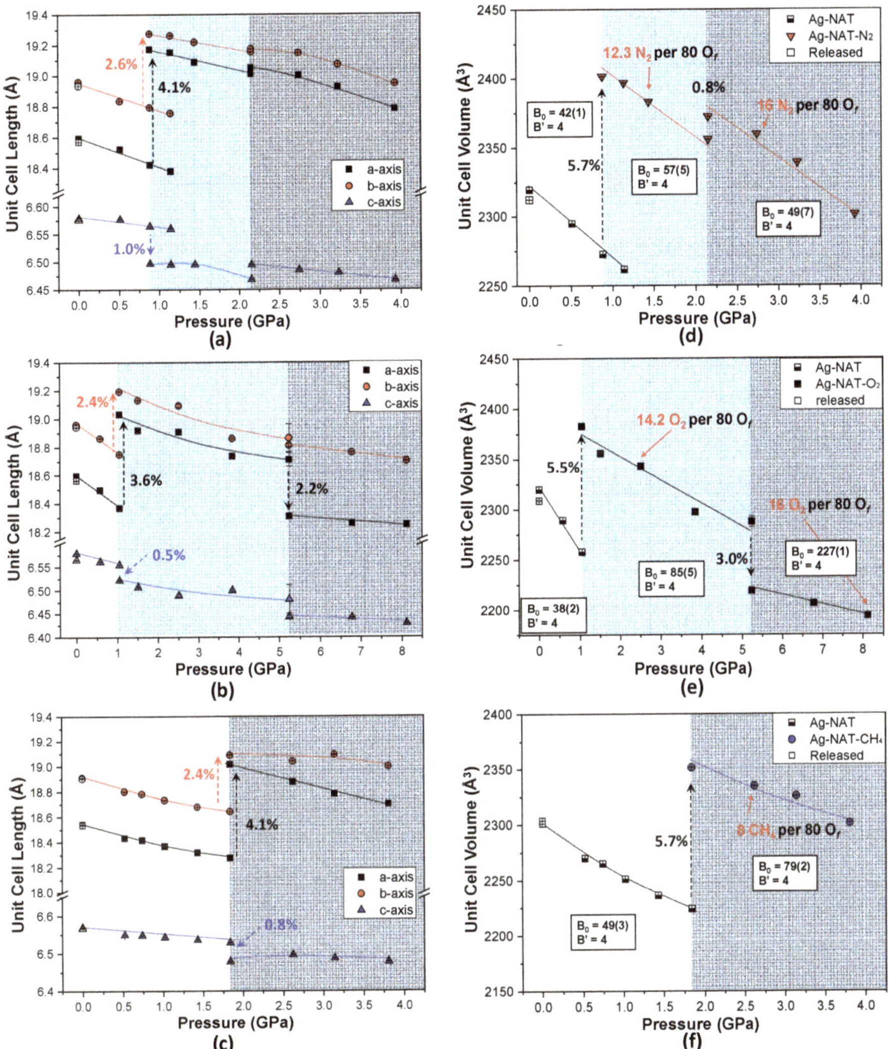

Figure 2. Pressure- and temperature-dependent axial changes of Ag-NAT in presence of (**a**) N_2, (**b**) O_2 and (**c**) CH_4 and changes of unit cell volumes in presence of (**d**) N_2, (**e**) O_2 and (**f**) CH_4. Each open symbol represents volume after pressure released.

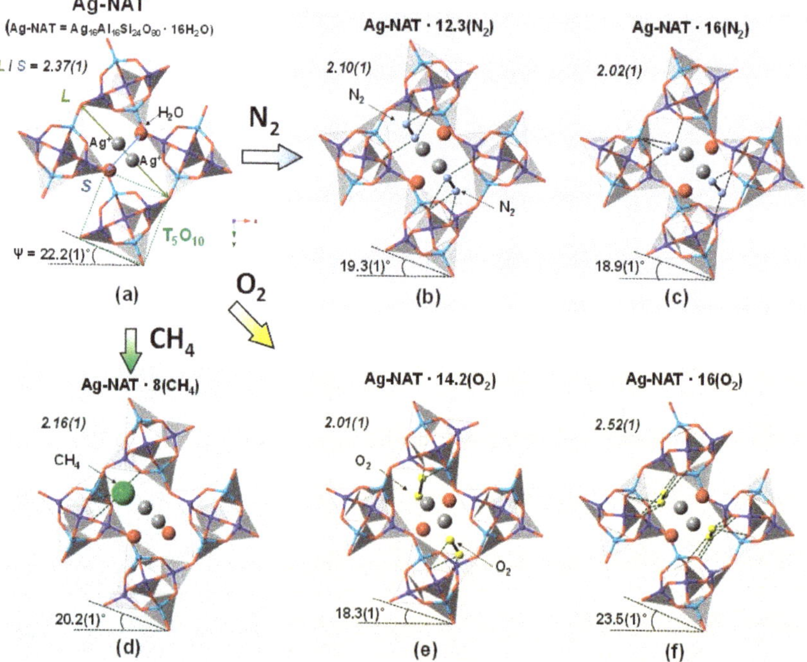

Figure 3. Polyhedral representations of (**a**) Ag-NAT and (**b**) Ag-NAT in N_2 at 1.44(5) GPa, (**c**) Ag-NAT in N_2 at 2.74(5) GPa, (**d**) Ag-NAT in CH_4 at 2.62(5) GPa, (**e**) Ag-NAT in O_2 at 2.51(5) GPa, (**f**) Ag-NAT in O_2 at 8.12(5) GPa. Blue balls in tetrahedra illustrate ordered distributions of Al (Si) atoms in the framework. Yellow, dark blue and green balls represent O_2, N_2 and CH_4, respectively. Dashed lines inside channel represent coordinate bonds of gas molecule and framework oxygen.

Concomitantly with our previous high-pressure studies of the Ag-NAT, we found that the onset pressure of the first PII exponentially increases as a function of the kinetic diameter of molecules as PTM (Figure 4a) [9,13]. In terms of the kinetic diameter, the water molecule can be most permeable in the angstrom scale channel and therefore, the pressure required for an over-hydrated state is as low as 0.4(1) GPa. In the case of the Xe molecule, however, greater mechanical forces, such as pressure, are required in order to open the channel window of the Ag-NAT and subsequently transfer it into the channel. All our experimental results regarding the unit cell volume of the Ag-NAT obtained using the applied pressure are summarized in Figure 4b. We determined that the Ag-NAT has the limitation of maximum expansion (or contraction) rate of the unit cell volume. Adopting a linear function by the least-square fitting method, the relationship of expanded phases (red symbols) after PII and the as-prepared phases before PII (black symbols) follow Equations (1) and (2), respectively.

$$y = -0.014(1)x + 1.054(3) \tag{1}$$

$$y = -0.016(1)x + 1.00(2) \tag{2}$$

where y is the normalized unit cell volume and x is the pressure.

The comparison of normalized unit cell volumes shows that their maximum degrees of expansion and contraction are approximately 5% and 6%, respectively, in the pressure range from ambient to 8 GPa. The two values of slope in these functions show that the expanded phase would be more incompressible by the evaluated pressure because the inserted molecules sustain the channel to prevent collapse. These accumulated results can provide guidance for similar experiments using the Ag-NAT in the future. For example, we can expect PII will be almost complete if the volume is increased up to ~5% and follow Equation (1). In other words, the Ag-NAT can absorb more gas if the volume contraction trend follows Equation (2). We expect this approach to also be adapted to other porous materials under non-ambient conditions.

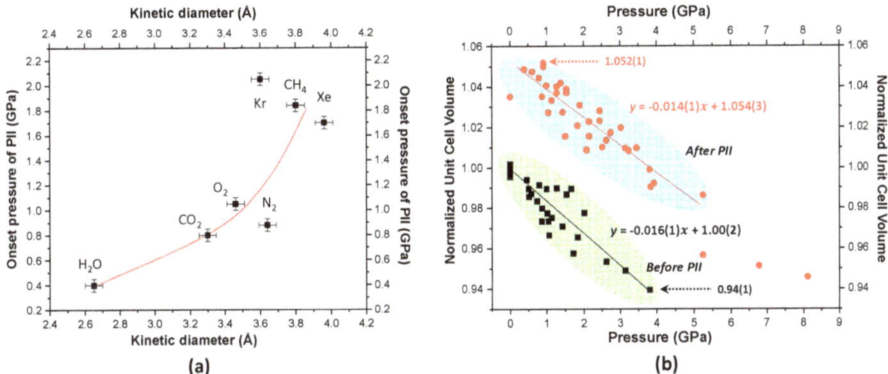

Figure 4. (a) Onset pressure of pressure-induced insertion vs. kinetic diameter of molecule as PTM; (b) normalized volume plot in a presence of different PTMs (N_2, O_2, CH_4, CO_2, H_2O, Xe and Kr) [9,13].

4. Conclusions

We established thermo-compressional behaviors of the Ag-NAT in the presence of N_2, O_2 and CH_4 by pressure and temperature conditions with comparative structural changes as the Ag-NAT takes up molecules. The PII occurs twice in the case of N_2 and O_2 and once for CH_4. The unit cell parameters linearly decrease with the evaluated pressure, except when pressure-induced insertion occurs, in which case it abruptly decreases. The pressure required to insert PII into the channel depends on the molecule size, and the incompressibility increases after molecule insertion. We investigated the reversible thermo-compressional interactions of Ag-NAT and guest molecules that are systematically dependent on the ionic diameter of the guest molecules. We expect that our results and experimental approach can be used as fundamental data for application in pressure-induced capture and storage of noble gases or even radioactive species, using microporous materials.

Supplementary Materials: The following are available online at http://www.mdpi.com/1996-1944/13/18/4096/s1, Figure S1: Graphic result of thermogravimetric analysis of the Ag-NAT. Table S1: Chemical composition of the Ag-NAT calculated from energy dispersive spectroscopy (EDS) method. Table S2: Refined cell parameters and atomic coordinates of Ag-NAT in O_2, N_2 and CH_4 under pressure. Table S3: Selected interatomic distances (Å) and angles (°) of Ag-NAT in O_2, N_2 and CH_4 under pressure.

Author Contributions: Conceptualization, Y.L.; methodology, H.K. and P.K.; formal analysis, H.K. and P.K.; data curation, H.K. and P.K.; writing—original draft preparation, Y.L. and D.S.; writing—review and editing, Y.L. and D.S.; visualization, Y.L.; supervision, D.S.; project administration, Y.L.; funding acquisition, Y.L. and D.S. All authors have read and agreed to the published version of the manuscript.

Funding: Please add: This research was funded by National Research Foundation of the Ministry of Science and ICT of Korean Government, Grant Number NRF-2019R1F1A106258, NRF-2017R1D1A1B03035418, NRF-2019K1A3A7A09101574, NRF-2020R1C1C1013642 and Pusan National University Research Grant, 201902840001 and Chonnam National University Research Grant, 2019-0215.

Acknowledgments: Experiments using synchrotron radiation were supported by Pohang Light Source II (PLS-II) at Pohang Accelerator Laboratory (PAL). We thank T. Jeon and H.-H. Lee for the support at beamline 3D and 5A.

Conflicts of Interest: The authors declare no conflict of interest.

References

1. Ackley, M.W. Application of natural zeolites in the purification and separation of gases. *Microporous Mesoporous Mater.* **2003**, *61*, 25–42. [CrossRef]
2. Barrer, R.M. *Zeolites and Clay Minerals as Sorbents and Molecular Sieves*; Academic Press: London, UK; New York, NY, USA, 1978. [CrossRef]
3. Martins, L.; Boldo, R.T.; Cardoso, D. Ion exchange and catalytic properties of methylammonium FAU zeolite. *Microporous Mesoporous Mater.* **2007**, *98*, 166–173. [CrossRef]
4. Mumpton, F.A. La roca magica: Uses of natural zeolites in agriculture and industry. *Proc. Natl. Acad. Sci. USA* **1999**, *96*, 3463–3470. [CrossRef] [PubMed]
5. Gatta, G.D.; Lee, Y. Zeolites at high pressure: A review. *Miner. Mag.* **2014**, *78*, 267–291. [CrossRef]
6. Seoung, D.; Lee, Y.; Kao, C.-C.; Vogt, T.; Lee, Y. Super-Hydrated Zeolites: Pressure-Induced Hydration in Natrolites. *Chem. A Eur. J.* **2013**, *19*, 10876–10883. [CrossRef] [PubMed]
7. Lee, Y.; Hriljac, J.A.; Parise, J.B.; Vogt, T. Pressure-induced stabilization of ordered paranatrolite: A new insight into the paranatrolite controversy. *Am. Miner.* **2005**, *90*, 252–257. [CrossRef]
8. Lee, Y.; Liu, D.; Seoung, D.; Liu, Z.; Kao, C.-C.; Vogt, T. Pressure- and heat-induced insertion of CO2 into an auxetic small-pore zeolite. *J. Am. Chem. Soc.* **2011**, *133*, 1674–1677. [CrossRef] [PubMed]
9. Lee, Y.; Seoung, D.; Jang, Y.-N.; Lee, Y.; Vogt, T. Pressure-induced hydration and insertion of CO2 into Ag-natrolite. *Chem. A Eur. J.* **2013**, *19*, 5806–5811. [CrossRef] [PubMed]
10. Lee, Y.; Lee, Y.; Seoung, D.; Im, J.; Hwang, H.-J.; Kim, T.-H.; Liu, D.; Liu, Z.; Lee, S.Y.; Kao, C.-C.; et al. Immobilization of Large, Aliovalent Cations in the Small-Pore Zeolite K-Natrolite by Means of Pressure. *Angew. Chem. Int. Ed.* **2012**, *51*, 4848–4851. [CrossRef] [PubMed]
11. Im, J.; Seoung, D.; Lee, S.Y.; Blom, D.A.; Vogt, T.; Kao, C.-C.; Lee, Y. Pressure-Induced Metathesis Reaction to Sequester Cs. *Environ. Sci. Technol.* **2014**, *49*, 513–519. [CrossRef] [PubMed]
12. Im, J.; Lee, Y.; Blom, D.A.; Vogt, T.; Lee, Y. High-pressure and high-temperature transformation of Pb(ii)-natrolite to Pb(ii)-lawsonite. *Dalton Trans.* **2016**, *45*, 1622–1630. [CrossRef] [PubMed]
13. Seoung, D.; Lee, Y.; Cynn, H.; Park, C.; Choi, K.-Y.; Blom, D.A.; Evans, W.J.; Kao, C.-C.; Vogt, T.; Lee, Y. Irreversible xenon insertion into a small-pore zeolite at moderate pressures and temperatures. *Nat. Chem.* **2014**, *6*, 835–839. [CrossRef] [PubMed]
14. Gatt, R.; Zammit, V.; Caruana, C.; Grima, J. On the atomic level deformations in the auxetic zeolite natrolite. *Phys. Status Solidi b* **2008**, *245*, 502–510. [CrossRef]
15. International Zeolite Association. Available online: http://www.iza-structure.org/databases (accessed on 26 August 2020).
16. Baur, W.H.; Kassner, D.; Kim, C.-H.; Sieber, N.H.W. Flexibility and distortion of the framework of natrolite: Crystal structures of ion-exchanged natrolites. *Eur. J. Miner.* **1990**, *2*, 761–770. [CrossRef]
17. Lee, Y.; Hriljac, J.A.; Vogt, T.; Parise, J.B.; Artioli, G. First Structural Investigation of a Super-Hydrated Zeolite. *J. Am. Chem. Soc.* **2001**, *123*, 12732–12733. [CrossRef] [PubMed]
18. Lee, Y.; Vogt, T.; Hriljac, J.A.; Parise, J.B.; Artioli, G. Pressure-Induced Volume Expansion of Zeolites in the Natrolite Family. *J. Am. Chem. Soc.* **2002**, *124*, 5466–5475. [CrossRef] [PubMed]
19. Lee, Y.; Seoung, D. Natrolite is not a "soda-stone" anymore: Structural study of alkali (Li+), alkaline-earth (Ca^{2+}, Sr^{2+}, Ba^{2+}) and heavy metal (Cd^{2+}, Pb^{2+}, Ag^+) cation-exchanged natrolites. *Am. Miner.* **2011**, *96*, 1718–1724. [CrossRef]
20. Toby, B.H. EXPGUI, a graphical user interface for GSAS. *J. Appl. Crystallogr.* **2001**, *34*, 210–213. [CrossRef]
21. Dollase, W.A. Correction of intensities for preferred orientation in powder diffractometry: Application of the March model. *J. Appl. Crystallogr.* **1986**, *19*, 267–272. [CrossRef]
22. Thompson, P.; Cox, D.E.; Hastings, J.B. Rietveld refinement of Debye–Scherrer synchrotron X-ray data from Al_2O_3. *J. Appl. Crystallogr.* **1987**, *20*, 79–83. [CrossRef]
23. Larson, A.C.; Von Dreele, R.B. *General Structure Analysis System Report (GSAS)*; LAUR: Baltimore, MD, USA, 1986; pp. 86–748.

Article

Energy Conversion Capacity of Barium Zirconate Titanate

Nawal Binhayeeniyi [1,*], Pisan Sukwisute [2], Safitree Nawae [1] and Nantakan Muensit [3,4]

[1] Faculty of Science and Technology, Princess of Naradhiwas University, Narathiwat 96000, Thailand; safitree.n@pnu.ac.th

[2] Department of Physics, Faculty of Science, King Mongkut's Institute of Technology Ladkrabang, Bangkok 10520, Thailand; pisan.su@kmitl.ac.th

[3] Department of Physics, Faculty of Science, Prince of Songkla University, Songkhla 90110, Thailand; nantakan.m@psu.ac.th

[4] Center of Excellence in Nanotechnology for Energy (CENE), Prince of Songkla University, Songkhla 90110, Thailand

* Correspondence: nawal.b@pnu.ac.th

Received: 6 November 2019; Accepted: 7 January 2020; Published: 9 January 2020

Abstract: In this study, we investigated the effect of zirconium content on lead-free barium zirconate titanate (BZT) (Ba(Zr$_x$Ti$_{1-x}$)O$_3$, with x = 0.00, 0.01, 0.03, 0.05, and 0.08), which was prepared by the sol–gel method. A single-phase perovskite BZT was obtained under calcination and sintering conditions at 1100 °C and 1300 °C. Ferroelectric measurements revealed that the Curie temperature of BaTiO$_3$ was 399 K, and the transition temperature decreased with increasing zirconium content. At the Curie temperature, Ba(Zr$_{0.03}$Ti$_{0.97}$)O$_3$ with a dielectric constant of 19,600 showed the best performance in converting supplied mechanical vibration into electrical power. The experiments focused on piezoelectric activity at a low vibrating frequency, and the output power that dissipated from the BZT system at 15 Hz was 2.47 nW (30 MΩ). The prepared lead-free sol–gel BZT is promising for energy-harvesting applications considering that the normal frequencies of ambient vibration sources are less than 100 Hz.

Keywords: lead-free ceramic; sol–gel process; barium zirconate titanate; dielectric property

---

1. Introduction

Lead zirconate titanate (PZT, Pb(Zr$_x$Ti$_{1-x}$)O$_3$), a lead-based material with a high piezoelectric coefficient and electromechanical coupling factor, is one of the most promising materials for use in fabricated energy-harvesting devices [1–3] because its perovskite structure exhibits dielectric, ferroelectric, and piezoelectric properties [2,3]. However, PZT is toxic to the environment. Therefore, innovative lead-free dielectric materials with piezoelectric properties have been formulated to address this environmental issue. Among these materials is BaTiO$_3$, which is a well-known material possessing a perovskite structure with high dielectric properties, a low dielectric loss tangent, and dielectric reliability [4–7].

BaTiO$_3$ can be modified by doping it with additives such as Sr^{2+}, Ca^{2+}, Sn^{4+}, and Zr^{4+} [4]. Doping BaTiO$_3$ with ZrO$_2$ can improve the dielectric and piezoelectric properties because the chemical stability of Zr^{4+} is greater than that of Ti^{4+} [4–7]. In addition, the Curie temperature also changes; that is, it decreases as the Zr content increases [5–8].

BaTiO$_3$ can be used in tunable capacitor devices and dynamic random-access memory applications. Moreover, it is also applied in actuators and energy storage devices because the strain that is induced by the electric field retains dipole moment behavior and energy storage properties [7,9,10]. Lui et al. prepared BaTi$_{0.7}$Zr$_{0.3}$O$_3$ ceramic by spark plasma sintering. The maximum energy storage density of the ceramic was determined to be 0.51 J/cm^3 [9]. Moreover, Puli et al. investigated the energy storage of barium calcium titanate (BCT) ceramic and obtained a high energy density (0.24 J/cm^3) [10].

There are various kinds of energy harvesters, including thermoelectric, electromagnetic, electrostatic, and piezoelectric. Of these methods, piezoelectric energy harvesting is very attractive for the system's small size, high output power, and ease of operation [11–13].

In this study, we investigated the crystal structure, dielectric properties, phase transition, and the degree of diffuseness of lead-free barium zirconate titanate (BZT) ceramics with various Zr contents. Additionally, the energy conversion behavior resulting from the modification of BZT was examined. These materials might lead to a reduction in the use of the lead-based bulky ceramics that are usually required in applications.

2. Materials and Methods

Ba(Zr$_x$Ti$_{1-x}$)O$_3$ (x = 0.00, 0.01, 0.03, 0.05, and 0.08) was prepared by the sol–gel method. Barium acetate (HIMEDIA®, Mumbai, MH, India, 99.0%), zirconium(IV) propoxide (Sigma-Aldrich®, St. Louis, MO, USA, 70 wt.% in 1-propanol), and titanium(IV) isopropoxide (Sigma-Aldrich®, St. Louis, MO, USA, were used as the reagents. Glacial acetic acid (Merck, Darmstadt, HE, Germany, 100%) and 2-methoxyethanol (Ajex Finechem Pty Ltd, Taren Point, NSW, Australia) were used as solvents in the sol–gel method following Jiwei et al. [14]. The procedure has been reported elsewhere [15,16]. The gels were dried in an oven for 24 h. All dried gels were calcined at 1100 °C for 2 h in alumina crucibles. The BZT powder was ball-milled in ethanol milling media (Merck, Ethanol Absolute, Darmstadt, HE, Germany) for 24 h (200 rpm) by using a high-energy planetary ball mill (Retsch PM100, Haan, NW, Germany). The milled powders were blended with a small amount of polyvinyl alcohol (PVA) to form discs (diameter 13 mm) at 100 MPa. All the green bodies were sintered at 1300 °C for 2 h in closed alumina crucibles. The upper and lower surfaces of the sintered ceramics were covered by silver paste and then calcined at 600 °C for 0.5 h for use as electrodes for the dielectric measurements.

The dielectric properties and ferroelectric phase transitions of all samples were characterized at 25–150 °C (at 1 kHz) by an LCR meter (Hewlett Packard 4263 B, Mississauga, ON, Canada). The crystalline structure of BZT was determined by X-ray diffraction (XRD, PHILLIPS X'pert MPD, Almelo, OV, Netherlands) with Ni-filtered CuK$_\alpha$ radiation. The XRD analysis was performed at room temperature (20° ≤ 2θ ≤ 77°) with a step size of 0.02°. The bulk densities of the sintered BZT discs were measured in accordance with the Archimedes method. Thermal analysis of the dried BZT gels was performed by differential thermal analysis (DTA, Perkin Elmer DTA7, Norwalk, CT, USA) and thermogravimetric analysis (TGA, Perkin Elmer TGA7, Norwalk, CT, USA). The thermal analysis results were collected from 50 °C to 1300 °C at a rate of 10° C/min. Surface microstructures were observed using scanning electron microscopy (SEM, quanta400, Thermo Fisher Scientific, Brno, JM, Czech Republic) with an accelerating voltage of 20 kV and 3000× magnification. The grain sizes were analyzed by averaging over the total number of grains in the SEM images.

3. Results and Discussion

The TGA and DTA results in Figure 1 show three mechanisms. First, the endothermic reaction observed in the temperature range of 25–200 °C is associated with the dehydration of the dried BZT gels, as observed by the mass loss of about 20%. Second, in the temperature range of 200–650 °C, a major mass loss occurs with the emission of CO$_2$, solvents, and organic compounds because of the thermal disintegration of the polymeric dried gels and primary synthesis of Ba(Zr$_x$Ti$_{1-x}$)O$_3$ via BaCO$_3$–TiO$_2$ and BaCO$_3$–ZrO$_2$ core–shell particles [17–19]. Third, the exothermic peak in the range of 650–1200 °C exhibits a slight weight loss that can be attributed to Ba(Zr$_x$Ti$_{1-x}$)O$_3$ crystallization and the subsequent formation of the perovskite structure. This final mechanism is due to the decarbonation of BaCO$_3$ to react with TiO$_2$ and ZrO$_2$. For these results, it is worth noting that although the calcination process is typically performed at temperatures as low as 650 °C, the calcination temperature used in this work was 1100 °C [15,16] to ensure the formation of the pure perovskite structure of Ba(Zr$_x$Ti$_{1-x}$)O$_3$ without secondary phases, as seen in the following XRD result (Figure 2). Calcination at a temperature above 1100 °C should not be undertaken, because of the agglomeration and enlargement of $Ba(Zr_xTi_{1-x})O_3$ particles. The compression of large calcined particles might result in a low bulk density of the sintered ceramics [20,21]. Table 1 presents the measurable bulk density of sintered $Ba(Zr_xTi_{1-x})O_3$. The relative density of all samples is 93.5% ± 0.21%. The addition of zirconium does not affect density [8]. A sintering temperature of 1300 °C is sufficient to fuse the as-calcined $Ba(Zr_xTi_{1-x})O_3$ powders, and a calcination temperature of 1100 °C has an insignificant effect on the bulk density. The XRD patterns of sintered $Ba(Zr_xTi_{1-x})O_3$ ceramics (with x = 0.00, 0.01, 0.03 0.05, and 0.08) are shown in Figure 2. The structure of all $Ba(Zr_xTi_{1-x})O_3$ ceramics is a pure perovskite phase without an impurity phase. With the addition of Zr, the peak shifts to a lower angle because the ionic radius of Zr^{4+} (0.079 nm) is larger than that of Ti^{4+} (0.068 nm) [5]. It is clear that the tetragonal phase of $BaTiO_3$ ceramic is characterized by the splitting of the (0 0 2) and (2 0 0) diffraction peaks at 44.93° and 45.40°, respectively (the calculated values of the cell parameters of $BaTiO_3$ are (a ~ 3.9906 Å, c ~ 4.0301 Å), respectively). As the zirconium content increases, the two diffraction peaks merge into one peak. This corresponds with the change in the structure of the BZT system from tetragonal to orthorhombic at room temperature, as previously reported by [4,6,22]. According to, the separation of (1 3 3) and (3 1 1) diffraction peaks of $Ba(Zr_{0.03}Ti_{0.07})O_3$ occurs at diffraction angles of 74.63° and 74.91°, respectively; upon the addition of 5 mol.% zirconium content, a single diffraction peak is observed. This is caused by the structure transforming from orthorhombic to rhombohedral [4,6]. It is concluded that the increased zirconium content changes the structure of BZT ceramic from tetragonal to rhombohedral, which is confirmed by the gradual merging of XRD peaks [6,8,21]. Finally, the dense ceramic discs exhibit large grains and a small proportion of fine grains with pores. The grains are irregular in shape, with an average grain size of 10–30 μm, because the initial size of the powder is changed by the ball milling process [20,21], as shown in Figure 3.

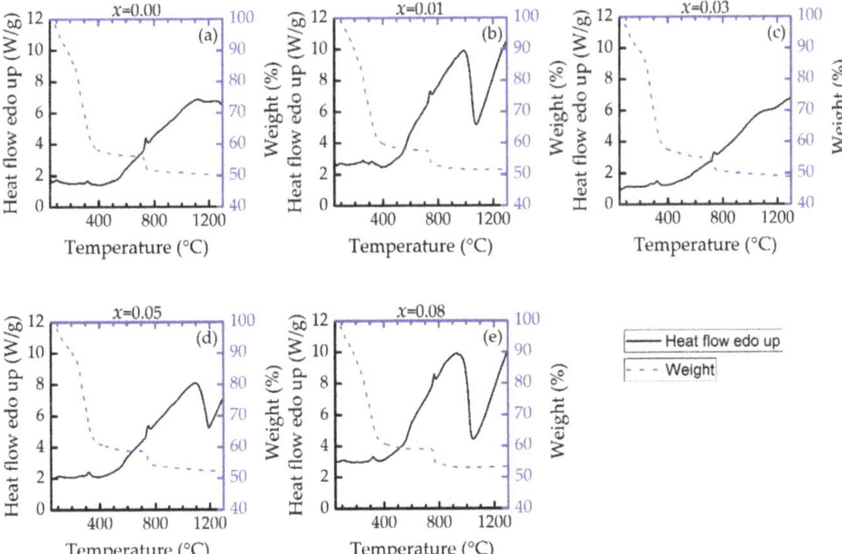

Figure 1. TGA and DTA plots for $Ba(Zr_xTi_{1-x})O_3$ samples with x composition of (**a**) 0.00, (**b**) 0.01, (**c**) 0.03, (**d**) 0.05 and (**e**) 0.08 mol.

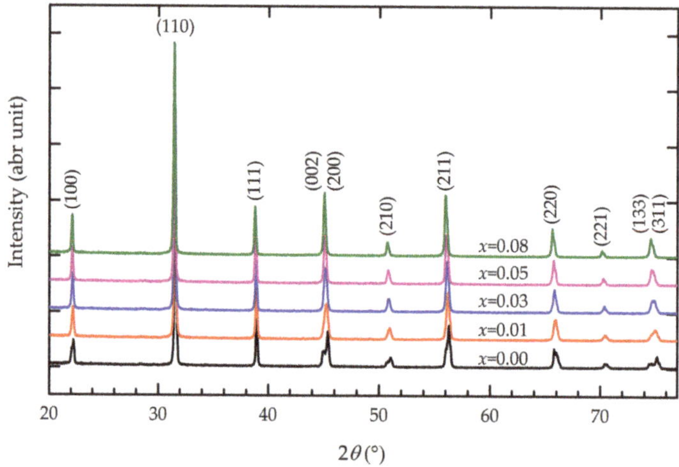

Figure 2. XRD patterns of the Ba(Zr$_x$Ti$_{1-x}$)O$_3$ ceramics.

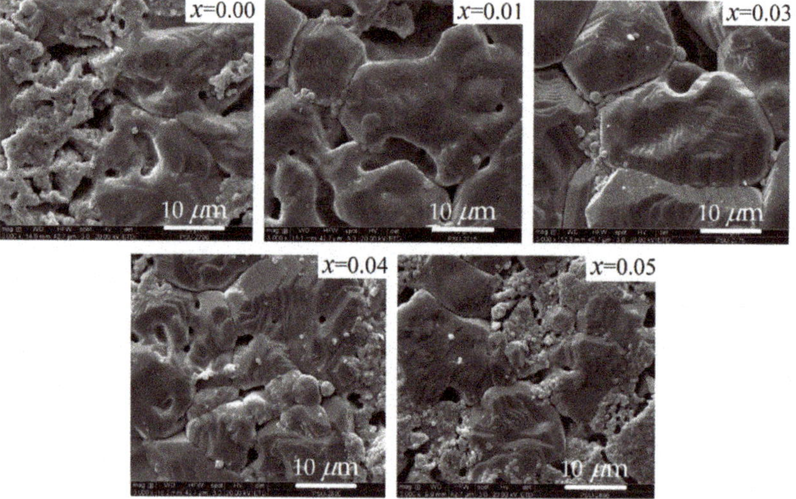

Figure 3. SEM images of the sintered Ba(Zr$_x$Ti$_{1-x}$)O$_3$ ceramics.

Table 1. Relative density, the values of the dielectric constant (ε_r) at T_m (1 kHz), dielectric loss (tanδ) at T_m (1 kHz), Curie–Weiss temperature (T_0), Curie–Weiss constant (C), Curie–Weiss law temperature (T_{cw}), T_m, ΔT_m, and γ for all x values of Ba(Zr$_x$Ti$_{1-x}$)O$_3$.

Ba(Zr$_x$Ti$_{1-x}$)O$_3$	Relative Density (%)	ε_r at T_m (1 kHz)	tanδ at T_m (1 kHz)	T_0 (K)	C (×10^5 K)	T_{cw} (K)	T_m (K)	ΔT_m (K)	γ
$x = 0.00$	93.26	9,496	0.0072	357	4.04	400	399	1	1.01
$x = 0.01$	93.66	15,702	0.0207	366	4.08	395	392	3	1.05
$x = 0.03$	93.76	19,698	0.0314	353	3.92	378	370	8	1.21
$x = 0.05$	93.49	16,891	0.0382	335	3.79	368	353	14	1.26
$x = 0.08$	93.32	11,294	0.0392	312	3.36	355	331	24	1.38

The relative permittivity (ε_r) or dielectric constant and dielectric loss (tan δ) at T_m (1 kHz) are listed in Table 1. The dielectric constant increases with zirconium content until it reaches 3 mol.%. Ba(Zr$_{0.03}$Ti$_{0.07}$)O$_3$ ceramic has the highest dielectric constant, which is reduced when zirconium reaches

5 mol.%. The dielectric loss of all BZT ceramics depends on the zirconium content and ranges from 0.072 to 0.0392, similar to the results of our previous work [15].

Figure 4 presents the values of the relative permittivity (ε_r) or dielectric constant measured at 1 kHz for the Ba(Zr$_x$Ti$_{1-x}$)O$_3$ samples. The position of each dielectric peak moves to a higher temperature with the addition of Zr, which ranges from 0 to 3 mol.%. For $x = 0.08$, the dielectric peak is broad because of the inhomogeneous distribution of Zr^{4+} ions in the Ti sites and the non-uniform stress in the grains [8,23]. The highest dielectric constant is 19,600 for Ba(Zr$_{0.03}$Ti$_{0.97}$)O$_3$. Further increases in Zr content cause a decrease in the temperature T_m with the maximum dielectric value (Table 1), as described in the literature [6,8]. This is the result of the increased substitution of the Zr^{4+} ion in the B sites of BaTiO$_3$, causing a change in the d-spacing of the Ba(Zr$_x$Ti$_{1-x}$)O$_3$ structure [6,16] and resulting in a decrease in the phase transition temperature or T_m [8,22]. For low Zr content ($x < 0.15$), at the apex of the dielectric curve, T_m can be considered the Curie temperature (T_c) [21]. A rapid increase in the ε_r value occurs near T_c because the BZT structure is thermally excited to a tetragonal–cubic intermediate phase (ferroelectric–paraelectric phase transition) when the temperature changes to T_m. This results in a large degree of unstable polarization, and consequently, an applied electric field can easily produce considerable variation in polarization [24]. The decrease in the dielectric constant above T_c is caused by the thermal detriment of polarization alignment [24,25].

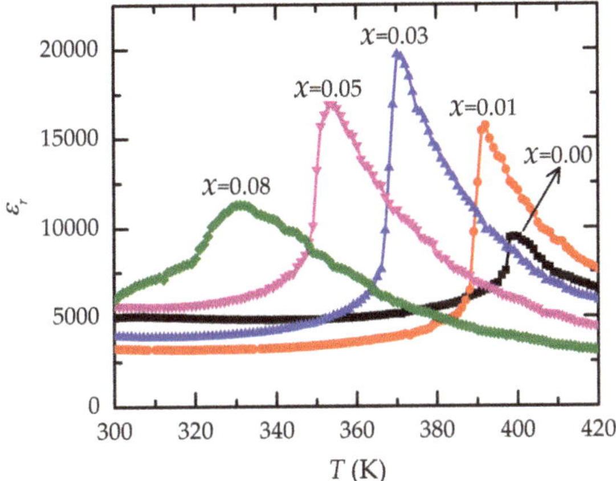

Figure 4. Relationships between the dielectric constant (ε_r) and temperature for all samples of Ba(Zr$_x$Ti$_{1-x}$)O$_3$ ceramics.

Because BZT ceramic is classified as a ferroelectric material, the dielectric characteristic of BZT above the Curie temperature corresponds to the Curie–Weiss law: $1/\varepsilon_r = (T - T_0)/C$ $(T > T_c)$, where T_0 and C are the Curie–Weiss temperature and constant, respectively. For all analyzed BZT compositions, the inverse ε_r versus temperature data were fitted to the Curie–Weiss law, as shown in Figure 5. The T_0 fitting parameters are listed in Table 1. It is clear that the reciprocal ε_r value follows the Curie–Weiss law for $T > T_m$. The divergence of the reciprocal ε_r value from the Curie–Weiss law is defined as $\Delta T_m = T_{cw} - T_m$, where T_{cw} is the temperature at which the value of the reciprocal ε_r value begins to diverge from the Curie–Weiss law. From the results in Table 1, the ΔT_m value increases with Zr content because of the influence of the Zr^{4+} ions on the shift in the ferroelectric–paraelectric transition temperature of BZT [5,6,26].

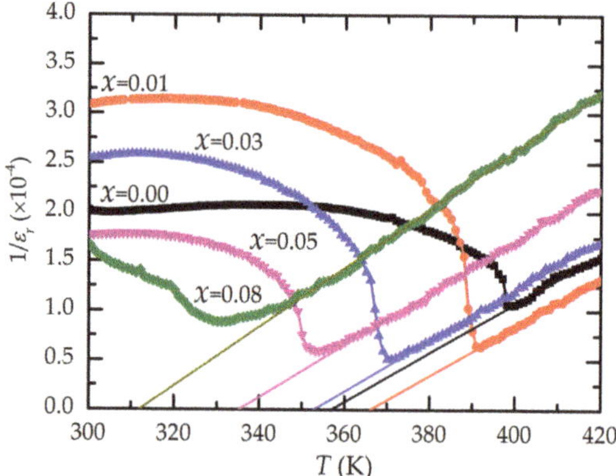

Figure 5. Relationships between the temperature and the inverse dielectric constant at 1 kHz for all the Ba(Zr$_x$Ti$_{1-x}$)O$_3$ ceramics.

The degree of diffuseness of the phase transition can be formulated by a modified Curie–Weiss law [27]:

$$\frac{1}{\varepsilon_r} - \frac{1}{\varepsilon_m} = \frac{(T - T_m)^{\gamma}}{C^*} (1 \leq \gamma \leq 2), \tag{1}$$

where γ and C^* are constants derived from fitting the experimental data. The γ value provides information about the behavior of ferroelectric materials. For a normal ferroelectric, $\gamma = 1$. For an ideal ferroelectric relaxor associated with quadratic dependence, $\gamma = 2$. Figure 6 shows the plot of $\ln(1/\varepsilon_r - 1/\varepsilon_m)$ against $\ln(T - T_m)$ at 1 kHz. The fitted γ values (Table 1) show that the higher the Zr content, the higher the diffuse phase transition, as reported in previous works [6,8]. Consequently, the inclusion of the diffusive Zr^{4+} ion in the octahedral site of the perovskite structure causes the common ferroelectric to transform into a ferroelectric relaxor [9,28]. The dielectric losses of all the BZT ceramics range from 1% to 5%. It is clear that the dielectric losses show the same trends with increasing temperature.

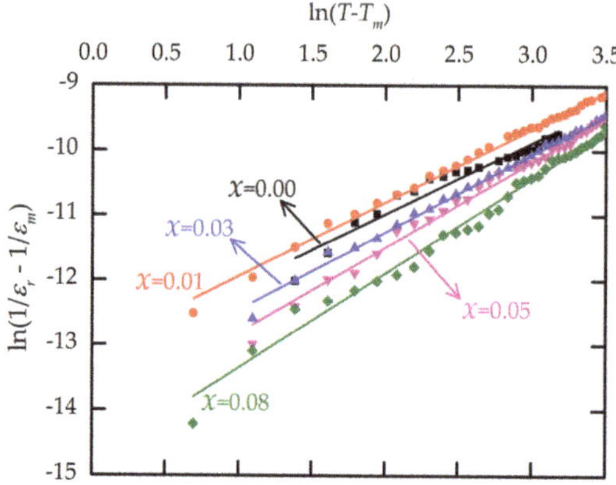

Figure 6. Linear relationships between $\ln(1/\varepsilon_r - 1/\varepsilon_m)$ and $\ln(T - T_m)$ for all x values.

Each sample was investigated for the capability of energy conversion, as described by Sukwisute et al. [1]. Each disc (thickness ~ 1 mm) was rigidly glued onto a vibrating structure at a constant operating frequency of 15 Hz. Varying resistors were connected to each disc, and the potential in the circuit was measured to calculate the output power according to P$_{ac}$ = V^2/R, where V is the potential and R is the resistance. The calculated values are summarized in Table 2. The Ba(Zr$_{0.03}$Ti$_{0.97}$)O$_3$ ceramic shows the capability of energy conversion of the supplied mechanical vibration to electrical power. This is attributed to the highest dielectric constants and the transformation of the common ferroelectric to a relaxor ferroelectric, as reported previously [9,28]. In previous work, Rukbamrung et al. used the standard harvesting approach to determine the energy-harvesting ability of PZT + 1 mol.% Mn and PMN-25PT, and they obtained a power of 1.7 and 4.5 µW [2]. The BZT ceramics in our study can be operated at the low frequencies used in daily activities, such as walking and running. In addition, the normal frequencies of ambient vibration sources are much less than 100 Hz [11,12]. From this practical viewpoint, BZT ceramic can be very useful in low-frequency energy-harvesting devices.

Table 2. Output power dissipated from the barium zirconate titanate (BZT) system.

Ba(Zr$_x$Ti$_{1-x}$)O$_3$	V ($\pm$0.05 V)	R (MΩ)	P$_{ac}$ (nW)
$x = 0.00$	0.24	132	0.044
$x = 0.01$	0.26	90	0.075
$x = 0.03$	0.86	30	2.47
$x = 0.05$	0.68	50	0.92
$x = 0.08$	0.28	90	0.09

4. Conclusions

Ba(Zr$_x$Ti$_{1-x}$)O$_3$ ceramics with various zirconium contents (x = 0.00, 0.01, 0.03, 0.05, and 0.08) were produced by the sol–gel method. A single-phase perovskite BZT was obtained under calcination and sintering conditions at 1100 °C and 1300 °C. All BZT samples had a pure perovskite structure without a secondary phase. The crystal structure changed with the zirconium content. Ferroelectric measurements of the ceramics showed that the Curie temperature of BaTiO$_3$ was 399 K, and further increases in the zirconium content decreased the Curie temperature to 331 K. At the phase transition, Ba(Zr$_{0.03}$Ti$_{0.97}$)O$_3$ had the highest dielectric constant of 19,600 and exhibited good performance in converting supplied mechanical vibration to electrical power. Thus, Ba(Zr$_{0.03}$Ti$_{0.97}$)O$_3$ is promising for mechanical energy-to-electrical energy coupling at low frequencies, with no damage observed at high temperatures.

Author Contributions: N.B. performed the experiments, characterization, data analysis, research discussion and manuscript preparation. P.S. and N.M. provided the supervision and research discussion. N.B., P.S., N.M. and S.N. reviewed the manuscript. All authors have read and agreed to the published version of the manuscript.

Funding: This research was supported by the Faculty of Science and Technology, Princess of Naradhiwas University, Thailand.

Acknowledgments: The authors are profoundly grateful to the Faculty of Science and Technology, Princess of Naradhiwas University, the Faculty of Science, King Mongkut's Institute of Technology Ladkrabang, and the Department of Physics and the Center of Excellence in Nanotechnology for Energy at the Prince of Songkla University for equipment and other support.

Conflicts of Interest: The authors declare no conflict of interest.

References

1. Sukwisute, P.; Muensit, N.; Soontaranon, S.; Rugmai, S. Micropower energy harvesting using poly (vinylidene fluoride hexafluoropropylene). *Appl. Phys. Lett.* **2013**, *103*, 063905. [CrossRef]
2. Rakbamrung, P.; Lallart, M.; Guyomar, D.; Muensit, N.; Thanachayanont, C.; Lucat, C.; Guiffard, B.; Petit, L.; Sukwisut, P. Performance comparison of PZT and PMN-PT piezoceramics for vibration energy harvesting using standard or nonlinear approach. *Sens. Actuators A Phys.* **2010**, *163*, 493–500. [CrossRef]

3. Lü, C.; Zhang, Y.; Zhang, H.; Zhang, Z.; Shen, M.; Chen, Y. Generalized optimization method for energy conversion and storage efficiency of nanoscale flexible piezoelectric energy harvesters. *Energy Convers. Manag.* **2019**, *182*, 34–40.

4. Acosta, M.; Novak, N.; Rojas, V.; Patel, S.; Vaish, R.; Koruza, J.; Rossetti, G.A.; Rödel, J. $BaTiO_3$-based piezoelectrics: Fundamentals, current status, and perspectives. *Appl. Phys. Rev.* **2017**, *4*, 041305. [CrossRef]

5. Shen, B.; Zhang, Q.; Zhai, J.; Xu, Z. DC field effect on dielectric property of Ba (Zr_xTi_{1-x}) O_3 ceramics. *Ceram. Int.* **2013**, *39*, S9–S13. [CrossRef]

6. Kuang, S.J.; Tang, X.G.; Li, L.Y.; Jiang, Y.P.; Liu, Q.X. Influence of Zr dopant on the dielectric properties and Curie temperatures of $Ba(Zr_xTi_{1-x})O_3$ ($0 \leq x \leq 0.12$) ceramics. *Scr. Mater.* **2009**, *61*, 68–71. [CrossRef]

7. Yang, L.; Kong, X.; Li, F.; Hao, H.; Cheng, Z.; Liu, H.; Li, J.-F.; Zhang, S. Perovskite lead-free dielectrics for energy storage applications. *Prog. Mater. Sci.* **2019**, *102*, 72–108. [CrossRef]

8. Julphunthong, P.; Chootin, S.; Bongkarn, T. Phase formation and electrical properties of Ba (Zr_xTi_{1-x}) O_3 ceramics synthesized through a novel combustion technique. *Ceram. Int.* **2013**, *39*, S415–S419. [CrossRef]

9. Liu, B.; Wu, Y.; Huang, Y.H.; Song, K.X.; Wu, Y.J. Enhanced dielectric strength and energy storage density in $BaTi_{0.7}Zr_{0.3}$ O_3 ceramics via spark plasma sintering. *J. Mater. Sci.* **2019**, *54*, 4511–4517. [CrossRef]

10. Puli, V.S.; Pradhan, D.K.; Riggs, B.C.; Chrisey, D.B.; Katiyar, R.S. Investigations on structure, ferroelectric, piezoelectric and energy storage properties of barium calcium titanate (BCT) ceramics. *J. Alloys Compd.* **2014**, *584*, 369–373. [CrossRef]

11. Dhakar, L.; Liu, H.; Tay, F.E.H.; Lee, C. A new energy harvester design for high power output at low frequencies. *Sens. Actuators A Phys.* **2013**, *199*, 344–352. [CrossRef]

12. Selvan, K.V.; Muhammad, M.S. Micro-scale energy harvesting devices: Review of methodological performances in the last decade. *Renew. Sust. Energ. Rev.* **2016**, *54*, 1035–1047. [CrossRef]

13. Zhang, X.; Gao, S.; Li, D.; Jin, L.; Wu, Q.; Liu, F. Frequency up-converted piezoelectric energy harvester for ultralow-frequency and ultrawide-frequency-range operation. *Appl. Phys. Lett.* **2018**, *112*, 163902. [CrossRef]

14. Jiwei, Z.; Xi, Y.; Liangying, Z.; Bo, S.; Chen, H. Orientation control and dielectric properties of sol–gel deposited Ba(Ti, Zr)O_3 thin films. *J. Cryst. Growth* **2004**, *262*, 341–347. [CrossRef]

15. Binhayeeniyi, N.; Sukvisut, P.; Thanachayanont, C.; Muensit, S. Physical and electromechanical properties of barium zirconium titanate synthesized at low-sintering temperature. *Mater. Lett.* **2010**, *64*, 305–308. [CrossRef]

16. Thanachayanont, C.; Yordsri, V.; Kijamnajsuk, S.; Binhayeeniyi, N.; Muensit, N. Microstructural investigation of sol–gel BZT powders. *Mater. Lett.* **2012**, *82*, 205–207. [CrossRef]

17. Wang, Z.; Zhao, K.; Guo, X.; Sun, W.; Jiang, H.; Han, X.; Tao, X.; Cheng, Z.; Zhao, H.; Kimura, H.; et al. Crystallization, phase evolution and ferroelectric properties of sol-gel-synthesized $Ba(Ti_{0.8}Zr_{0.2})O_3$–$x(Ba_{0.7}Ca_{0.3})TiO_3$ thin films. *J. Mater. Chem. C* **2013**, *1*, 522–530. [CrossRef]

18. Buscaglia, M.T.; Buscaglia, V.; Alessio, R. Coating of $BaCO_3$ crystals with TiO_2: Versatile approach to the synthesis of $BaTiO_3$ tetragonal nanoparticles. *Chem. Mater.* **2007**, *19*, 711–718. [CrossRef]

19. Mochizuki, Y.; Tsubouchi, N.; Sugawara, K. Synthesis of $BaTiO_3$ nanoparticles from TiO_2-coated $BaCO_3$ particles derived using a wet-chemical method. *J. Asian Ceram. Soc.* **2014**, *2*, 68–76. [CrossRef]

20. Maiwa, H. Electromechanical properties of Ba $(Zr_{0.2}Ti_{0.8})$ O_3 ceramics prepared by spark plasma sintering. *Ceram. Int.* **2012**, *38*, S219–S223. [CrossRef]

21. Xu, Q.; Zhan, D.; Huang, D.P.; Liu, H.X.; Chen, W.; Zhang, F. Dielectric inspection of $BaZr_{0.2}Ti_{0.8}O_3$ ceramics under bias electric field: A survey of polar nano-regions. *Mater. Res. Bull.* **2012**, *47*, 1674–1679. [CrossRef]

22. Hemeda, O.M.; Salem, B.I.; Abdelfatah, H.; Abdelsatar, G.; Shihab, M. Dielectric and ferroelectric properties of barium zirconate titanate ceramics prepared by ceramic method. *Phys. B* **2019**, *574*, 411680. [CrossRef]

23. Tang, X.G.; Wang, J.; Wang, X.X.; Chan, H.L.W. Effects of grain size on the dielectric properties and tunabilities of sol-gel derived Ba $(Zr_{0.2}Ti_{0.8})$ O_3 ceramics. *Solid State Commun.* **2004**, *131*, 163–168. [CrossRef]

24. Xue, D.; Gao, J.; Zhou, Y.; Ding, X.; Sun, J.; Lookman, T.; Ren, X. Phase transitions and phase diagram of $Ba(Zr_{0.2}Ti_{0.8})O_3$–$x(Ba_{0.7}Ca_{0.3})TiO_3$ Pb-free system by anelastic measurement. *J. Appl. Phys.* **2015**, *117*, 124107. [CrossRef]

25. Trainer, M. Ferroelectrics and the Curie–Weiss law. *Eur. J. Phys.* **2000**, *21*, 459–464. [CrossRef]

26. Sun, Z.; Li, L.; Zheng, H.; Yu, S.; Xu, D. Effects of sintering temperature on the microstructure and dielectric properties of $BaZr_{0.2}Ti_{0.8}O_3$ ceramics. *Ceram. Int.* **2015**, *41*, 12158–12163. [CrossRef]

27. Uchino, K.; Nomura, S. Critical exponents of the dielectric constants in diffused-phase-transition crystals. *Ferroelectr. Lett. Sect.* **1982**, *44*, 55–61. [CrossRef]
28. Ahmad, M.M.; Alismail, L.; Alshoaibi, A.; Aljaafari, A.; Kotb, H.M.; Hassanien, R. Dielectric behavior of spark plasma sintered BaTi$_{0.7}$Zr$_{0.3}$O$_3$ relaxor ferroelectrics. *Results Phys.* **2019**, *15*, 102799. [CrossRef]

Article

Piezoelectric Properties of $Pb_{1-x}La_x(Zr_{0.52}Ti_{0.48})_{1-x/4}O_3$ Thin Films Studied by In Situ X-ray Diffraction

Thomas W. Cornelius [1,*], Cristian Mocuta [2], Stéphanie Escoubas [1], Luiz R. M. Lima [3,4], Eudes B. Araújo [4], Andrei L. Kholkin [5,6] and Olivier Thomas [1]

[1]   Aix Marseille Univ, Univ Toulon, CNRS, IM2NP, CEDEX 20, 13397 Marseille, France; stephanie.escoubas@im2np.fr (S.E.); olivier.thomas@im2np.fr (O.T.)

[2]   Synchrotron SOLEIL, L'Orme des Merisiers, Saint-Aubin-BP 48, 91192 Gif-sur-Yvette, France; cristian.mocuta@synchrotron-soleil.fr

[3]   Faculty of Mechanical Engineering, University of Rio Verde (UniRV), Rio Verde 75901-970, Brazil; luizrogerio@unirv.edu.br

[4]   School of Natural Sciences and Engineering, Department of Physics and Chemistry, São Paulo State University (UNESP), Ilha Solteira 15385-000, Brazil; eudes.borges@unesp.br

[5]   Department of Physics & CICECO—Aveiro Institute of Materials, University of Aveiro, 3810-193 Aveiro, Portugal; kholkin@ua.pt

[6]   Laboratory of Functional Low-Dimensional Structures, National University of Science and Technology MISiS, 119049 Moscow, Russia

*   Correspondence: thomas.cornelius@im2np.fr

Received: 1 July 2020; Accepted: 24 July 2020; Published: 27 July 2020

Abstract: The piezoelectric properties of lanthanum-modified lead zirconate titanate $Pb_{1-x}La_x(Zr_{0.52}Ti_{0.48})_{1-x/4}O_3$ thin films, with x = 0, 3 and 12 mol% La, were studied by in situ synchrotron X-ray diffraction under direct (DC) and alternating (AC) electric fields, with AC frequencies covering more than four orders of magnitude. The Bragg reflections for thin films with low lanthanum concentration exhibit a double-peak structure, indicating two contributions, whereas thin films with 12% La possess a well-defined Bragg peak with a single component. In addition, built-in electric fields are revealed for low La concentrations, while they are absent for thin films with 12% of La. For static and low frequency AC electric fields, all lanthanum-modified lead zirconate titanate thin films exhibit butterfly loops, whereas linear piezoelectric behavior is found for AC frequencies larger than 1 Hz.

Keywords: X-ray diffraction; piezoelectric properties; lanthanum-modified lead zirconate titanate (PLZT)

---

1. Introduction

Lead zirconate titanate $Pb(Zr_{1-x}Ti_x)O_3$ (PZT) has been perhaps one of the most studied ferroelectric materials, due to its excellent piezoelectric, pyroelectric, ferroelectric and dielectric properties [1]. Due to its remarkable technological importance, a wide range of Zr/Ti ratios has been investigated from both a scientific and a technological point of view. The special interest is in compositions close to the morphotropic phase boundary (MPB), occurring at around $Zr/Ti \sim 52/48$, because the arising monoclinic C_m phase enhances the piezoelectric response [2,3]. By doping PZT with different ions, its physical properties can be radically modified, or new properties appear with potential for use in different technological devices.

Lanthanum-modified lead zirconate titanate, $Pb_{1-x}La_x(Zr_{1-y}Ti_y)_{1-x/4}O_3$ (PLZT $x/1-y/y$), is a material with peculiar properties, with a clear potential for technological applications [4,5]. The addition of La^{3+} to the conventional PZT system distorts the unit cell and decreases the concentration of oxygen vacancies, leading to transitions from normal ferroelectric to relaxor behavior with the increase in La content [6]. Since the discovery of PLZT in 1971, relaxor ferroelectric compositions with higher relative permittivity, spontaneous polarization and lower leakage current [7] have been widely investigated as thin films for a variety of applications, including non-volatile memory and piezoelectric devices. In contrast to ordinary ferroelectrics, whose physical properties—including ferroelectric, dielectric and piezoelectric responses—are adequately described by the Landau–Ginzburg–Devonshire theory [8], relaxor ferroelectrics escape a simple physical description, and possess a number of unique features that make them promising candidates for technological applications, such as piezoelectric transducers and light shutters. Despite several experiments and proposed theoretical models [9] to explain the origin of the relaxor phenomenon in terms of polar nanoregions—relaxor materials exhibit a diffuse phase transition, a large temperature and frequency-dependent dielectric maximum—there is no consensus between different theoretical models. Its real nature is still a subject of discussion, and the fundamental physics of the relaxors remain a fascinating puzzle [10,11]. With their complex structures and intriguing properties, relaxors represent truly a frontier of research in ferroelectrics and related materials, offering great opportunities both for fundamental breakthroughs and for technological applications. In addition to the classic relaxor PLZT 9/65/35 composition, the PLZT 12/52/48 also exhibits a relaxor behavior.

A photovoltaic effect has been reported in several ferroelectric perovskite oxides, including PZT and $BaTiO_3$. It is noteworthy that the maximum photocurrent and photovoltage are also observed at different PLZT compositions [12]. Regarding compositions at around the MPB, the maximum photovoltaic properties have been reported at PLZT 3/52/48 [13]. PLZT was demonstrated to be photovoltaically active, with efficiencies in the range of 0.28%. Theoretical analyses predict that an extremely high efficiency may exist in high-quality ferroelectric ultrathin films, reaching 18.7% for 8-nm thin PLZT films, which is comparable with, or even higher than, semiconductor-based photovoltaics [13].

To study the piezoelectric properties of ferroelectric thin films, various techniques, such as piezoresponse force microscopy (PFM), laser interferometry [14] and in situ X-ray diffraction (XRD) during the application of electric fields, exist [15–18]. While PFM is a surface-sensitive tool, XRD allows for measuring the piezoelectrically-induced strain within the complete depth of a thin film, with strain resolution of 10^{-4}. In addition, X-ray diffraction provides access to the piezoelectric anisotropy of the material by measuring different Bragg reflections, and thus probing grains of different orientations. This in situ technique has been used on PZT thin films to study domain switching [19], the imprint effect [20,21] and the structural evolution during imprint [22].

The present work focuses on the piezoelectric properties of PLZT x/52/48 thin films, with $x = 0, 3$ and 12 mol% La, studied via electrical measurements of the dielectric permittivity and the polarization as a function of applied electric fields, as well as by in situ synchrotron X-ray diffraction during the application of DC and AC electric fields with frequencies covering more than four orders of magnitude. The interrelation between PLZT compositions, characterized by their piezoelectric, photovoltaic and relaxor properties and their in situ piezoelectric response, is investigated.

2. Materials and Methods

Lanthanum-modified lead zirconate titanate $Pb_{1-x}La_x(Zr_{0.52}Ti_{0.48})_{1-x/4}O_3$ (PLZT x/52/48) thin films with $x = 0, 3$ and 12 mol% La were prepared via the acetate solution route. For the preparation of the solution, stoichiometric zirconium butoxide $Zr(OC_4H_9)_4$ (Aldrich 80%), titanium isopropoxide $C_{12}H_{28}O_4Ti$ (Fluka), lead acetate $(CH_3COO)_2Pb.3H_2O$ (Dinâmica) and lanthanum oxide La_2O_3 (Aldrich) precursors were added sequentially. The titanium isopropoxide was put into a beaker on a hot plate containing zirconium butoxide at room temperature under stirring for 5 min. Subsequently, 1 mL of glacial acetic acid was added to the solution at room temperature under stirring for an additional 5 min. Finally, lead acetate and lanthanum oxide were added to the solution, increasing the temperature in the sequence to 80 °C. After completing the solubilization by adding 2 mL of acetic acid and 1 mL of distilled water (~20 min), the hot plate was switched off, keeping the solution under stirring until it cooled down to room temperature. The solution was stirred throughout the process to obtain a final solution (0.4 M) that was completely transparent and stable.

For the preparation of PLZT thin films, the precursor solutions were initially deposited on $Pt/TiO_2/SiO_2/Si(100)$ substrates by spin coating at 5000 rpm for 30 s, then placed on a hot plate at ~200 °C for 5 min to remove water and, in a final step, pyrolyzed in an electric furnace (Furnace EDG 1800, The Mellen Company Inc., Concord, NH, USA) at 300 °C for 10 min. The same procedure was repeated on the previously annealed film to increase the film thickness. Finally, the films were crystallized in an electric furnace (in air) at 700 °C for 30 min. The thickness of the final films was ~500 nm. The obtained films are denoted as PLZT0 (0% La), PLZT3 (3% La) and PLZT12 (12% La). Scanning electron microscopy images and energy dispersive X-ray spectra of the PLZT3 and the PLZT12 thin films can be found in Figures S1 and S2 in the Supplementary Material. For electrical measurements, circular Au top electrodes of 500 μm diameter were sputtered on the film surfaces using a homemade shadow mask. The DC electric field dependence of the dielectric permittivity $\varepsilon(E)$ was measured using an Agilent 4284A LCR meter (Keysight Technologies, Barueri, Brazil) at 100 kHz. A modified Sawyer-Tower circuit at 10 Hz was employed to measure the polarization vs. electric field (*P-E*) hysteresis loops.

in situ X-ray diffraction experiments were performed at the DiffAbs beamline at SOLEIL synchrotron (St Aubin, France) [23,24]. The incident monochromatic 10 keV X-ray beam was collimated to a size of 50 μm using a pinhole placed about 20 cm upstream of the sample. This spot size ensured that the footprint of the incident X-ray beam was always smaller than the electrode diameter at a fixed incident angle of 10° used during the diffraction measurements. Thus, this configuration guarantees experiments in the central part of the electrodes where homogeneous electric fields are expected, hence avoiding any edge effects. The in situ measurements were performed with co-planar vertical diffraction geometry with the sample mounted horizontally on a *xyz* translation stage for precise (better than 1 μm) sample positioning. The diffracted X-rays were recorded using a two-dimensional hybrid pixel area detector (XPAD, ImXPAD, La Ciotat, France) with 560×960 pixels and a pixel size of 130 μm (details about such detectors and data conversion can be found in [25]). It was installed at a distance of 650 mm downstream from the sample, with its short dimension along the vertical direction, thus covering an angular range of about 6.5° in 2θ. In order to apply an electric field, one of the gold electrodes was contacted electrically using a thin gold wire with a diameter of 50 μm while another contact was taken at the Pt back electrode. The sample was then laterally scanned, generating *XY* maps with Au diffraction contrast that allow for a precise positioning of the particular (contacted) Au electrode with respect to the incident X-ray beam. Static and alternating electric fields were applied, employing an Analog Output card (model PXI 3U from ADLINK) that allows for the generation of a bipolar tension in the range of ±10 V (5 mA maximum current). For AC measurements of the piezoelectric hysteresis loops, the X-ray diffraction signal was acquired by accumulating (internally) several thousand images, each of them with very short exposure (typically 1 μs) and synchronously taken at the very same voltage during the AC cycle [24]. Once the counting statistics were sufficient, the synchronization was shifted to the next voltage point in order to describe the hysteresis loop.

3. Results

3.1. Electric Measurements

Hysteresis loops of the polarization P, as a function of the applied electric field E of the studied PLZT thin films with different compositions measured at room temperature, are presented in Figure 1. The *P-E* hysteresis loops show the signature of a normal ferroelectric for PLZT0 ($x = 0$ mol% La), or PZT 52/48, while a slim hysteresis loop is observed for PLZT12, which is a typical characteristic of a relaxor material. The remanent polarizations (P_r) amount to 18.7, 10.5 and 8.1 $\mu C/cm^2$ for the PLZT0, PLZT3 and PLZT12 thin films, respectively, while the coercive fields (E_c) are 125.8, 142.0 and 94.7 kV/cm for the same films. As expected, increasing lanthanum doping decreases the remanent polarization, such that the smallest P_r is observed for the relaxor composition PLZT12 ($x = 12$ mol% La). In addition, the hysteresis loops are asymmetric, and shifted towards positive electric fields, in particular for the PLZT0 and PLZT3 thin films. These asymmetries suggest a macroscopic self-polarization effect in the studied films. The asymmetries are quantified by the differences $\Delta P_r = P_r^+ - P_r^-$ and $\Delta E_c = E_c^+ - E_c^-$, which are given in Table 1 for the different compositions. The difference in polarization is negative, with the highest absolute value for the PLZT0 thin film, and the absolute value decreases to zero for PLZT12. The offset in the electric field is also the largest for the pure PZT thin film, and absent for the relaxor thin film. These results demonstrate that the self-polarization effect is more pronounced in PLZT0 (PZT 52/48), and tends to vanish in the relaxor composition PLZT12. The positive offset in the electric field ($\Delta E_c > 0$) for both PLZT0 and PLZT3 compositions indicates the presence of an internal electric field pointing towards the bottom electrode.

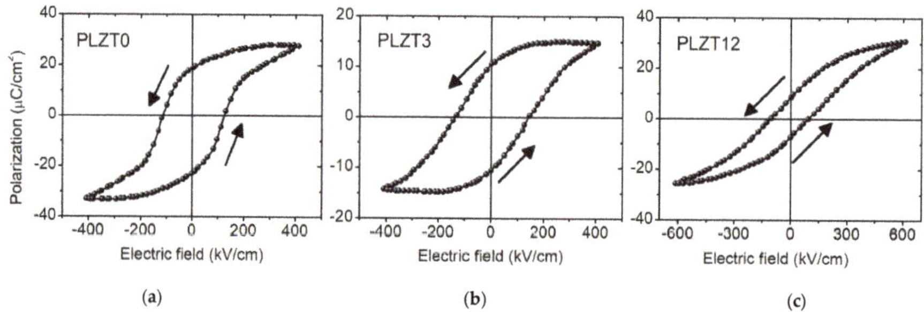

Figure 1. *P-E* hysteresis loops at a frequency of $f = 10$ Hz for the studied PLZT0 (**a**), PLZT3 (**b**) and PLZT12 (**c**) thin films.

Table 1. Summary of remanent polarization (P_r^+), electric coercive field (E_c^+) and differences in remanent polarization (ΔP_r) and electric coercive field (ΔE_c).

Sample	P_r^+ (μC/cm^2)	E_c^+ (kV/cm)	$\Delta P_r = P_r^+ - P_r^-$ (μC/cm^2)	$\Delta E_c = E_c^+ - E_c^-$ (kV/cm)
PLZT0	18.7	125.8	−4.2	+6.3
PLZT3	10.5	142.0	−0.1	+4.7
PLZT12	8.1	94.7	0.0	0.0

Besides *P-E* hysteresis loops, capacitance, as a function of electric field curves, displayed as dielectric permittivity ε vs. electric field E curves in Figure 2, is an alternative way to reveal the hysteresis in the studied PLZT thin films. The butterfly-like shapes of the ε-E curves also signify the polarization switching in ferroelectrics. Although the peaks in the ε-E curves reflect the double coercive electric field, there is some incompatibility in the values compared to the coercive fields inferred from the *P-E* hysteresis loops shown in Figure 1, because the ε-E curves were recorded at higher frequencies (100 kHz). An asymmetry is also observed in Figure 2 for PLZT0 and PLZT3, in contrast to the symmetric curve for the PLZT12. These data confirm the existence of macroscopic self-polarization in the studied PLZT films, which is more pronounced in the PLZT0 film and non-existent in the PLZT12 film.

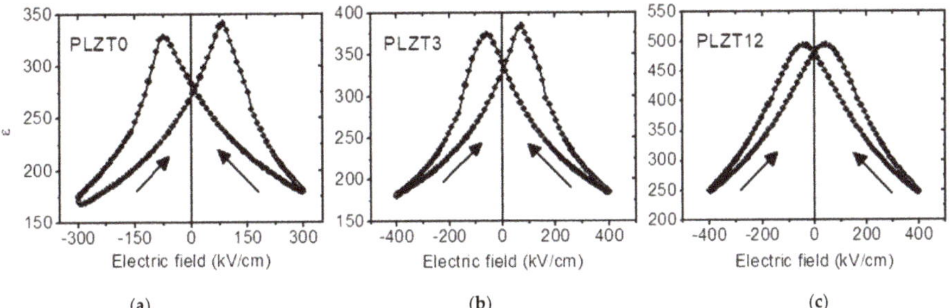

Figure 2. Dielectric permittivity–electric field (ε-E) curves for PLZT0 (**a**), PLZT3 (**b**) and PLZT12 (**c**) films with $x = 0, 3$ and 12 mol% La recorded at 100 kHz.

3.2. In situ X-ray Diffraction

The profiles of the PLZT 110 Bragg peaks of the three different compositions of the PLZT thin films, for various voltages applied during a DC voltage sweep ranging from −9 V to +9 V, are presented in Figure 3. The profiles for the different applied voltages are vertically shifted to improve their visibility. While the Bragg reflection of the PLZT0 and PLZT3 thin films consists of at least two contributions, the PLZT12 thin film exhibits a single well-defined Bragg peak.

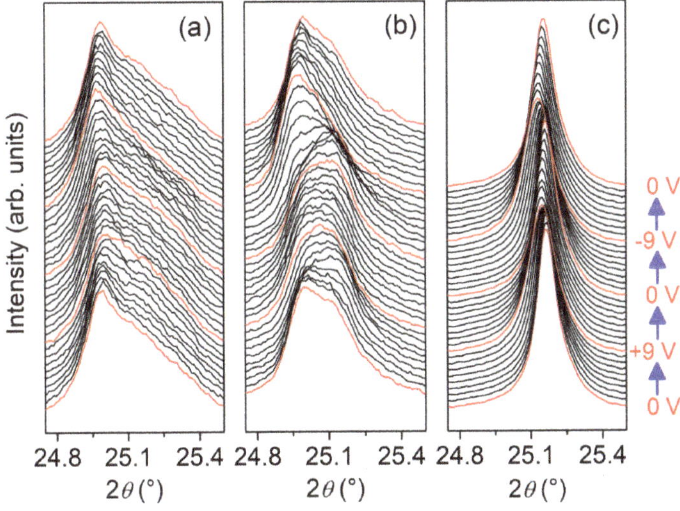

Figure 3. Profiles of the PLZT 110 Bragg peak for (**a**) PLZT0, (**b**) PLZT3 and (**c**) PLZT12 thin film. The profiles for different applied voltages, ranging from −9 V to +9 V, are shifted vertically for clarity. One can note the presence of double peak features for the PLZT0 and PLZT3 samples, and a simple Bragg peak for the PLZT12 thin film.

For high-resolution X-ray diffraction, the conventional interpretation of the double peak structure observed in Figure 3a,b would lead to the erroneous assignment of secondary phases. In the case of PZT systems, a secondary peak can be associated to the usually reported pyrochlore phase. The presence of the pyrochlore phase in PZT system is often associated to stoichiometry deviation or uncomplete perovskite growth during synthesis. However, the PLZT thin films studied in the present work were prepared under strict stoichiometric control and rigorous thin film growth protocols, leading to a very low concentration of pyrochlore phase (as demonstrated by the diffractogram shown in Figure S3 in the Supplementary Material). In addition, the main Bragg peak of the pyrochlore phase in the PZT system is observed at $2\theta \sim 24.2°$ (not shown here). Hence, the observed double peak structure cannot be associated to the pyrochlore phase, but can be explained in terms of an adaptive diffraction phenomenon of the PZT monoclinic phases at the MPB, as will be discussed further below [26]. When applying an electrical potential ranging from −9 V to +9 V, the 2θ angle of the Bragg reflections varies.

The piezoelectrically-induced strain, inferred from the 2θ position of the center of mass of the Bragg reflections for the three PLZT thin films, is presented in Figure 4a–c. For all three compositions, the strain vs. electric field curve shows butterfly loops. The largest strain is obtained for the PLZT thin film with a La concentration of 3%. The cross-point where the two wings of the butterfly cross each other shifts from 0 kV/cm for a La concentration of 12% to 60 kV/cm for 3% La, and to 100 kV/cm for the pure (0% La) PZT thin film (as displayed in Figure 4d). The strain at these electric field cross-points is compressive for the pure PZT and PLZT3 thin films, amounting to −0.5% and −0.6%, respectively, while the PLZT12 thin film is almost strain-free at the cross-point of the butterfly loops (see Figure 4e). These differences go along with the asymmetry of the two wings of the butterfly loops, as also illustrated by the ratio of the areas of the two wings presented in Figure 4f, which decreases with the increasing La concentration. Asymmetric loops are generally caused by an internal bias field within the thin films, whose magnitude and distribution depend on the composition/microstructure and on the thermal/electrical history of the system.

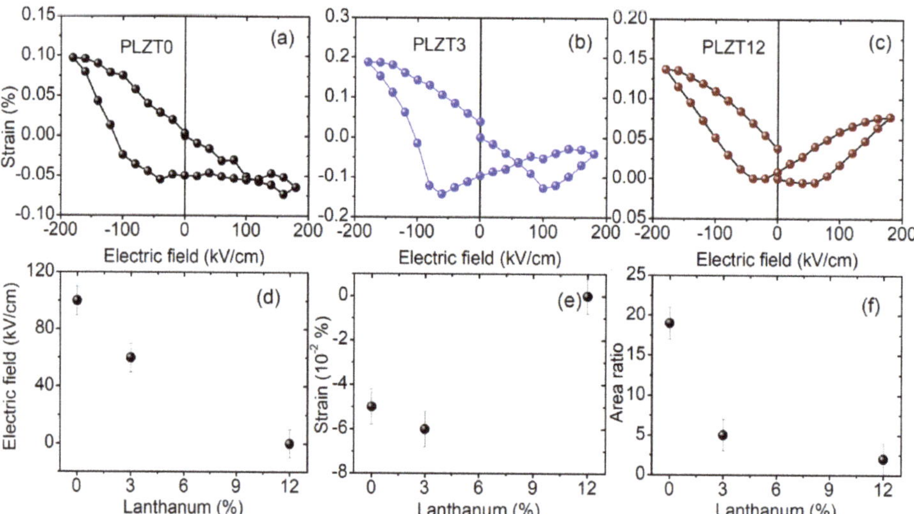

Figure 4. Piezoelectrically-induced strain inferred from the 2θ-displacement of the center of mass of the PLZT 110 Bragg reflection for PLZT thin films shown in Figure 3, as a function of the applied voltage: (**a**) PLZT0, (**b**) PLZT3 and (**c**) PLZT12. (**d**) Electric field and (**e**) strain at which the two wings of the butterfly loops presented in parts (**a–c**) cross each other as a function of the La concentration in the PLZT thin films. (**f**) Ratio of the areas of the left and right 'wings' of the butterfly loop, as a function of the La concentration.

To further analyze the two contributions of the Bragg reflection for the pure PZT thin film (PLZT0) and the PLZT3 thin film, the profiles of the PLZT 110 Bragg peaks were fitted using two pseudo-Voigt functions. The strains induced in the PLZT0 and the PLZT3 thin films by the applied electric field are presented in Figure 5a,d, respectively. While for the pure PZT thin film, only the main contribution of the Bragg reflection seems to be actually sensitive to the applied electric field, for the PLZT3 thin film both contributions show similar behaviors. However, the noise, and thus the uncertainty regarding the contribution, at larger 2θ angles is much larger, which might indicate that the seemingly piezoelectrically-induced strain is actually an artefact. The full width at half maximum (FWHM) for both pseudo-Voigt functions used to fit the Bragg reflections shows the same behavior as the piezoelectrically-induced strain (Figure 5b,e). Note that the FWHM is very different (factor of 2 to 3) for the two contributions. In the case of the PLZT3 thin film, the FWHM first increases when reducing the electric field to negative values, indicating an increased disorder, followed by a decreasing peak width for $E < -100$ kV/cm. This reducing peak width goes along with increasing strain, changing from compressive to tensile strain, implying the switching of domains. This particular evolution of the integrated XRD peak intensity (area) (Figure 5c,f) and its corresponding FWHM, not necessarily in phase with each other, makes the butterfly loop effect even more visible in the scattered intensity distribution (e.g., the maximum intensity of this Bragg peak, see Figure S4 of the Supplementary Material). In the case of the PLZT3 thin film, it is easily seen that both of the peaks (sharp and broad contributions) exhibit changes vs. the applied voltage.

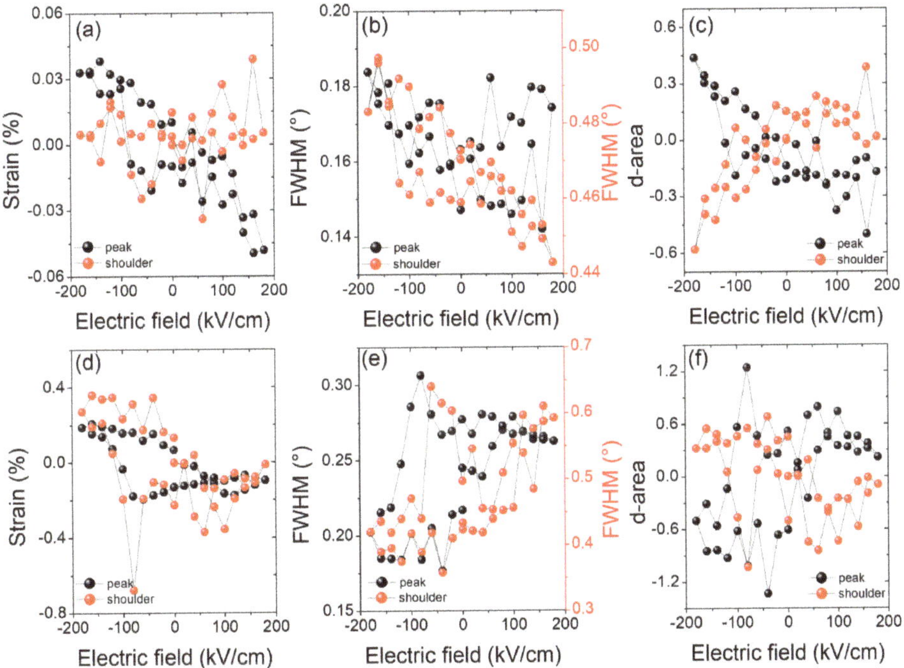

Figure 5. (**a**) Piezoelectrically-induced strain, (**b**) full width at half maximum (FWHM), and (**c**) variation of the area below the two pseudo-Voigt functions, used for fitting the profiles of the Bragg peaks shown in Figure 1a for the PLZT0 thin film (with respect to $E = 0$ kV/cm). (**d**) Piezoelectrically-induced strain, (**e**) full width at half maximum, and (**f**) variation of the area below the two pseudo-Voigt functions used for fitting the profiles of the Bragg peaks shown in Figure 1b for the PLZT3 thin film (with respect to $E = 0$ kV/cm).

The piezoelectrically-induced strain in the thin films, as a function of the applied AC electric field with frequencies ranging from 0.05 Hz to 1250 Hz, is displayed in Figure 6. Due to the low intensity of the diffracted X-rays during the AC measurements, the Bragg reflections could not be reliably fitted using two pseudo-Voigt functions, and thus could not separate the two contributions. Therefore, the strain (one single component) was inferred here from the center of mass of the Bragg reflections. For the thin film with low La concentration (PLZT0 and PLZT3), a quasi-linear piezoelectric behavior is observed for AC frequencies > 1 Hz, while for very low frequencies (< 1 Hz) the shape of the hysteresis loops resembles, again, butterfly loops, and thus the piezoelectric behavior of the thin films approaches the same behavior as seen for the DC electric fields. The piezoelectrically-induced strain for the PLZT12 thin film (Figure 6c), on the other hand, does not show this linear behavior, but is closer to a typical piezoelectric hysteresis curve over the four orders of magnitude in frequency tested in the present work. As illustrated in the inset of Figure 6c, the wings of the butterfly loops for the PLZT12 thin film, at AC frequencies of 62.5 and 1250 Hz, are not apparent.

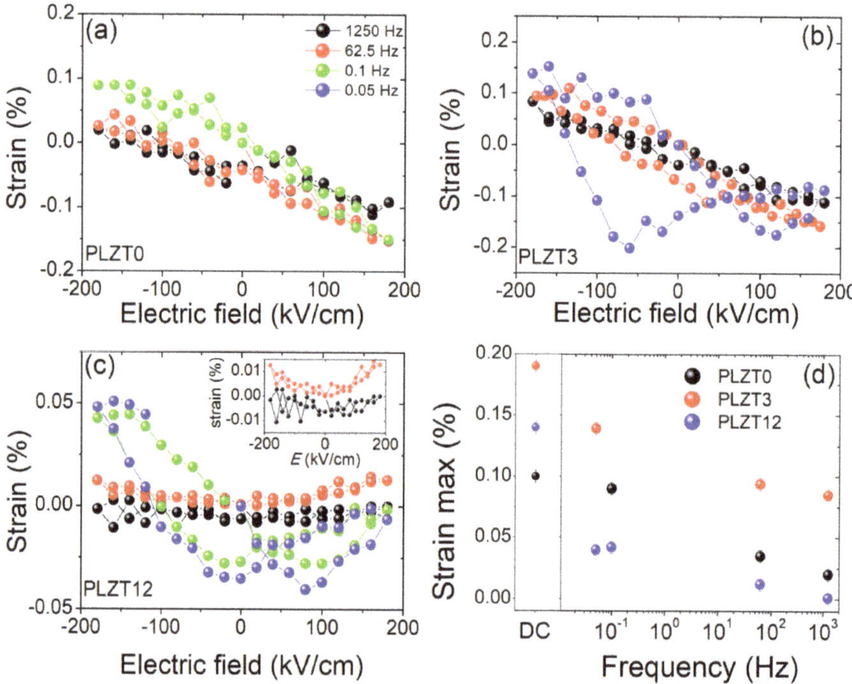

Figure 6. Piezoelectrically-induced strain as a function of electric field for different AC frequencies for (**a**) the PLZT0, (**b**) the PLZT3 and (**c**) the PLZT12 thin films. (**d**) Maximum piezoelectrically-induced strain at $E = -180$ kV/cm as a function of the applied AC frequency.

As illustrated by Figure 6d, for all three PLZT thin films studied in this work, the maximum strain measured at -180 kV/cm decreases for AC electric fields compared to DC electric fields. This decrease is particularly pronounced for the PLZT12 sample, where the strain decreases by a factor of about 3, compared to a factor of about 2 for PLZT3 and 1.2 for PLZT0. In addition, it decreases with the increasing AC frequency.

4. Discussion

Studies of the monoclinic phases in the PZT system, usually conducted by high-resolution X-ray diffraction, remain a fascinating topic. Three types of monoclinic phases (M_A, M_B and M_C) have been reported to exist in this system around the MPB [27], as well as in similar ferroelectric systems where the presence of a MPB depends on the composition as well as on the thermal and electric histories [28]. Due to the complex phases in the PZT system, factors like internal stress, small domain sizes and the coexistence of multiple phases easily affect the diffraction patterns, and thus strongly complicate their analysis regarding the determination of the precise phases, as well as their lattice parameters and other related effects. Although the intermediate phases may be associated with rotational polarization instabilities [29], an explanation for these phases around the MPB using the classical Landau–Ginzburg–Devonshire theory is rather difficult, since the homogeneous ferroelectric phases can be satisfactorily described only if high order terms are included in the free energy expansion [30]. In addition to these difficulties, the polarization rotation model indicates that the crystallographic anisotropy of the polarization direction disappears around the MPB as a consequence of the polarization instabilities, although the polarization rotation leads to a giant piezoelectric response in these ferroelectric systems [29]. In this scenario, an adaptive ferroelectric phase model was developed for intermediate monoclinic phases near the morphotropic phase boundaries in the ferroelectrics of complex oxides, based on the conformal miniaturization of stress-accommodating tetragonal and rhombohedral microdomains that coincide with the M_C and M_A monoclinic phases [31]. In the adaptive diffraction phenomenon, contributions to the Bragg reflections are determined by the coherent superposition of waves scattered from individual twin-related nanocrystals, and they are adaptively shifted along the twin peak splitting vectors as a consequence of changes in the twin variant volume fraction [26]. Applied to the tetragonal phase, this theory could explain the intrinsic lattice parameter relationships of the monoclinic M_C phase also observed in lead magnesium niobate–lead titanate solid solutions (PMN-PT). On the other hand, the (001) and (110) twin planes diffract incident waves just like the monoclinic M_A and M_B phases when applied to the rhombohedral phase of nanotwin superlattices, indicating that one or more extra diffraction peaks would appear when nanodomains coexist with coarse domains [32].

Based on the above discussion, since the presence of the pyrochlore phase is discarded in the studied PLZT thin films, the double peak structure found for low lanthanum concentrations (Figure 3a,b) suggests an adaptive diffraction phenomenon, which disappears with high lanthanum content (as shown in Figure 3c). This double peak structure is thus probably composed by the inconsistent sum of the conventional peak and the corresponding new adaptive peak originating from the coexistence of nanodomains and coarse domains. With increasing lanthanum content, the tetragonality in the PZT system decreases, thus leading to a reduction of the nanodomain and coarse domain coexistence, associated with a decrease in the spontaneous polarization in the unit cell. This reasonably explains the disappearance of the double peak structure with the increasing lanthanum content. Another possibility is that the diffraction pattern of the lanthanum-doped PZT structure behaves like an adaptive diffraction of nanotwins, where scattered waves from several nanodomains overlap coherently to form a single Bragg peak [32]. However, further studies on this system are needed to corroborate or refute this assumption.

Electrical measurements, both of *P-E* hysteresis curves and *ε-E* curves, evidenced a built-in electric field in lanthanum-modified lead zirconate titianate thin films, which is the largest for pure PZT and diminishes with increasing lanthanum content. These findings were confirmed by in situ synchrotron X-ray diffraction, revealing asymmetric butterfly loops for DC electric fields. The butterfly shape of the hysteresis loops disappears for AC electric fields with frequencies larger than 1 Hz. Similarly, differences in the coercive fields inferred from the *P-E* hysteresis loops and from *ε–E* curves, which were recorded at 10 Hz and 100 kHz, respectively, were found. This frequency dependence may originate from internal fields within the thin films that have to be overcome, and thus higher electric fields would have to be applied to eventually switch the domains. The same effect may also be the origin of the reduced piezoelectrically-induced strain for AC electric fields, compared to DC electric fields. Considering that both the butterfly shape and the absolute value of the strain are recovered at very low frequencies, i.e., for quasi-static electric fields, this indicates that the piezoelectric domains are actually hindered from switching, or that the switching process is significantly slowed down. It is well-known that quenched electric fields in ferroelectric capacitors may lead to the significant slowing down of polarization switching, due to pinning and creep effects [33,34]. A transformation of the domain structure, from micrometric 90° domains to a polar nanodomain configuration, was demonstrated for bulk lanthanum-modified tetragonal PZT with increasing La content [35]. Similarly, transmission electron microscopy studies on PLZT 10/20/80 films also revealed the absence of the normal 90° domain configuration, while it was present for $Pb_{0.9}(Zr_{0.2}Ti_{0.8})O_3$ films [36]. This transformation of the domain structure may be the origin of the frequency-dependence reported in the present work. In addition, the interaction of the 90° domain walls with point defects may contribute to higher activation voltages. Thus, future and ongoing works focus on the behavior of PLZT thin films with regard to higher electric fields, as well as the function of time, in order to elucidate the origin of the aforementioned phenomena.

5. Conclusions

In conclusion, the piezoelectric properties of lanthanum-modified lead zirconate titanate thin films around the morphotropic phase boundary of pure PZT, with Zr/Ti = 52/48 and with different lanthanum concentrations ranging from 0 to 12 mol%, were investigated by in situ synchrotron X-ray diffraction during the application of DC and AC electric fields with frequencies covering more than four orders of magnitude. A double-peak Bragg reflection was found for thin films with low lanthanum concentrations, which disappears for PLZT films with 12 mol% La content. This double-peak structure is attributed to the inconsistent sum of the conventional Bragg peak and the corresponding new adaptive peak, originating from the coexistence of nanodomains and coarse domains. The piezoelectrically-induced strain describes butterfly loops for all compositions, which are symmetric for $x = 12$ mol% and asymmetric for lower La concentration. These asymmetries are due to internal electric fields in the PLZT thin films, which increase with decreasing x and which were also evidenced by electrical measurements. While the PLZT12 thin film shows butterfly-like loops for DC as well as AC electric fields, thin films with lower La concentrations exhibit linear piezoelectric behavior for AC frequencies larger than 1 Hz, whereas butterfly loops are apparent for low AC frequencies of about 0.1 Hz. The maximum piezoelectrically-induced strain was found to decrease for AC electric fields compared to DC electric fields, and it diminishes further with increasing AC frequency.

Supplementary Materials: The following are available online at http://www.mdpi.com/1996-1944/13/15/3338/s1, Figure S1: Scanning electron microscopy images of a PLZT3 and a PLZT12 thin film, Figure S2: Energy dispersive X-ray spectrum of a PLZT3 and a PLZT12 thin film, Figure S3: Diffractograms of the three PLZT thin films studied in the present work, Figure S4: Representation of the XRD profiles measured for the different applied DC voltages for (a) the PLZT0 and (b) the PLZT3 thin film.

Author Contributions: L.R.M.L. and E.B.A. fabricated the samples and performed the electrical measurements on them. T.W.C., C.M., S.E. and O.T. performed the synchrotron experiments. T.W.C. and C.M. analyzed the X-ray diffraction data. E.B.A., A.L.K. and T.W.C performed the formal analysis. T.W.C. prepared the original draft. All authors participated to the reviewing and editing of the manuscript. All authors have read and agreed to the published version of the manuscript.

Funding: This work was supported by the Fundação de Amparo à Pesquisa do Estado de São Paulo-FAPESP (Project: 2017/13769-1); Conselho Nacional de Desenvolvimento Científico e Tecnológico-CNPq (Grant: 304604/2015-1 and Project No. 400677/2014-8); and Coordenação de Aperfeiçoamento de Pessoal de Nível Superior-CAPES (CAPES-PRINT Project: 88881.310513/2018-01 and CAPES-COFECUB Project No. 801-14).

Acknowledgments: The authors gratefully acknowledge the SOLEIL Synchrotron for allocating the beam time. Detector and Electronics Support Groups as well as Philippe Joly are acknowledged for excellent technical support during the experimental campaign at SOLEIL Synchrotron on DiffAbs beamline. The authors further acknowledge Florian Ruske from Helmholtz Zentrum Berlin for Materials and Energy for SEM and EDX analysis. We would like to thank the Brazilian agencies FAPESP, CNPq and CAPES for their financial support. Part of this work was developed within the scope of the project CICECO-Aveiro Institute of Materials, UIDB/50011/2020 & UIDP/50011/2020, financed by national funds through the FCT/MEC and, when appropriate, cofinanced by FEDER under the PT2020 Partnership Agreement. ALK acknowledges Ministry of Education and Science of the Russian Federation for the support in the framework of the Increase Competitiveness Program of NUST MISiS under Grant K2-2019-015.

Conflicts of Interest: The authors declare no conflict of interest.

References

1. Jaffe, B.; Cook, W.R.; Jaffe, H. *Piezoelectric Ceramics*; Elsevier: New York, NY, USA, 1971; p. 146.
2. Noheda, B.; Cox, D.E.; Shirane, G.; Gonzalo, J.A.; Cross, L.E.; Park, S.-E. A monoclinic ferroelectric phase in the Pb(Zr$_{1-x}$Ti$_x$)O$_3$ solid solution. *Appl. Phys. Lett.* **1999**, *74*, 2059. [CrossRef]
3. Noheda, B.; Gonzalo, J.A.; Cross, L.E.; Guo, R.; Park, S.E.; Cox, D.E.; Shirane, G. Tetragonal-to-monoclinic phase transition in a ferroelectric perovskite: The structure of PbZr$_{0.52}$Ti$_{0.48}$O$_3$. *Phys. Rev. B* **2000**, *61*, 8687. [CrossRef]
4. Haertling, G.H. Electro-optic ceramics and devices. In *Electronic Ceramics*; Levinson, L.M., Ed.; Marcel Dekker: New York, NY, USA, 1988.
5. Haertling, G.H. PLZT electrooptic materials and applications—A review. *Ferroelectrics* **1987**, *75*, 25. [CrossRef]
6. Zou, Q.; Ruda, H.; Yacobi, B.; Farrell, M. Microstructural characterization of donor-doped lead zirconate titanate films prepared by sol-gel processing. *Thin Solid Films* **2002**, *402*, 65–70. [CrossRef]
7. Aggarwal, S.; Ramesh, R. Point defect chemistry of metal oxide heterostructures. *Ann. Rev. Mater. Sci.* **1998**, *28*, 463. [CrossRef]
8. Lines, M.E.; Glass, A.M. *Principles and Applications of Ferroelectrics and Related Materials*; Oxford University Press: Clarendon, UK, 2001.
9. Kholkin, A.; Morozovska, A.; Kiselev, D.; Bdikin, I.; Rodriguez, B.; Wu, P.; Bokov, A.; Ye, Z.-G.; Dkhil, B.; Chen, L.-Q.; et al. Surface domain structure and mesoscopic phase transition in relaxor ferroelectrics. *Adv. Funct. Mater.* **2011**, *21*, 1977–1987. [CrossRef]
10. Kleemann, W. The relaxor enigma—Charge disorder and random fields in ferroelectrics. *J. Mater. Sci.* **2006**, *41*, 129–136. [CrossRef]
11. Shvartsman, V.; Dkhil, B.; Kholkin, A.L. Mesoscale domains and nature of the relaxor state by piezoresponse force microscopy. *Ann. Rev. Mater. Res.* **2013**, *43*, 423–449. [CrossRef]
12. Poosanaas, P.; Uchino, K. Photostrictive effect in lanthanum-modified lead zirconate titanate ceramics near the morphotropic phase boundary. *Mater. Chem. Phys.* **1999**, *61*, 36–41. [CrossRef]
13. Qin, M.; Yao, K.; Liang, Y.C. High efficient photovoltaics in nanoscaled ferroelectric thin films. *Appl. Phys. Lett.* **2008**, *93*, 122904. [CrossRef]
14. Kholkin, A.L.; Wütchrich, C.; Taylor, D.V.; Setter, N. Interferometric measurements of electric field-induced displacements in piezoelectric thin films. *Rev. Sci. Instrum.* **1996**, *67*, 1935–1941. [CrossRef]
15. Young, J.; Chen, P.; Sichel, R.J.; Callori, S.J.; Sinsheimer, J.; Dufresne, E.M.; Dawber, M.; Evans, P.G. Nanosecond dynamics of ferroelectric/dielectric superlattices. *Phys. Rev. Lett.* **2011**, *107*, 055501.
16. Gorfman, S.; Simons, H.; Iamsasri, T.; Prasertpalichat, S.; Cann, D.P.; Choe, H.; Pietsch, U.; Watter, Y.; Jones, J.L. Simultaneous resonant x-ray diffraction measurement of polarization inversion and lattice strain in polycrystalline ferroelectrics. *Sci. Rep.* **2016**, *6*, 20829. [CrossRef]
17. Pramanick, A.; Damjanovic, D.; Daniels, J.E.; Nino, J.C.; Jones, J.L. Origins of electro-mechanical coupling in polycrystalline ferroelectrics during subcoercive electrical loading. *J. Am. Ceram. Soc.* **2011**, *94*, 293–309. [CrossRef]
18. Gorfman, S.; Schmidt, O.; Tsirelson, V.; Ziolkowski, M.; Pietsch, U. Crystallography under external electric field. *Z. Anorg. Allg. Chem.* **2013**, *639*, 1953–1962. [CrossRef]
19. Lee, K.S.; Kim, Y.K.; Baik, S.; Kim, J.; Jung, I.S. In situ observation of ferroelectric 90°-domain switching in epitaxial Pb(Tr,Ti)O$_3$ thin films by synchrotron X-ray diffraction. *Appl. Phys. Lett.* **2001**, *79*, 2444–2446. [CrossRef]
20. Baturin, I.; Menou, N.; Shur, V.; Muller, C.; Kuznetsov, D.; Hodeau, J.L.; Sternberg, A. Influence of irradiation on the switching behavior in PZT thin films. *Mater. Sci. Eng. B* **2005**, *120*, 141–145. [CrossRef]
21. Iamsasri, T.; Esteves, G.; Choe, H.; Vogt, M.; Prasertpalichat, S.; Cann, D.P.; Gorfman, S.; Jones, J.L. Time and frequency-dependence of the electric field-induced phase transition in BaTiO$_3$-BiZn$_{1/2}$Ti$_{1/2}$O$_3$. *J. Appl. Phys.* **2017**, *122*, 064104. [CrossRef]
22. Cao, J.L.; Zhang, K.; Solbach, A.; Yue, Z.; Wang, H.H.; Chen, Y.; Klemradt, U. In situ X-ray reflectivity study of imprint in ferroelectric thin films. *Mater. Sci. Forum* **2011**, *687*, 292–296. [CrossRef]
23. Davydok, A.; Cornelius, T.W.; Mocuta, C.; Lima, E.C.; Araùjo, E.B.; Thomas, O. In situ X-ray diffraction studies on the piezoelectric response of PZT thin films. *Thin Solid Films* **2016**, *603*, 29–33. [CrossRef]

24. Cornelius, T.W.; Mocuta, C.; Escoubas, S.; Merabet, A.; Texier, M.; Lima, E.C.; Araùjo, E.B.; Kholkin, V.; Thomas, O. Piezoelectric response and electrical properties of $Pb(Zr_{1-x}Ti_x)O_3$ thin films: The role of imprint and composition. *J. Appl. Phys.* **2017**, *122*, 164104. [CrossRef]

25. Mocuta, C.; Richard, M.-I.; Fouet, J.; Stanescu, S.; Barbier, A.; Guichet, C.; Thomas, O.; Hustache, S.; Zozulya, A.; Thiaudière, D. Fast pole figure acquisition using area detectors at the DiffAbs beamline-Synchrotron SOLEIL. *J. Appl. Crystallogr.* **2013**, *46*, 1842–1853. [CrossRef]

26. Wang, Y.U. Diffraction theory of nanotwin superlattices with low symmetry phase. *Phys. Rev. B* **2006**, *74*, 104109. [CrossRef]

27. Noheda, B.; Cox, D.E. Bridging phases at the morphotropic boundaries of lead oxide solid solutions. *Phase Transit.* **2006**, *79*, 5–20. [CrossRef]

28. Araújo, E.B. Recent advances in processing, structural and dielectric properties of PMN-PT ferroelectric ceramics at compositions around the MPB. In *Advances in Ceramics–Electric and Magnetic Ceramics, Bioceramics, Ceramics and Environment*; Sikalidis, C., Ed.; IntechOpen: London, UK, 2011; pp. 43–60.

29. Fu, H.; Cohen, R. Polarization rotation mechanism for ultrahigh electromechanical response in single-crystal piezoelectrics. *Nature* **2000**, *403*, 281–283. [CrossRef]

30. Vanderbilt, D.; Cohen, M. Monoclinic and triclinic phases in higher-order Devonshire theory. *Phys. Rev. B* **2001**, *63*, 094108. [CrossRef]

31. Jin, Y.M.; Wang, Y.U.; Khachaturyan, A.G.; Li, J.F.; Viehland, D. Conformal miniaturization of domains with low domain-wall energy: Monoclinic ferroelectric states near the morphotropic phase boundaries. *Phys. Rev. Lett.* **2003**, *91*, 197601. [CrossRef]

32. Wang, Y. Diffraction theory of nanotwin superlattices with low symmetry phase: Application to rhombohedral nanotwins and monoclinic M_A and M_B phases. *Phys. Rev. B* **2007**, *76*, 024108. [CrossRef]

33. Yang, S.M.; Jo, J.Y.; Kim, T.H.; Yoon, J.-G.; Song, T.K.; Lee, H.N.; Marton, Z.; Park, S.; Jo, Y.; Noh, T.W. Ac dynamics of ferroelectric domains from an investigation of the frequency dependence of hysteresis loops. *Phys. Rev. B* **2010**, *82*, 174125. [CrossRef]

34. Jo, J.Y.; Yang, S.M.; Kim, T.H.; Lee, H.N.; Yoon, J.-G.; Park, S.; Jo, Y.; Jung, M.H.; Noh, T.W. Nonlinear dynamics of domain-wall propagation in epitaxial ferroelectric thin films. *Phys. Rev. Lett.* **2009**, *102*, 045701. [CrossRef]

35. Dai, X.; Xu, Z.; Viehland, D. Normal to relaxor ferroelectric tran sformations in lanthanum-modified tetragonal-structured lead zirconate titanate ceramics. *J. Appl. Phys.* **1996**, *79*, 1021–1026. [CrossRef]

36. Ramesh, R.; Keramidas, V.G. Metal-oxide heterostructures. *Ann. Rev. Mater. Sci.* **1995**, *25*, 647–678. [CrossRef]

MDPI
St. Alban-Anlage 66
4052 Basel
Switzerland
Tel. +41 61 683 77 34
Fax +41 61 302 89 18
www.mdpi.com

Materials Editorial Office
E-mail: materials@mdpi.com
www.mdpi.com/journal/materials